T0295295

Time-Variant and Quasi-separable Systems

Matrix theory is the lingua franca of everyone who deals with dynamically evolving systems, and familiarity with efficient matrix computations is an essential part of the modern curriculum in dynamical systems and associated computation.

This is a master's-level textbook on dynamical systems and computational matrix algebra. It is based on the remarkable identity of these two disciplines in the context of linear, time-variant, discrete-time systems and their algebraic equivalent, quasi-separable systems. The authors' approach provides a single transparent framework that yields simple derivations of basic notions, as well as new and fundamental results such as constrained model reduction, matrix interpolation theory and scattering theory. This book outlines all the fundamental concepts that allow readers to develop the resulting recursive computational schemes needed to solve practical problems.

An ideal treatment for graduate students and academics in electrical and computer engineering, computer science and applied mathematics.

Patrick Dewilde was a professor at the Delft University of Technology for 31 years and previously Director of the Delft Institute for Microelectronics, Chairman of a major Dutch research funding agency and Director of the Institute of Advanced Study of the Technical University of Munich. He is an Institute of Electrical and Electronics Engineers (IEEE) Fellow, a winner of the IEEE Belevitch Award and an elected member of the Dutch Royal Academy of Arts and Science.

Klaus Diepold is a professor at the Technical University of Munich. During his time in industry, he was the chief architect of award-winning software tools MotionPerfect and SteadyHand and is a coauthor of *Understanding MPEG-4: Technology and Business Insights* (2004). He is a board member of the Center for Digital Technology and Management and Fellow for Innovation in University Education. He was awarded the Start-Up Mentor of Excellence Award in 2021 by the Technical University of Munich. In 2023 he received the Unipreneurs Award by the Federal Ministry for Education and Research for his entrepreneurial activities.

Alle-Jan Van der Veen is Professor and Chair of the Signals and Systems group at Delft University of Technology. He is an IEEE Fellow and IEEE SPS Vice President – Technical Directions. Previous IEEE positions include: Editor-in-Chief of *Transactions on Signal Processing*, Chairman of the Signal Processing Society, elected member of the SPS Board of Governors and Chair of the SPS Signal Processing Theory and Methods Technical Committee and the Kilby Medal selection committee.

Time-Variant and Quasi-separable Systems

Matrix Theory, Recursions and Computations

PATRICK DEWILDE
Technical University of Munich

KLAUS DIEPOLD
Technical University of Munich

ALLE-JAN VAN DER VEEN
Delft University of Technology

CAMBRIDGE
UNIVERSITY PRESS

Shaftesbury Road, Cambridge CB2 8EA, United Kingdom

One Liberty Plaza, 20th Floor, New York, NY 10006, USA

477 Williamstown Road, Port Melbourne, VIC 3207, Australia

314–321, 3rd Floor, Plot 3, Splendor Forum, Jasola District Centre,
New Delhi – 110025, India

103 Penang Road, #05–06/07, Visioncrest Commercial, Singapore 238467

Cambridge University Press is part of Cambridge University Press & Assessment,
a department of the University of Cambridge.

We share the University's mission to contribute to society through the pursuit of
education, learning and research at the highest international levels of excellence.

www.cambridge.org
Information on this title: www.cambridge.org/9781009455626

DOI: 10.1017/9781009455640

When citing this work, please include a reference to the DOI 10.1017/9781009455640

First published 2025

A catalogue record for this publication is available from the British Library

Library of Congress Cataloging-in-Publication Data
Names: Dewilde, Patrick, author. | Diepold, Klaus, author. |
Veen, Alle-Jan van der, author.
Title: Time-variant and quasi-separable systems :
matrix theory, recursions and computations / Patrick Dewilde,
Klaus Diepold, Alle-Jan Van der Veen.
Description: Cambridge ; New York, NY : Cambridge University Press, 2025. |
Includes bibliographical references and index.
Identifiers: LCCN 2023051167 | ISBN 9781009455626 (hardback) |
ISBN 9781009455640 (ebook)
Subjects: LCSH: Matrices. | Linear time invariant systems. | Separable
algebras. | Mathematical optimization. | Computer algorithms.
Classification: LCC QA188 .D49 2025 | DDC 512.9/434–dc23/eng/20240404
LC record available at https://lccn.loc.gov/2023051167

ISBN 978-1-009-45562-6 Hardback

Contents

Preface

This book develops the intimate connection between linear time-variant dynamical systems and matrix calculus.

Both dynamical systems and computers implementing matrix calculus are ubiquitous in our present technical world. Dynamical systems are often steered or controlled by an embedded computer that executes an algorithm. Conversely, a computer executing an algorithm may be viewed as a dynamical system in its own right. This correspondence can be exploited to great benefit for the understanding of the behavior of a dynamical system and, by the same token, the construction of accurate and efficient computations. The present book develops this connection for the most simple and important class, linear systems, and, equivalently, matrix computations.

Traditionally, system theory and computational algebra have developed and evolved in very different ways. We show in this book that bringing the fields together in one body of knowledge has great advantages both from a theoretical and a practical point of view. In doing so, we show that only a few amazingly simple basic concepts and numerical methods are needed to cover the whole field, provided one is willing to look at the key issues in an unadorned and fresh way.

The insight that such an approach was possible arose in the seminal work of Bellman and Kalman, who ventured into the new environment of time-variant systems in order to solve estimation and control problems related to the Apollo spaceflight mission. Classically, such problems are treated with Laplace and z-transform theory and have an algebraic basis in complex function analysis and module theory. It soon appeared that an effective time-variant algebra was lacking. The latter was developed over time, on the mathematical side as nest algebras, on the system theory side as time-variant systems, and on the numerical side, in a limited way, as semi-separable systems. It took time to harmonize the various approaches, but the net result is that discrete-time, time-variant linear systems are now fully equivalent to a generalization of semi-separable matrix systems, which has been given the name *quasi-separable systems*, to our knowledge by Israel Gohberg. The present book offers a systematic but also fully didactical introduction to the unified theory.

The basis for the transformation or rejuvenation of system theory presented in this book is both traditional and remarkably simple. The central idea may be informally formulated as "the state of a system is what the system remembers of its past at any moment," or, equivalently, "the minimal information needed at any moment to produce its future development, given future inputs." This information presents itself as Nerode equivalent classes and, in the traditional thinking, as an *ideal* or a *submodule* in an overall algebraic setting – that is, strong structures that incorporate the time invariance. It was thought for a long time that these notions could not extend beyond a time-invariant framework. We show that, on the contrary, a systematic exploiting of the notion of state in a time-variant setting is not only possible but also produces most if not all essential system operations and properties. Time invariance is not essential, but a new, alternative, and as it turns out, more elementary but also more powerful algebraic framework is needed.

At the numerical algebra side, the power of orthogonal transformations gradually became evident, exemplified by the present day resurgence of QR factorization and the singular value decomposition (SVD) as the most ubiquitous numerical methods, insights going way back to Jacobi and Gauss, but then resurrected by numerical analysts such as Givens and Householder. It turns out that the only central mathematical concept needed in dynamical system theory is that of "range of an operator," often represented by a suitable orthonormal basis, and computed recursively using QR- or LQ-type transformations (its dual), or perhaps SVD for even more accuracy.

In the present book, we treat operations on matrices as a dynamical system in its own right, that is, as respecting the recursive order viewed as evolution in time. General matrix computations can just be viewed as evolution of a numerical system and, conversely, the evolution of time-variant dynamical systems consists of matrix operations. The notions "matrix" plus "linear progression in time" or "indexing order" coincide with the notion "discrete-time dynamical system" in this approach.

Basic matrix operations are then operations on systems and vice versa. For the development of the whole theory, only the most elementary matrix operations are needed, namely additions and multiplications as they occur in QR and LQ factorizations. The majority of results ranging from system identification to system optimization, estimation and model reduction can remarkably be obtained without much more than elementary matrix operations. Conversely, this approach leads to novel matrix methods and new results in matrix theory, especially in the areas of matrix inversion and matrix approximation, interpolation and factorization.

To realize this plan, an extension of the matrix theory framework is needed in order to accommodate dynamically changing dimensions. A typical time-variant system starts at some time index, its evolving state, inputs and outputs have variable dimensions during its existence, and it dies out or stops operating at some time. Variable dimensions force an adaptation of the classical treatment of matrices, including the introduction of zero dimensions (i.e., empty matrices), matrices whose elements are matrices themselves and the systemic use of block diagonal matrices as building blocks for system representations. These extensions of classical matrix theory turn out to be pretty natural and to conform without difficulty to computer science practices.

The result is that time-variant system theory and elementary matrix calculus largely coincide. This is a major didactical advantage of the approach. No complex function calculus, transform theory, module theory or whatever complicated algebraic structures are needed for most, if not all of the results, provided one stays within the context of systems evolving in discrete time. With some extra effort, many time-invariant results can be derived from the time-variant basics, but they often require the solution of an additional fixed-point problem. Historically, the time-invariant way was thought to be simpler and more insightful. The opposite is true: invariance complicates matters considerably and may even be seen as "unnatural." By going time variant, one can sidestep most of the classical, transform-based literature in favor of straight and simple matrix theory, or, as a modern saying goes, *array processing* – a great logical and didactical simplification.

Motivation for the Method

System theory has mostly been restricted traditionally to linear, time-invariant systems. The consequence is that matrices that describe the global behavior of the system have a special structure, namely (block) Toeplitz or (block) Hankel, and the theory is geared toward representing such structures algebraically, mainly using transform theory, which tends to become a goal in itself. However, many instances of (linear) recursive computations deal with unstructured or weakly structured global system matrices. We show that the more general time-variant system theory developed in this book succeeds surprisingly easily in remedying the lack of correspondence between system theory and general matrix algebra of the traditional approach. This move, which at first seems to make things more complex, turns out to produce a great theoretical simplification, thereby using and even highlighting all the main principles and leading to all the main basic results.

With reference to the immense literature on system theory, control theory, optimization and filtering, it is perhaps hard to believe that there exists a simple and easily accessible computational approach to all these topics, given the variety of mathematical methods people have developed to attack the various problems they encountered, often using ad hoc or brute-force methods. But *there is* a unifying approach, and the present book shows how even *profound* system-theoretic problems can be approached with elementary matrix algebra.

The ability to compute effectively *and* to connect computations to fundamental issues in systems engineering is a major challenge limiting the ability of our future engineers, irrespective of whether they are more active in engineering or in data numerics; see Figure 1. One of the most appealing properties of the proposed approach is the fact that matrix computations and system theory are two sides of the same coin: methods of one field are directly relevant to the other.

The challenge to students and researchers in systems engineering appears to be the mastery of elementary matrix algebra, especially of the *geometry* connected to it. A matrix defines a linear operator, and various related subspaces (range, co-range, kernel and co-kernel) play the central geometric role. Most algebraic operations aim at characterizing these spaces, mostly by deriving orthonormal bases for them, an operation that in signal processing terms is called *orthogonal filtering*. Depending on the application at hand, whether realization theory, optimal control or estimation theory,

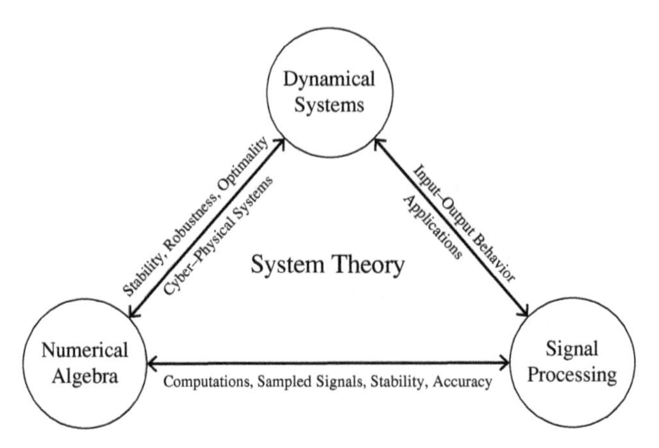

Figure 1 Our main disciplines are closely related.

the filtering procedure has an idiosyncratic name and is called, for example, dynamic programming, a Wiener filter, a Kalman filter or a selective filter. They all have similar operations that combine system-theoretic recursions with recursive numerical methods in common.

Conversely, the study we propose equips the students with a wealth of examples through which they can refine their intimate knowledge of matrix algebra as the main mathematical vehicle to achieve key results in numerical analysis and signal processing. Familiarity with matrix algebra is a *to be or not be* question for a modern engineer who has to handle data in a variety of situations, and the detailed study of system theory offers an ideal environment to develop that familiarity. Such an achievement can be seen as the main end term of the course we propose.

Adjoining Website

There is an adjoining website to this book containing:

- worked examples in concrete applications;
- computer programs of algorithms;
- PowerPoint presentations;
- advanced topics; and
- related papers from the open literature.

Prerequisites

We assume knowledge of elementary (undergraduate-level) matrix algebra (real and complex numbers, vectors, matrices, matrix products, echelon forms, matrix inversion, eigenvectors and eigenvalues, QR decomposition and SVD) and basic analysis (functions, limits, ordinary and partial differentials, integration, series). Matrices are our lingua franca, and our students and readers will develop familiarity with them by working through the book. More advanced mathematical properties (in particular, some elementary properties of Hilbert spaces) are introduced where they are used, and this does not happen in core theory, only in places where connections are made with other approaches. All these topics are covered by many textbooks on matrix or linear algebra, for example [40, 63], and textbooks in analysis and Hilbert space theory, for example [59].

A Summary of the Chapters

The book starts out with a motivating chapter to answer the question *why is it worthwhile to develop system theory?* To do so, we jump fearlessly into the very center of our methods, using a simple and straight example of *optimal control*. Although optimization is not our main subject – that is system theory – it provides for one of the main application areas, namely the optimization of the performance of a dynamical system in a time-variant environment (e.g., driving a car or sending a rocket to the moon). The chapter starts out with a review of the *Moore–Penrose pseudo-inverse*, which is a central concept of matrix algebra, used throughout the book. Next it describes a simple case of optimal control, which is first solved in a global way and then in a much more attractive recursive way called *dynamic programming*. The chapter then ends by showing how the method generalizes to linear, discrete-time, time-variant models.

Chapter 2 moves into a more philosophical mood to introduce *the basic notions on which system theory is based*. In this endeavor, it follows the insights first provided by Rudy Kalman and his coauthors in [49], and it does so as a narrative, laying the foundations for what will become the mathematical framework in the remainder of the book. A few fundamental and very appealing notions such as "state," "behavior," "reachability" and "observability" provide a sufficient basis for the whole theory (sufficient in the sense that no further fundamental notions are needed). In further chapters, we gradually develop the remarkable mathematical consequences of the concepts introduced here, exclusively using elementary matrix algebra and a necessary organizational extension of it. The chapter ends by introducing the reader to the main types of system they may encounter in practice, thereby illustrating how the basic concepts provide unity in diversity.

Chapter 3 starts developing our central *Linear Time-Variant (LTV) prototype environment*, a class that coincides perfectly with the linear algebra or matrix algebra context, making the correspondence systems-matrix computations a mutually productive reality. People familiar with the classical approach, in which the z-transform or other types of transform are used, will easily recognize the notational or graphic resemblance, but there is a major difference: everything remains elementary, no complex function calculus is involved and only the simplest matrix operations addition and multiplication are needed. Appealing expressions for the state-space realization of a system appear, as well as the global representation of the input–output operator in terms of four block-diagonal matrices $\{A, B, C, D\}$ and the (now non-commutative but elementary) *causal shift* Z. The consequences for and relation to linear time-invariant (LTI) systems and infinitely indexed systems are fully documented in the sections marked with $*$, which can be skipped by students or readers more interested in numerical linear algebra than in LTI system control or estimation.

From this point on, the main issues in system theory are tackled. The very first, considered in Chapter 4, is the all-important question of *system identification*. This is perhaps the most basic question in system theory and related linear algebra, with a strong pedigree starting from Kronecker's characterization of rational functions to its elegant solution for time-variant systems. Identification, often also called "realization," is the problem of deriving the internal system's equations (called state-space equations) from input–output data. In this chapter, we consider only the causal, or lower block triangular case, although the theory applies just as well to an anti-causal system, for which one lets the time run backwards, applying the same theory in a dual form.

In Chapter 5, we consider the central issue of *minimality* of the state-space system representation, as well as *equivalences of such representations*. The question introduces important new basic operators and spaces related to the state-space description. In our time-variant context, what we call the *Hankel operator* plays the central role, via a minimal composition (i.e., product) of a *reachability operator* and an *observability operator*. Corresponding results for LTI systems (a special case) follow readily from the LTV case as well as the theory for infinitely indexed systems, but they entail some extra complications, which are not essential for the main treatment offered and can be skipped on first reading.

Chapter 6 is a straightforward but essential chapter. It shows how the recursive structure of the state-space representations is exploited to make *elementary matrix operations* such as matrix addition and multiplication efficient. Also, elementary matrix inversion is treated for cases where the elementary inverse exists. The notions of *outer operator* and *inner operator* are introduced as basic types of matrices playing a central role in various specific matrix decompositions and factorizations to be treated in further chapters.

Several types of factorization solve the main problems of system theory (identification, estimation, system inversion, system approximation and optimal control). The factorization type depends on what kind of operator is factorized and what form the factors should have. Chapters 7 and 8 are therefore devoted to the two main types of factorization: Chapter 7 treats what is traditionally called *coprime factorization*, while Chapter 8 is devoted to *inner–outer factorization*. Coprime factorization, here called *external factorization* for more generality, characterizes the system's dynamics and plays a central role in system characterization and control issues. A remarkable result of our approach is the derivation of Bezout equations for time-variant and quasi-separable systems, made possible without the use of Euclidean divisibility theory. From a numerical point of view, all these factorizations reduce to recursively applied QR or LQ factorizations, applied on appropriately chosen operators.

Chapter 8 then considers the likely most important operation in system theory: *inner–outer (and its dual, outer–inner) factorization.* This factorization plays a different role than the previously treated external or coprime factorization, in that it characterizes properties of the *inverse or pseudo-inverse* of the system under consideration, rather than the system itself. An important point is that such factorizations are computed on the state-space representation of the original data. Inner–outer consists in nothing else but recursive QR factorization, as was already observed in our motivational Chapter 1, and outer–inner in recursive LQ factorization, in the somewhat unorthodox terminology used in this book for consistency reasons: QR for "orthogonal Q with right factor R" and LQ for "left factor L with orthogonal Q." These types of factorization play the central role in a variety of applications (such as optimal control and tracking, state estimation, system pseudo-inversion and spectral factorization). We conclude the chapter showing briefly how the time-variant linear results generalize to the nonlinear case.

The set of basic topics then continues with a major application domain of our theory: linear least-squares estimation (LLSE) of the state of an evolving system (also known as Kalman filtering – Chapter 9), which turns out to be an immediate application of the outer–inner factorization theory developed in Chapter 8. To complete this discussion, we also show how the theory extends naturally to cover the smoothing case (which is often considered "difficult").

Two chapters conclude the basic course.

Chapter 10 presents an alternative theory of external and coprime factorization, using polynomial denominators in the time-variant shift Z rather than inner denominators, as is done in Chapter 8. "Polynomials in the shift Z" are equivalent to block lower matrices with a support defined by a (block) staircase and are essentially different from the classical matrix polynomials of module theory, although the net effect on system analysis is similar. The polynomial method differs substantially and in a complementary way from the inner method. It is computationally much simpler but does not use orthogonal transformations. It offers the possibility of treating highly unstable systems using unilateral series. Also, this approach leads to famous Bezout equations, which, again, can be derived without the benefit of Euclidean divisibility methods.

Chapter 11 considers the *Moore–Penrose inversion of full matrices with quasi-separable specifications*, that is, matrices that decompose into the sum of a lower block triangular and a block upper triangular matrix, whereby each have a state-space realization given. We show that the Moore–Penrose inverse of such a system has, again, a quasi-separable specification globally of the same complexity as the original, and show how this representation can be recursively computed with three intertwined recursions. The procedure is illustrated on a 4×4 (block) example.

Chapters 12–17 exhibit further contributions of the theory of time-variant and quasi-separable systems to matrix algebra.

Chapter 12 considers another type of factorization of a quasi-separable system, namely *LU, or, equivalently, spectral factorization.* This type of factorization does not necessarily exist, and, when it exists, does not traditionally enjoy stable computation methods. Here we present necessary and sufficient existence conditions for the general quasi-separable case and a stable numerical algorithm to compute the factorization based on the quasi-separable representation. The algorithm uses orthogonal transformations and scalar normalizations exclusively, in contrast to classical Gaussian elimination.

Chapter 13 introduces a different kind of problem, namely *direct constrained matrix approximation via interpolation*, the constraint being positive definiteness. It is the problem of completing a positive definite matrix for which only a well-ordered partial set of entries is given (and also giving necessary and sufficient conditions for existence of the completion), or, alternatively, the problem of parametrizing positive definite matrices. This problem can be solved elegantly when the specified entries contain the main diagonal and further entries crowded along the main diagonal with a staircase boundary. This problem turns out to be equivalent to a constrained interpolation problem defined for a causal contractive matrix, with staircase entries again specified as before. The recursive solution calls for the development of a machinery known as *scattering theory*, which involves the introduction of non-positive metrics and the use of J-unitary transformations where J is a sign matrix.

Chapter 14 then develops the *scattering formalism*, whose usefulness for interpolation has been demonstrated in Chapter 13, for the case of systems described by state-space realizations, in preparation for the next three chapters, which use it to solve various further interpolation and embedding problems.

Chapter 15 shows how *classical interpolation problems of various types (Schur, Nevanlinna–Pick, Hermite–Fejer)* carry over to the time-variant and/or matrix situation. We show that they all reduce to a single generalized constrained interpolation problem, elegantly solved by time-variant scattering theory. An essential ingredient is the definition of the notion of *valuation* for time-variant systems, in generalization of the valuation in the complex plane provided by the classical z-transform.

Chapter 16 then provides for a further extension of constrained interpolation that is capable of solving the *constrained model reduction problem*, namely the generalization of Schur–Takagi-type interpolation to the time-variant setting. This remarkable result demonstrates the full power of time-variant system theory as developed in this book.

Chapter 17 completes the scattering theory with an elementary approach to inner embedding of a contractive, quasi-separable causal system (in engineering terms: the embedding of a *lossy* system in a *lossless* system, often called *Darlington synthesis*). Such an embedding is always possible in the finitely indexed case, but does not generalize to infinitely indexed matrices (this issue requires more advanced mathematical methods and lies beyond the subject matter of the book).

The appendix on the data model used throughout the book describes what can best be called an *algorithmic design specification*, that is, the functional and graphical characterization of an algorithm, chosen so that it can be translated to a computer architecture (be it in software or hardware). We follow hereby a powerful "data flow model" that generalizes the classical signal flow graphs and which can be further formalized to generate the information necessary for the subsequent computer system design at the architectural level (i.e., the assignment of operations, data transfer and memory usage.) The model provides for a natural link between mathematical operations and architectural representations. It is, by the same token, well adapted to the generation of parallel processing architectures.

Notation

We mostly use standard usage in linear algebra, but with some systematic differences, induced by keeping the notation close to the practice exercised by MATLAB. We do, however, use special notations for mathematical objects that occur often in our developments and try to avoid annoying overloads of symbols (which is sometimes impossible). Notation is an important issue, both in algebra and in computer science, and we try to be as straight and consistent as possible. Notations that deviate from common practice in either linear algebra or MATLAB are defined wherever they are introduced (this is especially true for "empty" objects, which play an important role in making the theory fully consistent). Also, we introduce the most important numerical methods in the chapters where they are intensely used. Here is a short summary of nonstandard notations used in this book:

- Many of our objects are matrices, and we often have to consider either special indexing conventions or take out submatrices from a given matrix. As such operations can become unwieldy, we adopt systematically a MATLAB-like annotation of index ranges. Suppose A is a matrix, then $A_{k:\ell,m:n}$ is a submatrix of A consisting of a selection of rows from index k to ℓ (inclusive) and columns from index m to n inclusive (if no confusion is expected, we may also write $A_{k:\ell}^{m:n}$: row indices at the bottom and column indices at the top). If rows or columns are not there originally, they are just considered empty (see the next item). We also systematically economize the notation (in contrast to many textbooks): A, a and $\mathbf{A}$ are different matrices, with $A_{i,j}$, $a_{i,j}$ and $\mathbf{A}_{i,j}$ being the different elements of each respectively.
- An index is actually also a function, namely from a subset of the natural numbers to the set of objects under consideration. So a_k can also be written as $a(k)$. We do this only when we want to emphasize or discuss this functional character. Often, either the functional arguments or the indices are omitted in a discussion or proof: these are then inferred from the context. This common practice enhances readability at the cost of precision (so we typically write A instead of $A_{1:n}^{1:m}$ for a simple $n \times m$ matrix.)
- Very often block matrices appear, that is, matrices whose entries are themselves matrices (called "blocks"). Blocks may consist of blocks themselves, but they are indexed in the usual fashion and have to be commensurate (dimensions in rows or columns must match throughout). For example: $A_{k,\ell}$ may be a block in a block matrix A, and $[A_{k,\ell}]_{m,n}$ is then a block entry at the position (m,n) in that original block. We do allow blocks with zero dimensions: they are just empty, but do have index numbers and dimensions. We introduce special notation for empty objects: an entry of dimensions $(0,1)$ is denoted by "–," an entry of dimensions $(1,0)$ by

"|" and one of dimensions (0,0) by "·" (these are actually place holders: they have indices but no entries). Special (logical) computational rules apply for such entries and are introduced in the text where this type of notation is first used. This extension of matrix algebra appears to be very useful: in many matrix operations (in particular, reductions and approximations), one cannot say beforehand whether certain entries survive the operation (e.g., deleting rows or columns of zeros in a matrix). Just as the introduction of the empty set $\emptyset$ and the number 0 prove extremely useful in set theory and algebra, so do indexed empty entries. They also correspond to the empty symbol ($\perp$) often used in computer science.

- In the literature, many notations are used to indicate the transpose of a real matrix or the Hermitian transpose (conjugate transpose) of a complex matrix. In this book and as in MATLAB, we use only real or complex arithmetic and indicate these transpositions by a single symbol, namely a single accent (i.e., A' is in all cases the Hermitian transpose of A; in the real case, it is then also just the transpose). A motivation for this choice is (i) the overload of the symbols T or H, as we have special use for those (we do not like the notation T^T for the transpose of T nor H^H for the Hermitian transpose of the Hankel matrix H!) and (ii) consistency with MATLAB (with the understanding that the accent is "transpose conjugate" in the complex case). We reserve the star (e.g., with a an operator, a^*) for the dual of an object in a context where duality is defined (a dual is not necessarily a transpose, although it often will be). We use tildes and hats as normal typographical symbols – they do not have any other meaning than to characterize the object under discussion. However, there is one exception. In estimation theory, we often use the upper bar as shorthand for expectation: $\overline{X} = \mathbf{E}X$, and the hat as indicating an estimate: $\widehat{X}$ is an estimate of X.
- We use *constructors* systematically, following a good habit of computer science. Constructors are written in normal font. For example, "col" is the column constructor. It takes a sequence of elements (e.g., numbers or blocks with appropriate dimensions) and makes a column out of them. For example, $\mathrm{col}(u_1,u_2) = \begin{bmatrix} u_1 \\ u_2 \end{bmatrix}$ (compare this with the awkward $\begin{bmatrix} u_1^T & u_2^T \end{bmatrix}^T$). Further constructors are "row," "Toeplitz," "Hankel" and so on (often abbreviated).
- Names and formulas: The general convention is that names are written in a regular font. For example, "ker" or "ran" are the names of the functions that return the kernel or the range of an operator, respectively, while *ker* is the product of *k*, *e* and *r*. This convention allows us to use indexed names as well: so K7 is just a name, while the notation K_7 assumes K to be a vector whose element with index 7 is K_7. A new name occurring in the text, a theorem or a proof is introduced with the symbol ":="; for example, $T := D + C(I - ZA)^{-1}ZB$ defines T in terms of the right-hand side symbols (assumed to be already defined).
- Another convenient notation that we shall use extensively is a shorthand for system realizations. Z is systematically used as the "forward" or "causal" shift, with its conjugate Z' the "backward" or "anti-causal" shift. A causal, LTV system T may

have a "realization" $T = D + C(I - ZA)^{-1}ZB$ for which we use the shorthand

$$T \sim_c \begin{bmatrix} A & B \\ C & D \end{bmatrix} \tag{1}$$

to mean "is represented by the causal realization." Similarly, an anti-causal LTV system may have a realization $T = D + C(I - Z'A)^{-1}Z'B$, with the shorthand

$$T \sim_a \begin{bmatrix} A & B \\ C & D \end{bmatrix} \tag{2}$$

to mean "is represented by the anti-causal realization." Z is a generic operator or constructor. It just shifts the indexing scheme but stands by the same token for a generic collection of shift matrices, which have a different effect whether applied to the left or the right of an indexed object. For example, suppose u is an indexed column vector, then Zu is again an indexed column vector whose elements are $(Zu)_k = u_{k-1}$. In our formalism, the dimensions of the u_k may vary (and even become empty), and the dimensions of the entries in Z, interpreted as a matrix, have to adapt. This question is addressed at length in the chapter on LTV systems. Some constructors can be represented as matrices. What characterizes a constructor is that it entails an organizational operation rather than a numerical operation.
- We also need shifts along diagonals. We denote $A^{\langle +1 \rangle}$ as a "forward" diagonal shift on a matrix, that is, a shift in the southeast direction, with as its conjugate the "backward" diagonal shift denoted as $A^{\langle -1 \rangle}$, that is, a shift in the northwest direction.
- We shall also occasionally use "continuous products," defined for integers $i > k$ as $A^{>}_{i,k} = A_{i-1}A_{i-2}\cdots A_{k+1}$ with $A^{>}_{i,i} := I$ (NB: this convention may differ from what is done in the literature). Also, $A^{\geq}_{i,k} := A_i A_{i-1} \cdots A_k$ will be used when convenient. To avoid confusion with the use of upper indices for columns and lower ones for rows, we often use the abbreviated notation $A^{i>k} := A^{>}_{i:k}$ and $A^{i\geq k} := A^{\geq}_{i:k}$ as well, in which the upper index now refers to a product instead of a column.

A Note on the Cover Picture

In 1903, the mathematician Max Dehn investigated for the first time "a simple and yet quite general problem of geometry" [78]: a square is to be divided exactly into partial squares of different sizes. This decomposition appeared to be highly nontrivial. A square together with such a division is called a *perfect* square.

The search for solutions went on in vain for more than three decades. The problem attracted an unusually high amount of interest – probably because of its simple and generally understandable nature as a puzzle and because it can be brought under the rubric of "mathematical recreations" for the edification of a broad public. But mathematicians especially were interested because of the "hard" topological and combinatorial aspects.

In 1940, four English students, Brooks, Smith, Stone and Tutte, proposed a completely new method of solving the problem. The student quartet approached the problem based on a physical analogy of the problem to Kirchhoff nets or electrical systems. Based on this system model, they found perfect squares of order $n = 26$ and $n = 28$.

In the following decades, computer programs were used to create catalogs of perfect decompositions, but all attempts to find perfect squares of order $n < 25$ failed. In the early 1960s [79], at the Philips Computing Center in Eindhoven, researchers proved, by full enumeration, that no perfect squares exist up to order $n = 20$. The picture on the book cover shows the perfect square of minimal order $n = 21$. It is uniquely determined and was found on March 22, 1978, by A. J. W. Duijvestijn at the University of Twente, with the help of a DEC-10 computer. The solution was published in the *Journal of Combinatorial Theory* [80] and has since been shown as a signet on the cover of both series of the journal.

This example shows that system models and the associated views and intuitions can enhance the understanding of the structure of difficult mathematical problems. The example shown should make it clear that system models can do much more than simply illustrate mathematical facts in a circuit diagram familiar to the engineer.

In 2008, Rainer Pauli picked up the story of the perfect square of order $n = 21$ and produced an artistically appealing rendition of the solution – the *Perfect Square*. This image is shown on the cover of the book and is based on an inner connection between the color shade of the squares and the geometry. The color representation is based on a mixing ratio of red and yellow corresponding to the size of the squares.

Acknowledgments

This book would not have been possible without the collaboration with and influence of many colleagues, students, assistants and relatives. A full list would be prohibitively long, and the short list that we are including here may (or even will) overlook important contributors, a fate that appears unavoidable. Nonetheless, here are persons to whom we are grateful for either their contributions to our knowledge and insights, for the support they provided or, in many cases, both.

In the first place, we would like to mention the massive seminal contributions of the giants of our field who have passed away in recent times: Rudy Kalman, Jan Willems, Alfred Fettweis and Uwe Helmke.

From a methodological point of view, we have been most strongly influenced by Tom Kailath and quite a few of his students. Besides Tom, who pioneered the algebraic and computational approach to system theory, we have been principally influenced by his students Martin Morf, Sun-Yuan Kung, George Verghese, Hanoch Lev-Ari, Ali Sayed and Augusto Vieira. Our contacts with the Information Systems Laboratory group in Stanford have always been exceedingly productive, and we cannot be sufficiently grateful to Tom for mostly making them happen. We were graciously invited to heaven so many times.

Many colleagues contributed ideas and specificities to our work, besides those already mentioned. In particular, we wish to mention Daniel Alpay, Brian Anderson, Thanos Antoulas, Jo Ball, John Baras, Vitold Belevitch, Bob Brayton, Adhemar Bultheel, Francky Catthoor, Samarjit Chakraborty, Shiv Chandrasekaran, Prabhakar Chitrapu, Leon Chua, Ed Deprettere, Harry Dym, Yuli Eidelman, Paul Fuhrmann, Tryphon Georgiou, Keith Glover, Israel Gohberg, Nithin Govindarajan, Bill Helton, Jochen Jess, Rien Kaashoek, Geert Leus, Wolfgang Mathis, Mohammed Najim, Bob Newcomb, Ralph Otten, Rainer Pauli, Ralph Phillips, Justin Rice, Jacqueline Scherpen, Wil Schilders, Malcolm Smith, Lang Tong, Eugene Tyrtyshnikov, Joos Vandewalle, Marc Van Barel, Paul Van Dooren, Michel Verhaegen, Eric Verriest, Jan Zarzycki and many of their students.

Klaus Diepold likes to acknowledge, in addition, the inspiring contributions from colleagues who have accompanied him through many years with collaboration, inspiration, discussion and friendship, notably Rainer Pauli, Wolfgang Mathis, Albrecht Reibiger, Harald Martens and Sunil Tatavarti.

Patrick Dewilde wishes to acknowledge the intense support and patience of his spouse, Anne Renaer, which made the many additional hours of research and writing possible, in addition to being a permanent and loving muse. He also wishes to extend his gratitude first to Sarah Kailath, who unfortunately passed away after an unforgiving illness, and later to Anu Luther, for so often and so graciously being our hosts and allowing many years of collaboration with Tom Kailath to take place in the ideal environment of their home.

Part I

Lectures on Basics with Examples

1 A First Example

Optimal Quadratic Control

Our book starts with a motivating chapter to answer the question: *Why is it worthwhile to develop system theory?* To do so, we jump fearlessly into the very center of our methods, using a simple and straightforward example of *optimal control*. Although optimization is not our main subject – that is system theory – it provides for one of the main application areas, namely the optimization of the performance of a dynamical system in a time-variant environment (think of driving a car or sending a rocket to the moon). The chapter starts out with a review of the *Moore–Penrose pseudo-inverse*, which is a central concept of matrix algebra, used throughout the book. Next it describes a simple case of optimal control, which is first solved in a global way and then in the much more attractive recursive way called *dynamic programming*. The chapter then ends by showing how the method generalizes to linear, discrete-time, time-variant models.

Menu

Hors d'oeuvre
The Moore–Penrose Inverse

First Course
Discovering the Power of
Dynamic Programming by Rowing

Second Course
The Bellman Problem: Optimal Quadratic Control
of a Linear Dynamical System

Dessert
Notes

1.1 Matrix Algebra Preliminary: The Moore–Penrose Inverse

Solving a system of linear equations $Ax = b$ is perhaps the first motivation for studying linear algebra. Here, A is a square $m \times m$ matrix with scalar entries, b is a given vector of dimension m, and x is an unknown vector of the same dimension m. When A has independent columns, these columns span the full real m-dimensional vector space $\mathcal{R}^m$ (or in the complex case $\mathcal{C}^m$), and there exists a unique solution $x = A^{-1}b$. In this case, the *range* of A, as an operator acting on x, is the full space $\mathcal{R}^m$, and there is a unique linear combination of columns of A that generates the given b of dimension m. Many other situations are, of course, conceivable: there may be fewer equations than unknowns (A has dimensions $n \times m$ with $n < m$) or just more ($n > m$), and the equations given may turn out to be contradictory. The result is that an infinite number of solutions might exist or just no solution at all. Hence, a more general approach is needed, and it is provided by the Moore–Penrose inverse.

An overdetermined situation ($n > m$) often arises as a result of many (similar or different) measurements involving the same unknown quantities in x, and then one wonders what to do about the resulting incompatibilities (in measurement practice involving more than one unknown variable, one should use a variety of measurement methods to obtain a nonsingular system of equations with more equations n than unknowns m). Let us look at such an overdetermined case in more detail.

Typically, when there are too many equations for the unknown quantities, these equations will be contradictory, and no exact solution for $Ax = b$ will exist. Rather, for each trial x, there will be an associated error $e_x = b - Ax$, and, assuming all measurements to be equally important, one may want to find x's that minimize the quadratic error $e_x' e_x = \sum_{i=1:n} [e_x]_i^2$. More generally, one might give weight to the importance of individual measurements, particularly when they lead to quantities with different dimensions.

Therefore, we consider the error equation

$$\begin{bmatrix} A_{1,1} & \cdots & A_{1,m} \\ \vdots & \vdots & \vdots \\ A_{m,1} & \cdots & A_{m,m} \\ \vdots & \vdots & \vdots \\ A_{n,1} & \cdots & A_{n,m} \end{bmatrix} \begin{bmatrix} x_1 \\ \vdots \\ x_m \end{bmatrix} - \begin{bmatrix} b_1 \\ \vdots \\ b_m \\ \vdots \\ b_n \end{bmatrix} = \begin{bmatrix} [e_x]_1 \\ \vdots \\ [e_x]_m \\ \vdots \\ [e_x]_n \end{bmatrix} \tag{1.1}$$

and try to minimize the error e_x in the least-squares sense. Matrix A has dimensions $n \times m$ with $n \geq m$, and let us assume that the columns of A are linearly independent, but since $n \geq m$, they span only a subspace of dimension m, and not the whole space $\mathcal{R}^n$ to which b belongs, unless $n = m$.

Using the quadratic norm $\|a\|_2 = \sqrt{a'a}$ for any vector a, we may write

$$x_{\min} = \operatorname{argmin}_x \|b - Ax\|_2, \tag{1.2}$$

meaning $x_{\min}$ is an argument x that minimizes the expression (notice: the square root does not matter for the minimization). We show:

Proposition 1.1 *The solution to the minimization problem* $\operatorname{argmin}_x \|b - Ax\|_2$, *where* A *is an* $n \times m$ *matrix* ($n \geq m$) *with independent columns, is unique and is given by*

$$x_{\min} = A^{\dagger} b, \tag{1.3}$$

in which $A^{\dagger} := (A'A)^{-1}A'$.

Moreover, the minimal error vector $e_{\min}$ *is given by*

$$e_{\min} = (I - \Pi_A)b, \tag{1.4}$$

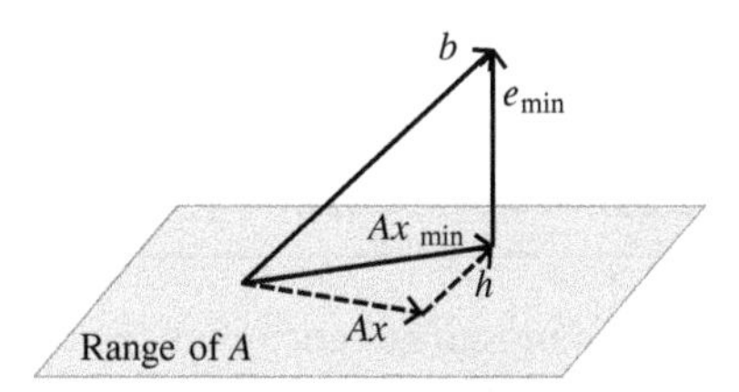

Figure 1.1 Best linear quadratic approximation.

in which $\Pi_A := AA^\dagger$ *is the orthogonal projection on the range of A in* $\mathcal{R}^n$, *and* $(I - AA^\dagger)$ *is the projection on the orthogonal complement of the range of A.*

Proof

(We follow the traditional orthogonality argument.) For any x of dimension m, Ax will lie in the linear subspace generated by the columns of A, that is, the *range* of A. The best $x_{\min}$ in a least-squares sense will then be such that the least-squares error $e_{\min} = b - Ax_{\min}$ is orthogonal on the range space of A. Expressing the orthogonality of the error vector on the columns of A, we require

$$A'(b - Ax_{\min}) = 0, \tag{1.5}$$

and hence $x_{\min} = (A'A)^{-1}A'b$ since $A'A$ is an $m \times m$ nonsingular matrix thanks to the assumed independence of the columns of A. The solution is unique, because for any x we have $b - Ax = (b - Ax_{\min}) + h$ with $h = A(x_{\min} - x) \perp e_{\min}$ since $e_{\min}$ is orthogonal to the range of A, see Fig. 1.1 for an illustration, and hence $\|b - Ax\|^2 = \|e_{\min}\|^2 + h^2 > \|e_{\min}\|^2$ when $h \neq 0$.

Next, one checks that $\Pi_A = AA^\dagger = A(A'A)^{-1}A'$ is indeed an orthogonal projection operator, for

1. it is a projection operator because $\Pi_A^2 = \Pi_A$, and
2. it is an orthogonal projection because $\Pi_A' = \Pi_A$

(these being the two necessary and sufficient properties for an operator to be an orthogonal projection), and, finally, the range of Π_A is the range of A as well, because, for any Ax whatever x may be, $\Pi_A Ax = A(A'A)^{-1}A'Ax = Ax$. $(I - \Pi_A)$ is then evidently the projection on the orthogonal complement of the range of A. □

Definition 1.2 Given a matrix A with independent columns, the matrix $A^\dagger = (A'A)^{-1}A'$ is called the Moore–Penrose inverse of A.

Example Suppose we have two measurements of a quantity x, the first giving $x = 9$ and the second $x = 11$. What is the "best" x in the least-squares sense? Writing the measurements in matrix form gives $b - Ax = e$ with $A = \begin{bmatrix} 1 \\ 1 \end{bmatrix}$ and $b = \begin{bmatrix} 9 \\ 11 \end{bmatrix}$. We find

$A'A = 2$ and $A^\dagger = \frac{1}{2}\begin{bmatrix} 1 & 1 \end{bmatrix}$, and hence $x_{\min} = 10$, with $e_{\min} = \begin{bmatrix} -1 \\ 1 \end{bmatrix}$ and the overall square-root error being $\sqrt{e'_{\min} e_{\min}} = \sqrt{2}$ as one would expect. □

This is the basic "geometric" result used in most quadratic optimization problems. Still, a number of remarks and/or refinements can be made:

1. A' is an $m \times n$ matrix, so the dimension of $A'b$ is the same as that of x. $\Pi_A := AA^\dagger = A(A'A)^{-1}A'$ is the orthogonal projection operator on the range of A, and we often write $\widehat{b} := \Pi_A b$. $\widehat{b}$ is the llse or *linear least-squares estimate of b in the range of A*.
2. Where the columns of A are not linearly independent, more work has to be done to solve the minimization problem, which typically will no longer have a unique solution. We shall treat such cases when they occur.

The QR Solution

The expression $A^\dagger = A(A'A)^{-1}A'$ is unwieldy and certainly not well suited to computations: not only is it largely inefficient, it is also computationally inaccurate – it is only mathematically satisfying because it is a closed-form solution. An adequate, first-hand, efficient and accurate solution is provided by the *upper QR algorithm* applied to A, which produces a factorization of the form

$$A = \begin{bmatrix} Q_1 & Q_2 \end{bmatrix} \begin{bmatrix} R \\ 0 \end{bmatrix}, \tag{1.6}$$

in which $Q = \begin{bmatrix} Q_1 & Q_2 \end{bmatrix}$ is an $n \times n$ orthogonal matrix and R a nonsingular $m \times m$ upper-triangular matrix. The columns of Q_1 form an orthonormal basis for the range of A, while the columns of Q_2 form an orthonormal basis for the kernel of A', also known as the *co-kernel* of A. When we dispose of such a QR factorization, then we can immediately write

$$A^\dagger = R^{-1}Q_1'. \tag{1.7}$$

Upper QR is not the only possibility for such a result; we could (and will) also use a *lower QR* version of the same type of algorithm, writing $A = \begin{bmatrix} Q_1 & Q_2 \end{bmatrix} \begin{bmatrix} 0 \\ L \end{bmatrix}$, in which Q (different from the previous version!) is also an orthogonal matrix and L is a nonsingular lower triangular matrix. In this latter case, we will still have $A^\dagger = L^{-1}Q_2'$. Both R and L can be seen as "compressed" versions of the rows of A with a special (upper or lower) structure.

Remark: upper or lower QR are not the only possibilities to obtain the range basis. A numerically more refined method is the *singular value decomposition – SVD*. We refer to the linear algebra literature for a more extensive explanation.

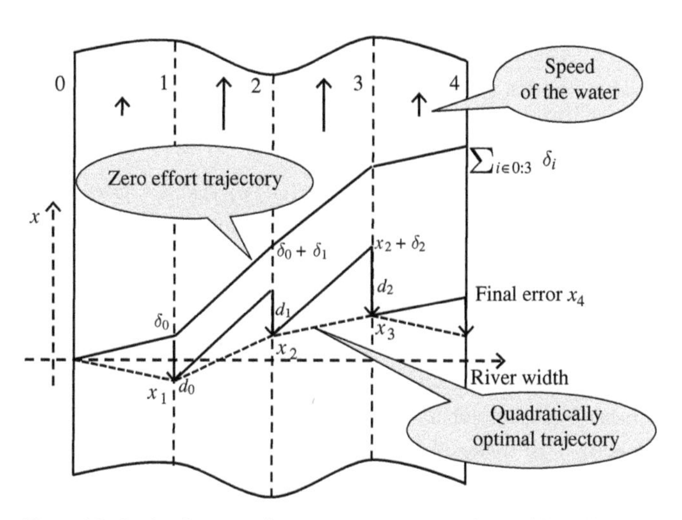

Figure 1.2 Optimal cost trajectory to row over a river with variable water speed.

Example In the previous example, we have $A = \begin{bmatrix} 1 \\ 1 \end{bmatrix} = \begin{bmatrix} \frac{1}{\sqrt{2}} & -\frac{1}{\sqrt{2}} \\ \frac{1}{\sqrt{2}} & \frac{1}{\sqrt{2}} \end{bmatrix} \begin{bmatrix} \sqrt{2} \\ 0 \end{bmatrix}$, giving orthonormal bases for both the range of A and the *co-kernel of* A, which is the kernel of A'.

1.2 A Toy Example of System Optimization: Row, Row, Row Your Boat

Suppose you want to cross a river in a rowing boat. The current in the river has variable velocities depending on the distance from the shore. You can let your boat drift, and with careful handling of the rudder or the oars, you can reach the other side without any effort on your part. However, you will drift too far downstream doing this, so instead, you would row against the current with the aim of reaching a point on the other side of the river that is close to the opposite of your starting point. You would try to do the best possible job by minimizing the effort you have to exert, while trying to get close to your intended destination.

The Modeling Phase

We start out by making a simplified model of the situation. Here are the assumptions (see Fig. 1.2):

- we subdivide the river into four segments enumerated 0:3, each segment having a uniform (actually average) speed of water $v_{0:3}$. We let the current flow in the (vertical) x-direction; the model will easily generalize to more segments;
- the "natural drift" in each segment (i.e., the drift of the boat with no rowing effort but keeping the boat going to the opposite shore as well as possible) is denoted by δ_i, $i \in 0{:}3$. For example, we assume the natural drift δ_i to be proportional to

the current flow v_i with some constant, which we have to specify further; we may assume that the water that pushes on the boat also pushes the boat to the other side when the rudder and/or oars are correctly set – we only need to assume that we know the no-effort drift a priori;

- rowing provides for an improvement on the drift of $d_i \geq 0$ in segment i, and the rowing effort is pegged at $N_i^2 d_i^2$ for some constant N_i solely dependent on v_i, motivated in the following paragraph.

A motivation to estimate the rowing effort in segment i to be proportional to d_i^2 is that two main effects combine to increase d_i, namely the fact that more force has to be used by the rower given the local push by the river, and, second, that that force has to be exercised over a longer relative distance due to the greater drift (energy = force times distance). That makes the effort in the first instance proportional to d_i^2 (an alternative argument is based on a power expansion, the observation that $d_i = 0$ means no effort, and any deviation requires effort.) The proportionality, in turn, is dependent on the local circumstances, and hence on v_i, perhaps proportional (this assumption is not used, but is not unreasonable). We write this constant, which is positive, as a square number N_i^2, for convenience, as will appear soon.

The total cost to be minimized hence becomes

$$C_4 = \sum_{i=0}^{3} N_i^2 d_i^2 + M^2 x_4^2, \tag{1.8}$$

in which the offset at destination x_4 is penalized as $M^2 x_4^2$ for some M, which one may choose: the larger the M is, the closer to the ultimate goal the rower will end up at. All the “modeling quantities” N_i and M are assumed known (this is the big “physics work” to be done before boarding!).

The dynamic model is very simple in this case. We take the position $x_i, i = 0{:}3$ of the boat as the state at position i, and its evolution is

$$x_{i+1} = x_i + \delta_i - d_i. \tag{1.9}$$

Notice that the model is not linear: it is *affine* because of the drift term δ_i, but we shall soon see that it can be handled with linear methods just as well.

Our *optimization strategy* now consists in writing down the complete *cost model* for this situation and then performing optimization on it. The cost model has to relate the control quantities that drive the dynamic model – the d_i – to their contribution in the cost function. It will soon appear that it is best to define the components in the cost model as squares of linear quantities, namely of $y_i = N_i d_i$ for $i = 0{:}3$ and $y_4 = M x_4$ – this will make the model linear or affine. Notice that these quantities are function of either the inputs (the d_i) or the states, in this case just x_4. x_4 can be expressed in terms of the input quantities by integrating the state equations $x_4 = \delta_t - \sum_{i=0:3} d_i$, where $\delta_t := \sum_{i=0:3} \delta_i$ is the total drift (assumed to be known).

Writing this out in matrix language and using the d_i as controlling inputs, we obtain the *global cost equation*

$$\left[\begin{array}{c} y_{0:3} \\ \hline y_4 \end{array}\right] = \left[\begin{array}{ccc} N_0 & & \\ & \ddots & \\ & & N_3 \\ \hline -M & \cdots & -M \end{array}\right] \begin{bmatrix} d_0 \\ \vdots \\ d_3 \end{bmatrix} + \left[\begin{array}{c} 0 \\ \vdots \\ 0 \\ \hline M\delta_t \end{array}\right]. \tag{1.10}$$

Defining $N := \text{diag}[N_i]$, $E = \text{col}\begin{bmatrix} 1 & \cdots & 1 \end{bmatrix}$ a column vector of 1's, and using vectors for the other quantities, the equations summarize as

$$y = \begin{bmatrix} N \\ -ME' \end{bmatrix} d + \begin{bmatrix} 0 \\ M\delta_t \end{bmatrix} \tag{1.11}$$

and the goal is to find the vector d that minimize $C = y'y$.

The Global Solution

As discussed in Section 1.1, the Moore–Penrose inverse produces the solution: in Eq. (1.11) and referring to the original Moore–Penrose equation $Ax - b = e_x$, d plays the role of x, y of e_x, $\begin{bmatrix} N \\ -ME' \end{bmatrix}$ of A and $-\begin{bmatrix} 0 \\ M\delta_t \end{bmatrix}$ of b. The Moore–Penrose inverse of the nonsingular *system matrix* $S := \begin{bmatrix} N \\ -ME' \end{bmatrix}$ is then

$$S^\dagger = (N^2 + M^2EE')^{-1}\begin{bmatrix} N & -ME \end{bmatrix} \tag{1.12}$$

and the solution of the optimization problem is given by

$$\widehat{d_{0:3}} := (N^2 + M^2EE')^{-1}M^2E\delta_t. \tag{1.13}$$

This expression can be computed explicitly, using the inversion rule for a low rank perturbation of a nonsingular matrix (sometimes called the "Sherman–Morrison formula": suppose that some low-dimensional (rectangular) matrices A and B of same dimensions are such that $I+B'A$ is nonsingular, then $(I+AB')^{-1} = I-A(I+B'A)^{-1}B'$ – proof is by direct verification; the simplest case is when A and B are just vectors – we leave details to the interested reader). The result is

$$\widehat{d_i} = \left(\frac{\frac{1}{N_i^2}}{\frac{1}{M^2} + \sum\left(\frac{1}{N_i^2}\right)}\right)\delta_t. \tag{1.14}$$

This result is a globally computed a priori control (not a *state-dependent* control), to be computed before boarding the boat. Notice that $\widehat{d_i} = K\frac{1}{N_i^2}$ with constant $K = \frac{\delta_t}{\left(\frac{1}{M^2}+\sum\frac{1}{N_i^2}\right)}$, so and assuming all N_i equal, the optimal efforts $N_i^2\widehat{d_i^2}$ to be spent at each step are equal (which is not unreasonable altogether: you distribute the energy to be exerted evenly over the sections – a pretty generally valid "principle" in optimization theory; notice also that in the limiting case $M \to \infty$, $\sum \widehat{d_i} = \delta_t$, forcing the rower to get at the destination point exactly).

Dynamic Programming

The global character of this solution can easily be seen as a problem: there is no adaptivity. Many things can happen when one is plodding in the river, and it pays to figure out a recursive solution that can adapt to the perspective from a local state of affairs, reached somewhere in the middle of the river. It turns out that the global solution can be converted to a local solution, by making the controls a function of the local state. But there is another advantage to a local solution (given the validity of the model of course): at any local position *only information on the cost of the next move* is needed to determine the optimal local move. The *reduction to minimal sufficient information* is what makes the recursive computation attractive and efficient. This we derive now. It is known as *dynamic programming* or *dynamic optimization.*

Let us therefore see how to do the local recursive optimization and derive the control law at stage k, which we shall see to be just a function of the local state x_k. The principle of dynamic optimization, or Bellman principle, is based on the observation that:

> once a state x_k has been reached, the cost must be optimal from that point on up to the final state, for if it were not so, there could be a lower total cost obtained by a modification of the final part of the trajectory. It follows that the optimal cost to reach the destination starting at a state x_k depends exclusively on that state x_k, that is, *all dependence on past history or controls* $d_{0:k-1}$ *go via the state* x_k*, which also determines what the optimal future controls are supposed to be.*

An important consequence of the principle is that local optimization can be done, provided one disposes of an expression for the cost of the trajectory following the current step, expressed in terms of the next state.

Concretely:

Suppose you have reached state x_k and you are ready to determine the optimal control $\widehat{d_k}$ to move to state x_{k+1} (using the state evolution equation, in this case $x_{k+1} = x_k + \delta_k - d_k$). What you need is the expression for the optimal cost of the trajectory starting at x_{k+1}, which by assumption depends only on x_{k+1}: $\widehat{C}_{k+1}(x_{k+1})$. You have to determine

$$\widehat{d_k} = \operatorname{argmin}_{d_k} \left(N_k^2 d_k^2 + \widehat{C}_{k+1}(x_k + \delta_k - d_k) \right), \tag{1.15}$$

which optimizes the cost from x_k on. Notice that $\widehat{d_k}$ depends solely on the state x_k, since x_{k+1} depends on x_k, so we should actually write $\widehat{d_k}(x_k)$. The minimum in the expression is the optimal cost from x_k on. It is

$$\widehat{C}_k(x_k) = N_k^2 \widehat{d}_k(x_k)^2 + \widehat{C}_{k+1}(x_k + \delta_k - \widehat{d}_k(x_k)). \tag{1.16}$$

So, knowing $\widehat{C}_{k+1}$, one can determine the optimal local control $\widehat{d}_k$, and the cost information needed for the step starting at x_{k-1}. This is the minimal sufficient information needed at step k: $\widehat{C}_{k+1}$, which has to be determined as a function of x_{k+1} by a backward recursion from the end point.

The key to dynamic optimization is therefore the recursive determination of the optimal cost $\widehat{C_k}(x_k)$ to reach the destination after having reached the state x_k and this to be done with a backward recursion, for all relevant k.

Let n be the index of the last stage for more generality (in this case, $n = 3$); then the dynamic programming equation (or Bellman equation) starts at n with $\widehat{C}_{n+1}(x_{n+1}) = M^2 x_{n+1}^2$, the cost of the deviation from the final goal, and then recurses back to $k = 0$, producing a control that is solely dependent on the x_k reached at each stage k, *provided the physical model does not change from the original assumptions*. If, after reaching x_k, one suddenly realizes that the model is not any more valid, then one would have to redo the backward recursive calculation using a new model, but of course only up to stage k, and derive a new control law, valid from stage k on.

Conclusion: for the optimization at stage k, what you need is

1. the optimal cost $\widehat{C}_{k+1}(x_{k+1})$ for each relevant x_{k+1}, and
2. the *cost model* at stage k, which is a function (still to be determined) of the effect produced by the choice of the optimal local control d_k and the optimal cost $\widehat{C}_{k+1}(x_{k+1})$ you will incur when transiting from x_{k+1} to x_{n+1}. Let us now see how this works out for our case.

The Local Cost Model

The main difficulty in dynamic programming is finding an expression for the cost $\widehat{C}_k(x_k)$, and this for each k. It turns out that, in the case of a linear or affine state cost model and a quadratic cost function, there is a simple solution to this problem. Concerning the optimal cost $\widehat{C}_k(x_k)$ for any k, we observe that

1. the cost is minimally zero when $x_k = -\delta_{t,k}$, where $\delta_{t,k} = \sum_{k=i}^{n} \delta_i$, because in that case we reach the ideal destination $x_{n+1} = 0$ with zero effort (all $d_{i:} = 0$, for $i \geq k$), and
2. the optimal cost expression in function of any x_i will likely be a quadratic expression (we shall prove this hypothesis recursively), has to be positive for all values of x_i, and has the correct minimum value zero when no cost is incurred. Assuming all this, the optimal cost then necessarily has the form

$$\widehat{C}_i(x_i) = Y_{i-1}^2 (x_i + \delta_{t,i})^2, \tag{1.17}$$

with Y_{i-1} being a new coefficient to be determined recursively (the choice for the index $i-1$ in Y_{i-1} instead of i is historical and motivated by the position of Y_i in the local cost model – see below). The proposed choice is quadratic in x_i and is zero for $x_i = -\delta_{t,i}$ as is satisfied by Eq. (1.17).

With this hypothesis, the (Bellman) *dynamic programming* equation at stage k becomes

$$\widehat{d}(x_k) = \operatorname{argmin}_{d_k} \left[N_k^2 d_k^2 + Y_k^2 (x_k + \delta_{t,k} - d_k)^2 \right]. \tag{1.18}$$

Noticing that $x_{k+1} + \delta_{t,k+1} = x_k + \delta_k - d_k + \delta_{t,k+1} = x_k + \delta_{t,k} - d_k$, where all δ_t are the properties of the river and hence known a priori, and introducing the optimal $\widehat{d}(x_k)$ found in the local cost expression, we should find

$$\widehat{C}_k(x_k) \mathrel{!=!} Y_{k-1}^2 (x_k + \delta_{t,k})^2 \tag{1.19}$$

for a new Y_{k-1}.

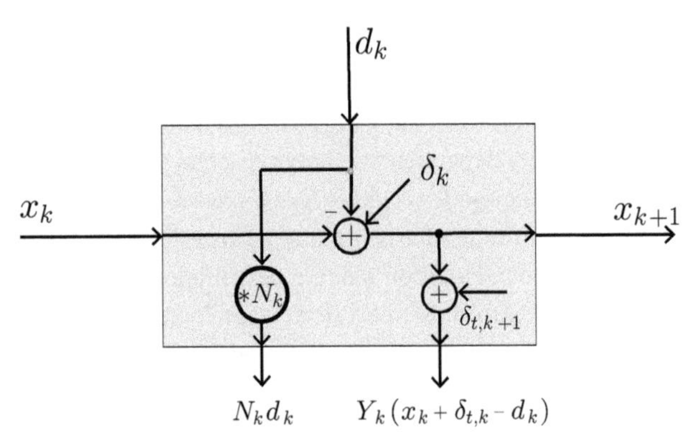

Figure 1.3 The local cost model for our rowing situation: the total cost starting with x_k is the quadratic norm of the output vector in the local model.

Equation (1.18) defines the optimization, and Eq. (1.19) says that the *expression* we guessed for $\widehat{C}_{k+1}(x_{k+1})$ is reproduced for any $\widehat{C}_k(x_k)$, thereby determining Y_{k-1} recursively (still to be proven: see the following remark).

Remark: The introduction of a recursive expression for $\widehat{C}_k$ is necessary. One cannot just add local quadratic costs, because it is not true that $(\sum(\delta_k - d_k))^2 = \sum(\delta_k - d_k)^2$! This illustrates the difficulty of general dynamic programming: one either has to guess the form of the recursive optimal cost function somehow, or else figure out some other method to find $\widehat{C}_k$ recursively in a meaningful way. In the quadratic cost case, as defined for an affine or linear model, it is easy to make the guess. One then shows correctness by recursive verification. This does not work for general norms, although even in such cases, dynamic programming remains interesting, at the cost of a more complex optimization strategy.

The local cost model at stage k, using the proposed cost expression, is shown in Figure 1.3. This cost model is affine, with steering (input) vector d_k and the cost vector expressed in terms of the present state x_k, the control d_k, and the known quantities, is then

$$\begin{bmatrix} Y_k(x_{k+1} + \delta_{t,k+1}) \\ N_k d_k \end{bmatrix} = \begin{bmatrix} Y_k(x_k + \delta_{t,k}) \\ 0 \end{bmatrix} + \begin{bmatrix} -Y_k \\ N_k \end{bmatrix} d_k. \tag{1.20}$$

Using the Moore–Penrose inverse, it follows that

$$\widehat{d}_k = -\begin{bmatrix} -Y_k \\ N_k \end{bmatrix}^{\dagger} \begin{bmatrix} Y_k(x_k + \delta_{t,k}) \\ 0 \end{bmatrix} = \frac{Y_k^2}{N_k^2 + Y_k^2}(x_k + \delta_{t,k}). \tag{1.21}$$

This is an affine control, partly proportional to x_k, with a constant *plus* an a priori known driving term $\delta_{t,k}$. For the cost, we find, after a small calculation,

$$\widehat{C}_k(x_k) = \frac{N_k^2 Y_k^2}{N_k^2 + Y_k^2}(x_k + \delta_{t,k})^2, \tag{1.22}$$

thereby proving the recursive hypothesis, with $Y_{k-1} = \frac{N_k Y_k}{\sqrt{N_k^2+Y_k^2}}$. Of course, the cost at n (which initializes the recursion) is simply $Y_n^2 x_{n+1}^2 = M^2 x_{n+1}^2$, and it cannot be optimized, since the other shore has been reached.

The (Lower) QR Way

A slightly different viewpoint, which will prove very effective in larger problems, works with orthogonalization and is, in the present case, particularly simple. For the cost vector (let us call it y_k, with the quadratic cost $y_k' y_k$), we have

$$y_k := \begin{bmatrix} Y_k(x_{k+1} + \delta_{t,k+1}) \\ N_k d_k \end{bmatrix} = \begin{bmatrix} Y_k & -Y_k \\ 0 & N_k \end{bmatrix} \begin{bmatrix} x_k + \delta_{t,k} \\ d_k \end{bmatrix}, \tag{1.23}$$

where d_k is the driving term. A lower QR factorization on the "system matrix" produces

$$\begin{bmatrix} Y_k & -Y_k \\ 0 & N_k \end{bmatrix} = Q_k \begin{bmatrix} \frac{Y_k N_k}{\sqrt{Y_k^2+N_k^2}} & 0 \\ \frac{-Y_k^2}{\sqrt{Y_k^2+N_k^2}} & \sqrt{Y_k^2 + N_k^2} \end{bmatrix}, \tag{1.24}$$

with $Q_k = \frac{1}{\sqrt{Y_k^2+N_k^2}} \begin{bmatrix} N_k & -Y_k \\ Y_k & N_k \end{bmatrix}$ being a single rotation in this case. (The lower QR factorization first starts with compressing the last column downward, in this case rotating $\begin{bmatrix} -Y_k \\ N_k \end{bmatrix}$ to $\begin{bmatrix} 0 \\ \sqrt{Y_k^2 + N_k^2} \end{bmatrix}$, and next moves to what remains of the next to last column, which in this case does not need any further compression as it is already a single scalar.)

Surprisingly perhaps, *the R-factor contains the result directly*! To see this, let $c_k := Q_k' y_k$ be the rotated cost vector over Q_k'; then $c_k' c_k = y_k' y_k$ (i.e., the quadratic cost is preserved after rotation), and

$$\begin{bmatrix} c_{k,1} \\ c_{k,2} \end{bmatrix} = \begin{bmatrix} \frac{Y_k N_k}{\sqrt{Y_k^2+N_k^2}} & 0 \\ \frac{-Y_k^2}{\sqrt{Y_k^2+N_k^2}} & \sqrt{Y_k^2 + N_k^2} \end{bmatrix} \begin{bmatrix} x_k + \delta_{t,k} \\ d_k \end{bmatrix}, \tag{1.25}$$

while the quadratic cost is now written as $c_{k,1}^2 + c_{k,2}^2$. Remembering that x_k is given at the beginning of each recursion, we see that $c_{k,1}$ is fixed, and the minimal cost is obtained by making $c_{k,2} = 0$, which as an optimal driving input gives

$$\widehat{d}_k = \frac{Y_k^2}{Y_k^2 + N_k^2}(x_k + \delta_{t,k}), \tag{1.26}$$

and the corresponding cost is $\widehat{C}_k(x_k) = c_{k,1}^2 = \frac{Y_k^2 N_k^2}{Y_k^2+N_k^2}(x_k + \delta_{t,k})^2$ as announced and as derived before.

Remarks

- The solution presented in Eq. (1.26) is a *feedback law*: it gives the required control in function of the state reached; see Fig. 1.4.
- The simplicity of the recursive solution should be obvious, but it requires some additional modeling effort to obtain it. It also has the great advantage that the optimization criteria may be modified adaptively as one proceeds (e.g., when one gets better estimates of the properties of the river, the boat, and/or of future costs). This is an issue we shall not address here but which may come up when we discuss optimization problems in later chapters.
- If one has to reach the exact destination ($x_{n+1} = 0$), then one may let M tend to infinity, with some care. This will not lead to much change in the derivation, except at the last stretch. It is a good exercise to perform!
- From the control formula, it is clear that the effort to be made at any stage k has N_k^2 in the denominator, meaning that the stronger the current, the less one should row against it. The wisdom to profit, mostly from the least resistance or the low hanging fruit, turns out to be a pretty general result. The general physical principle of "least action" has some of this flavor as well.
- It should be clear that the Moore–Penrose inverse method only works on linear or affine models and costs of quadratic type. With a different overall cost function or nonlinear model, the whole procedure becomes considerably more complex, but the principle of dynamic programming will still apply, at the cost of added numerical complexity.
- Our example was restricted to $n = 3$, but the treatment is sufficiently general (provided the model is valid, of course) and could also be utilized for a continuous-time

Figure 1.4 The optimal control feedback law.

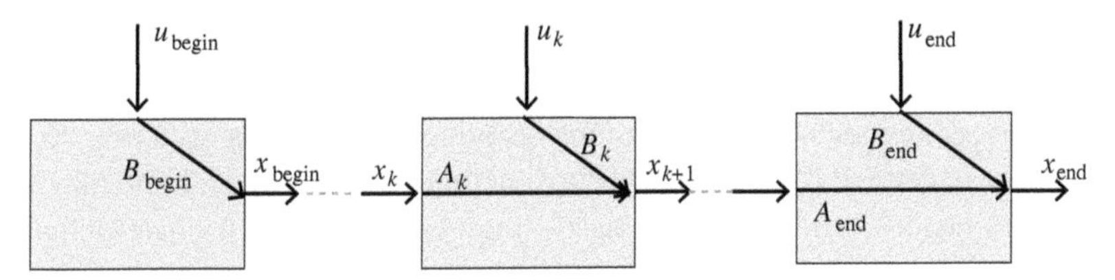

Figure 1.5 Signal flow model of the state evolution considered in the linear tracking model. At index k, $x_{k+1} = A_k x_k + B_k u_k$.

situation after discretization or, conversely, to derive the continuous-time treatment from the discretized, which is what is often done in the literature.

- It is easy to see that the recursive solution gives the same result as the global one. A good exercise!

1.3 The General Linear Model for Optimal Quadratic Control

Given the insights we have developed in our toy example, it is not difficult to derive the matrix algebra for the classical problem of *optimal quadratic state control* of a *linear dynamical system*, as originally proposed by Bellman [13], usually done for the continuous-time case, but done here for the discrete-time case. The goal is to show (1) how the conversion from a physical system environment to matrix algebra is done, and (2) how an upper or lower QR factorization greatly simplifies the procedure, thanks to orthogonalization. It also provides a good example of an important general method we shall develop later: inner–outer factorization.

We assume that the dynamics of our discrete-time system are represented by a real *state vector* x_k of dimension η_k, which is a function of a time index k (an integer) and characterizes the state of the system at the time instant k. Let the system be driven by an input u_k at index k, where u_k is a vector of dimension m_k, and let us assume (inspired by the example) that the evolution of the state of the system from the time point k to the time point $k+1$ is given by the linear (time-variant) state equations

$$x_{k+1} = A_k x_k + B_k u_k, \tag{1.27}$$

in which A_k is an $\eta_{k+1} \times \eta_k$ matrix and B_k is an $\eta_{k+1} \times m_k$ matrix. A simple *signal flow diagram* as shown in Fig. 1.5 is often used to represent such a model. (In Appendix, we present the "Functional Data Model" we typically use to represent computations graphically. It allows for an easy transformation of numerical computations into a hardware architecture or a computer program.)

Let us assume for illustrative purposes that our system starts at index 0 with a given initial state x_0 and that we wish to control the evolution of our system in the interval $[0, n+1]$, where n is the index of the final stage, so as to minimize a positive quadratic cost function keeping states and inputs small, by putting quadratically increasing costs on their values.

For simplicity and in order to represent a quadratic cost adequately, we write the cost of a state x_k as $x_k' M_k' M_k x_k$, in which M_k is a matrix of appropriate dimensions and the prime indicates real or complex conjugation, making x_k' a row vector. (Often the cost is defined by a strictly positive definite matrix C_k, and one may take $M_k = C_k^{1/2}$ or just a Cholesky factor, but C_k may be nonstrictly positive definite and M_k may be a rectangular factor accordingly.) Similarly, we write the cost of an input u_k as $u_k' N_k' N_k u_k$, and we assume N_k to be a square, nonsingular matrix (so that arbitrarily large inputs will not be possible).

Dynamic Programming

The problem of optimal quadratic control in the given set up is solved by *dynamic programming*. The basic idea of dynamic programming is: *once the system reaches the state x_k at the time index k, the trajectory has to be optimal from there on*; for, if that were not the case, there would be a better overall trajectory just by replacing the segment from the current position k to the end with a less costly path. This means, in particular, that all optimal inputs and costs, when started from x_k at the time index k, are only dependent on x_k and not directly on previous states; previous states will influence x_k, but can only influence later quantities via x_k, given the model, of course. To put it differently, all the future inputs have to be chosen so as to optimize the trajectory from the index k on and hence are only dependent on x_k and the model from that point in time on. Let us now prove the following recursive hypothesis for the quadratic cost case and linear system model:

> *The total optimal cost starting from any x_k in the remaining interval $[k, n+1]$ is given by a quadratic form $\widehat{C}_k(x_k) := x_k' Y_{k-1}' Y_{k-1} x_k$, in which Y_{k-1} is a (to be computed recursively) $\eta_k \times \eta_k$ matrix.*

The recursive hypothesis will be verified if (1) it is valid at the end point $k = n+1$ and (2) when valid for $k+1$ it is valid for k:

1. The first statement is true because at the end point $n+1$ the cost is $x_{n+1}' M_{n+1}' M_{n+1} x_{n+1}$, so $Y_n = M_{n+1}$.
2. The second statement is to be derived now.

Key to the derivation is the determination of the *cost model*: put the "square roots" $Y_k x_{k+1}$, $M_k x_k$ and $N_k u_k$ of the cost terms as outputs in the cost model, so that the cost equals their total square norm and the model itself is linear. Next, assume recursively that the optimal cost, from $k+1$ on, and for any x_{k+1}, is given by $x_{k+1}' Y_k' Y_k x_{k+1}$, the cost model at k (shown in Fig. 1.6) gives, after the multiplication of the first block row with Y_k,

$$y_k = \begin{bmatrix} Y_k x_{k+1} \\ \hline M_k x_k \\ N_k u_k \end{bmatrix} = \begin{bmatrix} Y_k A_k \\ \hline M_k \\ 0 \end{bmatrix} x_k + \begin{bmatrix} Y_k B_k \\ \hline 0 \\ N_k \end{bmatrix} u_k = \left[\begin{array}{c|c} Y_k A_k & Y_k B_k \\ \hline M_k & 0 \\ 0 & N_k \end{array}\right] \begin{bmatrix} x_k \\ u_k \end{bmatrix}, \quad (1.28)$$

and the optimization problem specializes to the following: *find the u_k that minimizes the cost $y_k' y_k = x_{k+1}' Y_k' Y_k x_{k+1} + x_k' M_k' M_k x_k + u_k' N_k' N_k u_k$, for given x_k.*

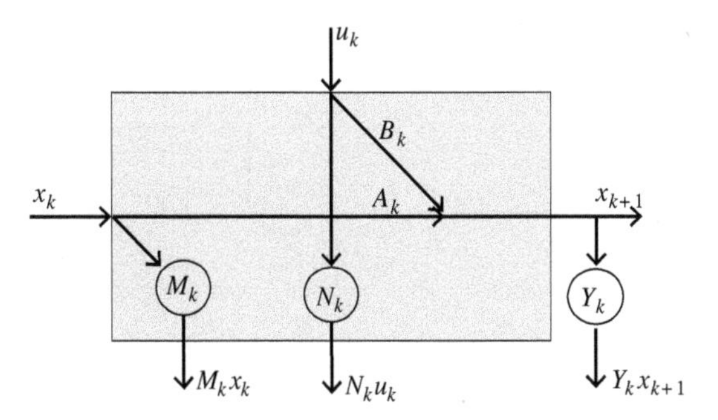

Figure 1.6 The full local cost model for optimal quadratic control, including cost outputs.

Notice: In this phase of the recursion, x_k is the only remaining "variable," whereas the others, namely u_k and x_{k+1}, will be "optimized out" as a function of x_k.

(Lower) QR Orthogonalization

The propagation of the cost function is now easily found using orthogonalization of the system matrix. Following the tradition in the literature, we use a lower QR factorization of the "system matrix" in Eq. (1.28), to produce the block decomposition

$$\left[\begin{array}{c|c} Y_k A_k & Y_k B_k \\ \hline M_k & 0 \\ 0 & N_k \end{array}\right] = Q_k \left[\begin{array}{c|c} 0 & 0 \\ \hline Y_{k-1} & 0 \\ C_{o,k} & D_{o,k} \end{array}\right], \tag{1.29}$$

in which

1. Q_k is an orthogonal (or, in the complex case, unitary) matrix that produces the block staircase in the right factor;
2. $D_{o,k}$ is square invertible, thanks to the assumption that N_k is square invertible (it is not hard to show this);
3. Y_{k-1} is, by construction, in the "row echelon form": it has independent rows and hence is right invertible (flat), which makes it of minimal dimension;
4. all $Q_k, Y_{k-1}, C_{o,k}$ and $D_{o,k}$ are new matrices computed from the entries in the system matrix and whose meaning will soon be clear.

The orthogonal transformation matrix Q_k' transforms the cost vector y_k to $c_k = Q_k' y_k$ in three orthogonal components, each with its own significance:

$$\begin{bmatrix} c_{k,1} \\ c_{k,2} \\ c_{k,3} \end{bmatrix} := Q_k' \begin{bmatrix} Y_k x_{k+1} \\ M_k x_k \\ N_k u_k \end{bmatrix} = \left[\begin{array}{c|c} 0 & 0 \\ \hline Y_{k-1} & 0 \\ C_{o,k} & D_{o,k} \end{array}\right] \begin{bmatrix} x_k \\ u_k \end{bmatrix} = \left[\begin{array}{c} 0 \\ \hline Y_{k-1} x_k \\ C_{o,k} x_k + D_{o,k} u_k \end{array}\right] \tag{1.30}$$

valid for any input u_k. It immediately follows from $Q_k Q_k' = I$ that the quadratic cost $y_k' y_k = c_k' c_k$ generated by x_k and u_k can now be expressed as

$$y_k' y_k = \|Y_{k-1} x_k\|^2 + \|(C_{o,k} x_k + D_{o,k} u_k)\|^2, \tag{1.31}$$

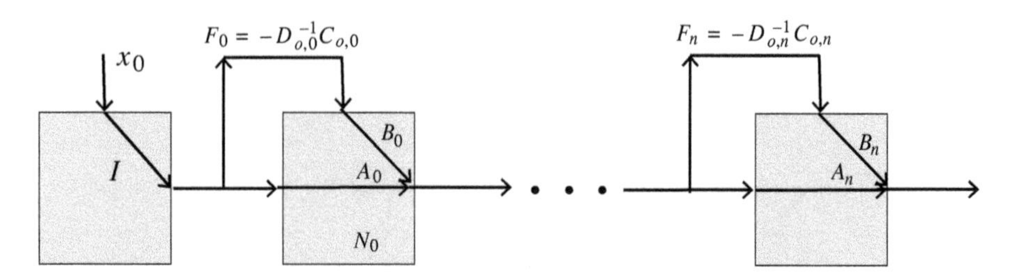

Figure 1.7 The optimal dynamic (i.e., state-dependent) Bellman controller.

in which $D_{o,k}$ is square invertible. It follows that, given x_k, the cost is minimized by choosing the optimal

$$\widehat{u}_k := -D_{o,k}^{-1}C_{o,k}x_k, \tag{1.32}$$

whereupon the minimal cost follows:

$$\widehat{C}_k(x_k) = \|Y_{k-1}x_k\|^2 = x_k'Y_{k-1}'Y_{k-1}x_k, \tag{1.33}$$

which had to be shown. □

The computation of new matrices from a given matrix, as is done in Eq. (1.29), has been termed *array processing* in the literature, because it is a computation on a given matrix that generates new matrices as a result. It is sometimes called a "square-root method" because it works directly on the data given rather than on the quadratic form that represents their cost.

In conclusion, we can state that a simple QR factorization gives the complete recursive solution of the quadratic tracking problem, whereby the orthogonal Q-factor filters the input data to produce both the optimal control and the new global cost function, all in function of the actual state x_k. Formula (1.32) shows that the optimal control consists of a simple state feedback. This realization is shown in Fig. 1.7.

1.4 A Question to Be Researched

Thinking about optimizing the behavior of a system evolving in time, under what conditions would dynamic programming be possible? Or, expressed negatively, when is dynamic programming definitely not possible?

1.5 Notes

1. One of the original motivations for considering a quadratic optimization model was the Apollo mission: How to get a rocket to the moon with minimal expenditure of fuel? It takes, of course, some work to reduce the Apollo mission problem to the simple model presented by discrete-time linear model in this book. One must first find a potentially optimal trajectory that respects gravity laws and that requires a minimal nominal amount of fuel to reach the goal within a domain of

feasibility defined by various limits in time and fuel needed. Once settled on such an optimal trajectory, the control problem is to keep the rocket close to the optimal trajectory with minimal expenditure of fuel, even though various inaccuracies may have occurred producing (stochastic) deviations. This is achieved by controlling the *deviation* of the state from the desired optimal and using the control to bring the rocket closer to the nominal optimal trajectory without spending too much fuel. When sufficiently small, the deviation of the state will satisfy a linear differential model derived from the optimal trajectory. After discretization, a model of the type given above is obtained. We leave the study of discretizations to the appropriate literature.

2. Dynamic programming is only one example of the use of system theory. There are many more examples, several of which will be discussed in the following chapters, and many more have to be relegated to further treatments. State estimation (Kalman filtering), control theory, system modeling and model reduction, approximation and interpolation of matrices, efficient computations with some types of structured matrix, and data filter design all rely on system-theoretic concepts. The value of dynamic programming as a prime example is the intimate connection it exhibits between the dynamic model and the recursion that leads to the optimal control. Further instructive examples of optimization problems on systems can be found in the book of Luenberger [50]. A classical textbook on linear optimization is due to Anderson and Moore [3].
3. There is a host of methods one can use to orthogonalize a set of commensurable vectors, to wit:

   1. global multidimensional rotations (to be preferred) or Householder transformations;
   2. Gram–Schmidt orthogonalization;
   3. bi-orthogonalization using hyperbolic transformations [18, 22].

   These methods are discussed in further chapters, when they come up, and in many textbooks on linear algebra.
4. The functional representation and data model used in this chapter, and described in the Appendix, was proposed and analyzed in [6]. It allows for an unbiased transformation of mathematical operations to a computer architecture at the functional and data transfer levels. It hides organizational details such as the conditional sequencing of functions, the partitioning, and storage and transfer of data in such a way that this information can easily be generated from the data model when architectural decisions like the localization of data in memories and the assignment of functions to processors has been made (such design phases will not concern us in this book). In particular, the model easily accommodates hierarchical representations and parallel processing. In this book, we shall only worry about practical implementation aspects as far as numerical properties (numerical accuracy and complexity) are concerned.
5. It is not easy to find a simple direct example for a deterministic quadratic optimization problem on a linear model. Most such problems are of the type "tracking a

nonlinear trajectory to counteract stochastic disturbances." In most practical examples of tracking with a linear model, the model has an extra disturbing noise term, which has to be controlled. Using the time-variant linear model found by differentiating a nonlinear trajectory, adding noise terms and using average quantities, one can, without too much effort, extend the model used in this chapter to handle the more general case.

6. Quite a few famous algorithms that go by specific names are, in fact, examples of dynamic programming. Take for example the famous Viterbi algorithm to decode a bitstream that has been coded with a convolutional code. This is an excellent example of the use of dynamic programming in a p-adic number system.

2 Dynamical Systems

What is a *system*? What is a *dynamical system*? Systems are characterized by a few central notions, such as their *state* and their defining behavior, and by some derived notions, such as *reachability* and *observability*. These notions pop up in many fields, so it is important to understand them in terms of their nontechnical terms. This chapter therefore introduces what people call a narrative that aims at describing the central ideas. In the remainder of the book, the ideas presented here are made mathematically precise. It turns out that a sharp understanding of just the notion of *state* suffices to develop most if not the whole mathematical machinery needed to solve the main engineering problems related to systems and their dynamics.

2.1 What Is a (Dynamical) System?

The *New Oxford American Dictionary* [57] defines the concept *system* as shown in Fig. 2.1.

We adopt the following definition:

Definition 2.1 A system is an assembly of interconnected and interacting entities that together achieve a behavior.

system |'sistem|

noun

1 a set of things working together as parts of a mechanism or an interconnecting network: *the state railroad system* | *fluid is pushed through a system of pipes or channels.*

- *Physiology* a set of organs in the body with a common structure or function: *the digestive system.*
- the human or animal body as a whole: *you need to get the cholesterol out of your system.*
- *Computing* a group of related hardware units or programs or both, especially when dedicated to a single application.
- *Geology* (in chronostratigraphy) a major range of strata that corresponds to a period in time, subdivided into series.
- *Astronomy* a group of celestial objects connected by their mutual attractive forces, especially moving in orbits about a center: *the system of bright stars known as the Gould Belt.*
- short for crystal system.

2 a set of principles or procedures according to which something is done; an organized scheme or method: *a multiparty system of government* | *the public school system.*

- a set of rules used in measurement or classification: *the metric system.*
- orderliness; method: *there was no system at all in the company.*
- a method of choosing one's procedure in gambling.

3 (the system) the prevailing political or social order, especially when regarded as oppressive and intransigent; *don't try bucking the system.*

4 *Music* a set of staves in a musical score joined by a brace.

Figure 2.1 The definition of *system* according to the *New Oxford American Dictionary.*

Hence, a system consists of

1. a characterizing *behavior*, giving each system instance an observable identity;
2. *entities*, with specific contributing functions;
3. *interconnections* that allow the entities to interact; and
4. *interactions* that realize the behavior.

Such a definition is very general and hence may include many types of systems. A discipline called *System Theory* would then attempt to find some commonality between all these types or at least discover classes of systems that can be described concretely. As it would be impossible to have one-size-fits-all descriptions, engineers and scientists who want to understand how a system functions, how to build one or how to control one, wisely restrict themselves to classes on which they have some grip, that is, types of systems they can manufacture and describe mathematically.

The combination of mathematical description and ability to realize is what allows engineers to *design* a system. As the quotation from the *New Oxford American Dictionary* in Fig. 2.1 shows, there are many types of extremely interesting systems that we cannot manufacture, but can be used in combination with man-made subsystems. A rocket that travels to Mars is an example of a man-made system (the rocket) that uses the planetary system to reach its goal. Luckily, the planetary system has a precise mathematical description, so that the rocket can be steered accurately to its destination.

Many natural systems have mathematical descriptions as well, but many are so complex or complicated that they defy unified mathematical treatment, for example, a cellular organism or the human brain. *Identifying* or *approximating* systems from their observed behavior is then a major issue.

In this book, we shall always work with mathematical descriptions or models of the systems we consider. It is the task of other disciplines, such as physics, biology or engineering, to develop adequate mathematical models for the systems they distinguish, but system theory can be helpful in the derivation of the models. In particular, we shall consider how to use measurements of a system to find potential models that are consistent with the measurements.

Most systems are not static but evolve as time progresses. So, often one shall be interested not only in how a system is put together (its structure) but also in how it evolves (its *dynamics*). When system evolution is the prime interest, one would call it a *dynamical system* instead of a purely static system like, for example, the architecture of a building. Our main interest in this book is dynamical systems.

The very first dynamical system ever described mathematically is the system of planets revolving around the sun. Isaac Newton discovered the basic dynamics of this system. In Newton's system, the sun and the planets are each given one static variable, their mass, and six characterizing dynamic variables, including three position parameters $\{x, y, z\}$ and three velocities $\{v_x, v_y, v_z\}$, of the respective center of gravity. The sun and the planets interact (according to Newton) with the gravitational forces that cause their velocities to change continuously, which in turn cause changes in the positions of the planetary bodies. To express this, Newton discovered the notion of "derivative" $v_x = \mathrm{d}x/\mathrm{d}t$, indicating the velocity of a body in a given direction as the time derivative of its position, while the time derivative of the velocity of each body is determined by the forces acting on it, which depend exclusively on the positions of the various masses in the system. It follows that the future evolution of the overall system is fully determined by initial positions and velocities of all the participating masses, hence the following definition.

Definition 2.2 The variables that characterize the dynamical entities of a system are called its state variables.

2.2 The State

Given the definition of "state" as the collection of dynamic variables that characterize the time evolution of a system, the question that arises immediately is: How do we know that a given parameter of a system we encounter is indeed a state variable? Over the years, the following insight has crystallized to answer this question (see [49]):

A state at a given time t is sufficient information on the system at t to determine its future evolution, given future inputs.

This means, among other things, that for a system with a finite-state vector, its whole past evolution up to time t is, as far as the future is concerned, fully accounted for in its state vector at time t. One does not need any additional information about the past of the system to assess its future evolution, given future inputs, or, to put it negatively, the system "forgets" everything from its past except what is contained in its state. This is schematically illustrated in Fig. 2.2.

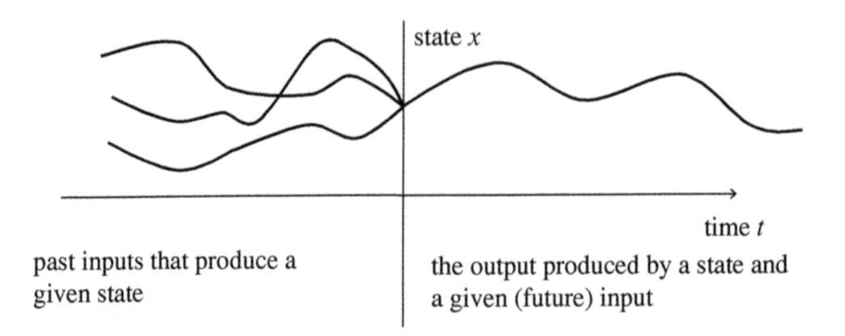

Figure 2.2 The state as the link between past and future.

For example, in Newton's planetary system, the state of a single planet is a time-dependent vector with six components

$$\mathrm{col}[x(t), y(t), z(t), v_x(t), v_y(t), v_z(t)],$$

where the state of the complete system is a column stack of the states of the member planets and the sun.

Newton's model of the planetary world is remarkable for both what it includes and what it does not include. Ancient Greek and Roman scientists and engineers (such as Aristotle and Archimedes) thought that "force changes position," and hence, they only included position (and not velocity) in their model. Thanks to improved mechanics and better insights into the planetary movements, Newton got the idea that it is the force on a body that indirectly changes its position. A force produces an acceleration instead. On the other hand, how the planets got to their position and velocities at a given time appears irrelevant for their future evolution: knowledge of these quantities suffices to determine their future evolution. We now know that this view is somewhat accurate, but not quite fully. But for the practical reason of getting a rocket to the moon, it is sufficient enough, although other effects, not accounted for by Newton, also act, such as relativistic effects and the influence of solar winds, which space engineers have to take into account.

Consider, as another example, the state of your PC. Surely, all the data that are stored in several memories (disk, cache and register files) belong to the state, but also to the state of various switches and controllers. To avoid ambiguities, your PC is equipped with a clock, and there will only be a well-defined state at some given point in the clock cycle, for example, when the clock comes up (which we may call a "clock tick"). As soon as the clock ticks, states are allowed to change, but this may be a pretty chaotic process, with races, spikes and disturbing influences, which the computer designer has to keep under control carefully, letting the system evolve to a new stable state before the next clock ticks. We, users, have to assume that these designers did a good job and that the next state is a well-defined function of the previous state and the inputs given to the system input ports. From the point of view of the user, your PC is a *discrete-time* machine that moves from one well-defined state to another at the tick of the clock, while for the designer, it is a fairly random machine, in which signals are propagated through gates and connections between clock ticks.

The given definition of "state" leaves a somewhat arbitrary choice dangling: Where is the "cut" between the past and the future with respect to a time point t? The traditional choice, in the wake of [49], is to let the past run to just before t (often written as t^-), and to let the "future" start with the present t (this definition is consistent with how your PC handles its state). In some fields that use system theory, most notably stochastic systems, often the opposite choice is made, and sometimes one sees mixed choices for inputs and outputs. The problem arises in the case of discontinuities, and it has to be considered carefully when it occurs.

Basic System Characteristics

State Minimality

In the state description of Section 2.2, the state is described as "sufficient" information. *Minimality* would require the information to be "necessary" as well. A computer memory is usually filled with all sorts of information that is not relevant to the problem at hand. A state is called *minimal* when no data in the state vector at any time t can be left out without affecting some future outputs, given appropriate future inputs. A more precise formulation is obtained as follows: two states at any time t may be considered equivalent, if the outputs produced after t are the same for both irrespective of which input is used from t onward; states are produced by past inputs up to and not including t, and we may say that two such inputs are equivalent in turn if they produce the same output irrespective of which input is used from t on. This leads to the notion of *Nerode equivalence*:

We say that two past inputs up to time t^- are *Nerode equivalent*, if all (future) inputs started at time t produce the same output from t on when applied to the system following (concatenated with) each of these two inputs.

It should be immediately noted from the definition that the notion of minimality depends on both the output *and* the input equations of the system. Whether a state is necessary depends on the future output generated by it together with future inputs. It may be that some internal parts of the system will never be visible from the observation of the output, in which case, they could be deemed superfluous.

Hence, the definition of state minimality is largely dependent on inputs and outputs of the system. This may be seen as a problem. A completely autonomous system has neither inputs nor outputs. Hence, a minimal description for it would simply be empty. This unpleasant situation can be remedied by the convention that in the case of an autonomous system, the state can be controlled by adequate outside influences, and one can also observe the state directly, so that the output equation simply becomes $\text{output}(t) = \text{state}(t)$. The definition given makes minimality "observer dependent" (the output is what can be observed) and "controller dependent" (perhaps imaginary agents that can influence). We shall build on this idea soon when we discuss observability and reachability.

An important further consequence of the notion of a minimal state is that its quantities (components of the state vector) are *algebraically independent*. They can be assigned values independently from each other (of course, within the number system used; in our case, they will be either real or complex numbers). Here also, there is a potential unwarranted generalization. It is quite possible that independent state variables may only take limited sets of values. That is, for example, the case in a computer, which does not allow any size of number. But also real-life systems are limited by ranges of relevant variables. The mathematical formalism often ignores these contingencies, hoping that good practice will be able to handle them when necessary.

Reachability

Returning to the characterization of a nonminimal state, it may be possible that some states at some time point t cannot be *reached*, that is, there is no past input up to time t that is able to produce that state. States that can be generated by at least one past input function are called *reachable*. It is not necessary that a state can be accessed directly to be reachable. For example, the position of a point mass cannot be accessed directly (say in a planetary system), but it can always be modified by having a force acting on it whereby the force influences the velocity that, in turn, influences the position.

Observability

It may also happen that some states at a given time point t will always produce equal output responses in the future, no matter what future input is applied (in other words, they remain undistinguishable, no matter what future input is applied). The distinction between such states is called *unobservable*: whether the system is in one or any of those states at time t cannot be determined by observing the evolution of the output from time t onward, using any future inputs. We call a state *observable* when its response to at least one future input is unique. Every *equivalence class* of states whose distinction is unobservable can be represented by just one state. In this case, it is also not necessary that states can be observed directly. Only necessary for observability is that their response to future inputs can be distinguished.

The Behavior of a System

We use the term "behavior of a person" to indicate how a person's actions and reactions appear to us, not how the person actually presents them physically. The notion involves both the person and how the environment perceives that person. In the case of a system, it would be *how the world outside the system perceives the system over time*. As the external world has access to the system solely via its inputs and outputs, one could say that the system behavior is "how the system attaches outputs to inputs over all time." In mathematical terms, this would be formulated as a *relation* between all inputs over all time and corresponding outputs over all time.

The point at issue could be whether a system behaves deterministically: one can conceive cases whereby hidden actors within cause the system to move one way or the other independently from the actual inputs, so that more outputs may correspond to a single input. Needless to say, such a behavior might greatly complicate formal descriptions. In the section on estimation theory, we shall allow such interferences but shall be careful to assign additional inputs to them, so that the erratic behavior is taken out of the the internal description of the system and put as an external influence, characterized by its statistics. That turns out to be a good strategy in many cases, at least where engineering is concerned. Therefore, many treatments of dynamical systems will assume that the system, given an initial state at some initial time t_0, defines *a map* from inputs starting at t_0 and running to some final time $t_1 > t_0$ to (1) the state at t_1 and (2) the output from t_0 to t_1 (a "map" is by definition univocal). This map is then the *functional behavior* or *transfer operator* of the system.

From the point of view of an observer of the system, the system may produce various types of behavior that are worth noting and that we would include under the term "behavior" as well. For example, the system may heat up, or it may age or it may develop some other physical characteristics that are not directly related to how it produces an output given an input, such as a change of size, color or form (these might even be taken as outputs). Certainly, the term "behavior" would always relate to something in the system that changes over time and can be observed. In engineering practice, one always selects a limited number of variables for which one wants to define the dynamics (i.e., the state variables), relegating all other (often long-term changes) to the evolution of the system's structure over time, calling it a (maybe slowly) "time-variant" system. Although one could actually assign states to such characteristics, it turns out to be more practical to leave them as system parameters that change over time – as one can see, the assignment of state variables is often, in some sense, arbitrary.

An Opposing Viewpoint

The late Jan Willems defines the *behavior* of a system differently as: *a mathematical model is a subset of a set of a priori possibilities* [75]. This subset is then defined as the behavior of the model. For a dynamical system, the behavior consists of the time trajectories that the model declares possible.

Willems attaches the notion "behavior" to an intrinsic mathematical model of a system. He thereby excludes inputs and outputs, in order to keep the definition untainted from outside influences. Such a definition has advantages and disadvantages. An important advantage is a precise definition of scope. A disadvantage is that the definition is "mathematical" and hence differs from the normal, nonmathematical usage of "behavior." The term "behavior" is also used in computer science with a different meaning, namely "how the system interacts with the outside world," not how it achieves its actions internally. We have defined behavior in a more common sense and have included inputs and outputs in the definition. The disadvantage of our definition is that it is more remote from the problem of mathematically modeling systems in a physically sound way than Willems's definition, but the advantage is that it meets the needs of signal processing, control theory and design engineering, where behavior is indeed seen as a goal to be achieved (namely a desired input–output behavior) rather than the characterization of how a given mathematical system model is thought to evolve internally. To conform with Willems's terminology, we could call our usage *input–output behavior*. Nonetheless, the methods we use on system models come very close to the treatment Willems gives on the basis of his notion of behavior. As we have already shown, the two notions can be brought together by a careful definition of what is considered input and output. Fully correct mathematical modeling of a system is never possible since every model involves abstractions and approximations, and correct modeling necessitates the observation of the system, hence connections to the outside. Willems recognizes this but reserves the term "behavior" for the properties of the mathematical model, while in our usage, it refers to the relation between the observer and the system. Whatever usage is made should be made explicit from the start.

All these concepts – state, behavior, reachability and observability – play an exceedingly important role in the development of dynamical system theory. It is remarkable that these few notions and their properties provide the basis for the mathematical theory needed to solve many problems in estimation theory, control theory and numerical linear algebra.

To conclude this introductory narrative, here is a quick short list of some typical dynamical systems and their state spaces.

Mechanical system:	positions and velocities
Computer:	various types of memory
Automaton:	control states, routing states
Airplane:	position, velocity, roll, yaw and pitch angles
Process plant:	pressure, temperature, concentrations
Brains:	states of synaptic connections

2.3 Inputs and Outputs

Inputs and outputs of a system are *quantities* that vary in *time*.[1] To start with time, it is customary to assume a unique time variable – often labeled t – valid for the full system, including all state variables. The simplest distinction was already given in Chapter 1: continuous time and discrete time. Continuous time would be a real number (often called t), and it would typically run from $-\infty$ to $+\infty$, by which we mean that it has no fixed beginning and no fixed end. If a system had a fixed beginning or end, this is embedded in the doubly infinite interval. Embedding is a strategy we shall often adopt, so we shall not consider "subcases" separately. It turns out that this will not lead to major difficulties. Next, *discrete time* would be just an index (an integer) also running from $-\infty$ to $+\infty$. Here the situation is a bit more delicate: (1) one could think of these indices to represent time points that are spread regularly over the

[1] In more advanced theories, there might be "space-time." Or various parts of a system may have different times, for example, various computers with different clocks communicating with each other. There are many more possibilities, but they will not concern us here.

time axis with constant intervals (this would be the normal clocked case) or (2) they could be representative of moments in time when something happens, not necessarily at regular intervals (like what happens in a waiting line), in which case we would talk of *discrete events*. There are even situations where in one given system, several time scales are present, for example, in a sampling system where the input would be continuous time and the output would be discrete time, or in a system in which different sampling rates are present. What also often happens in practice is that the system has an overall regular clock, while many of its subsystems have their own clocks. The strategy to be followed in such cases will probably be case dependent, but an effective strategy is to make all these various timing signals dependent on one general time (continuous or discrete when possible), so that one does not loose track of how the various events relate to each other. For example, when joining sequences produced by a video and an audio decoder into one sequence for rendering, one would time stamp the sequence produced by each decoder and then join the time-stamped sequences into one consistent and linearly ordered sequence that respects the overall timing.

In this book, we restrict ourselves to discrete time, presumably equally spaced timing points, but often a discrete-event system can be accommodated with the regular discrete-time framework. Although there are relatively easy ways to connect (regular) discrete time with continuous time algebraically, allowing the transposition of important results from one domain to another, we do not consider continuous-time systems in this book.

Concerning the *values* inputs and outputs may take, here also a great variety is possible. Inputs and outputs are functions of time, the point we just discussed. The next question is: to what range do they belong? In the simplest case, they would be one-dimensional real ($\mathcal{R}$) or complex valued ($\mathcal{C}$). Let us first consider the discrete-time case. What comes to mind is how a computer operates: at regular time intervals, it takes in new data from its input devices, does some computations on it, and then outputs data to output devices, after which it renews the cycle. In each such cycle, the computer may receive data as needed from a variety of sources and output data to various data repositories. So we would typically assume that (1) the data have a vectorial character and (2) the dimension of the vectors taken in or outputted may change from one event point to the next; hence, inputs would belong to $\mathcal{R}^{m_k}$ or $\mathcal{C}^{m_k}$, where m_k is the input dimension at the index point k and the output would have the form $\mathcal{R}^{n_k}$ or $\mathcal{C}^{n_k}$, with n_k being the dimension of the output data vector at k. This means that in such systems the input is an irregular sequence of vectors that are either real or complex, and the same applies to the output. *This is the point of view that we shall mostly adopt in this book.*

For continuous-time systems, the situation is more tricky. The traditional approach is to assume that the inputs and outputs are real or complex vectors of a fixed dimension (hence $\mathcal{R}^m$ and $\mathcal{R}^n$ for some fixed positive integers m and n, respectively, in the real case), which vary in time (functions of the type $t \mapsto u(t) \in \mathcal{R}^m$). Traditionally, one also assumes the continuous-time system to have fixed input, output and state dimensions, but that leads to a massive reduction in the field of interest, often justified, leaving out interesting time-variant systems (think, e.g., of an electrical power plant to which a new load is added or from which the load is removed by switching). When one is not satisfied with the restrictions mentioned, then the system under consideration automatically becomes a combination of discrete events and continuous-time evolution between events, often called a hybrid system. Although we do not consider such a case in this book, it may be mentioned that it can fairly easily be treated with regular time-variant theory, at least from an engineering point of view. In any event, we shall often assume that a continuous-time system is in fact (or can be approximated by) a discrete-time system with a small time step or, alternatively, that it is driven by step-wise input functions, and that its output gets sampled at regular intervals, so that there is an exact, discrete-time model of its behavior.

But why should the input, output, and state values be discrete (vectorial)? It is, of course, easily conceivable to have systems whose inputs and/or outputs are themselves functions of another variable besides time, for example, a space variable. Think, for example, of a boat in the waves (waves impinging on their full length) or, similarly, a near-field antenna in an electromagnetic wave. Such cases fall outside the theory we shall consider, but they can occasionally be treated by discretizing the input and output spaces as one does when one solves partial differential equations, so the theory we shall develop will at least be applicable to the discretized case.

Finally, why should an input or output be a real or complex number? In our computer age, we could just assume them to be bits, or bytes, or to belong to some other field, as happens in coding theory. Although we shall not consider the case of such *digital systems* in this book either, let us mention that all the basic notions that we are considering do apply to them as well (e.g., reachability and observability play an important role in digital optimization and testability of digital systems; see De Micheli [51]), but that the conversion of real and complex data to digital data requires careful consideration. Many problems can be dealt with by the development of algorithms that are robust for rounding – and this is an issue that we take at heart, concentrating on orthogonal transformations.

2.4 The Evolution of the State

The state of a system is also a function of time and hence will be subject to a similar taxonomy as the inputs and outputs. So, at a given point in time, it can be a real or complex vector (i.e., a function of an index) or a function of some continuous parameter (e.g., position and velocity of a point on the rod of a cantilever). In a discrete-time system, the dimension of the state may change from one time index to the next. In a continuous-time system, it is common to fix the dimension of the state over all time, unless one considers the already mentioned combination of discrete events and continuous-time evolution between events (after all a situation that can easily occur in practice). In most of the chapters of this book, we assume a state that evolves in discrete time and whose dimension can vary from one time point to the next. This is in line with what a state is in a computer memory, evolving stepwise in time, but whereby the size of the required memory may change.

There are, of course, many cases in which the state of a system is not a finite-dimensional vector. It may also be that it is finite dimensional, but that the vector structure is not obviously appropriate for it: its natural structure could, for example, be a matrix or a tensor or some other entity with structure (a network, a tree, ...). Such additional structures are very interesting, in particular as they may lead to efficient algorithms, but their extensive study would lead us too far afield for this introductory work, although we shall consider some interesting cases of states with structure, in particular matrices or tensors. The treatment of state variables that are continuously dependent on some parameters (e.g., space) is beyond the scope of this book – suffice it to say that in many cases a careful discretization of the continuous parameter may solve the problem.

As there are many possible types of states, there shall be many possible ways in which a state can evolve. However, since our main focus is on discrete-time systems (typically mapped on a computer), we shall generally assume the state evolution to be an index-dependent map at each time index point k, which maps the state vector $x(k) \in \mathcal{R}^{\eta_k}$ and the input vector $u(k)$ to the next state $x(k+1) \in \mathcal{R}^{\eta_{k+1}}$, where $\{\eta_k\}$ is the sequence of dimensions of the state vector $x(k)$:

$$\begin{cases} x(k+1) &= f_k(x(k),u(k)), \\ y(k) &= g_k(x(k),u(k)). \end{cases} \tag{2.1}$$

Here the state *transition function* f_k is a general nonlinear function, mapping the ordered pair $(x(k),u(k))$ of dimensions $\eta_k + m_k$ to $x(k+1)$ of dimensions η_{k+1}, and similarly for g_k.

Most of the theory we are going to develop makes a sweeping further assumption, namely that the transition function is actually linear, which means that both the state evolution and the output equations are described by a set of linear difference equations:

$$\begin{cases} x(k+1) &= A_k x(k) + B_k u(k), \\ y(k) &= C_k x(k) + D_k u(k), \end{cases} \tag{2.2}$$

in which A_k, B_k, C_k and D_k are the matrices of appropriate dimensions, which may vary with the index k. There are several motivations for this choice:

1. linear systems do occur a lot in engineering practice, since many technical systems have linear responses, at least ideally;
2. many properties of nonlinear systems can be derived from studying variations (i.e., differentials) of trajectories, and these turn out to be linear but time-variant;
3. the linear theory is by far the simplest, and most other theories have to use the results of the linear theory in some way.

Let us amplify the two last points by looking at how the state of a nonlinear system evolves (assuming sufficiently smooth evolution in a state space of fixed dimension). Such a state $x(k)$ follows a *trajectory* function of k, imposed by an initial state and an input sequence $u(k)$. Under some continuity requirements valid for many real-life systems, one may assume that a close-by initial state and a close-by input sequence generate close-by trajectories. Let the variational operator be denoted by Δ so that a new, close-by trajectory can be written as $x(k) + \Delta x(k)$, and the new input can be written as $u(k) + \Delta u(k)$. Assuming the differences to be very small, we shall have, to a first-order approximation

$$\begin{cases} \Delta x(k+1) &= A_k \Delta x(k) + B_k \Delta u(k), \\ \Delta y(k) &= C_k \Delta x(k) + D_k \Delta u(k), \end{cases} \tag{2.3}$$

where $A_k := \partial_x f_k$, $B_k := \partial_u f_k$, $C_k := \partial_x g_k$ and $D_k := \partial_u g_k$. The variation of the trajectory is a linear time-variant system. The difficulty with this approach is that the resulting linear variational system is trajectory dependent. To obtain the properties of the original system from this, an integration has to be performed from the differential system to the original. We shall discuss this point in Chapter 8, where we indicate how the linear result extends to nonlinear.

The same strategy can, of course, be followed in the continuous-time case, with the additional difficulty that it may interfere with discretization; discretization in both time and state space has to be done consistently. This can be achieved elegantly for an important class of physical systems, namely those described by a Lagrangian; but this topic falls outside our scope.

Some classes of systems have received much more attention historically than others, for example, linear time-invariant systems (called *LTI systems*) with a finite-dimensional real or complex state space. No doubt, they are important as they cover important application domains (e.g., elementary electrical circuits and elementary control systems), but, surprisingly, the time-invariant theory appears to be less elementary than the time-variant variety that forms the backbone of this book. One reason for this is that time invariance imposes global constraints that are often hard to fulfill. As computations on systems often lead to recursive equations, time invariance imposes a fixed-point solution that is much harder to obtain than just a recursive, time-variant solution.

In the continuous-time case, the state evolution of an LTI system with a finite-dimensional state space is often given by a set of ordinary differential equations:

$$\begin{cases} \dot{x}(t) &= Ax(t) + Bu(t), \\ y(t) &= Cx(t) + Du(t), \end{cases} \tag{2.4}$$

in which the dot indicates time derivation and the matrices A, B, C and D are constant. (In the time-variant version, they would be time dependent.) In the discrete-time variant, the state evolution is given by a difference equation, and the state-space equations become

$$\begin{cases} x(k+1) &= Ax(k) + Bu(k), \\ y(k) &= Cx(k) + Du(k). \end{cases} \tag{2.5}$$

There is a substantial distinction between the two cases. In the continuous-time case, matrix A is a *generator* of a state evolution "semi-group," and the zero-input state transition is properly described by an operator of the type e^{At}, while in the discrete-time case, matrix A properly transfers the state from the time point k to $k+1$, assuming the input $u(k) = 0$.

In the remaining chapters of this book we use the shorthand indexed notation for the state vector at index point k: x_k instead of $x(k)$. In this, more general, chapter we keep the functional notation to highlight the time dependence of the state.

In conclusion, a good balance for a textbook on the relation between dynamical system theory and computational algebra is provided by discrete-time, time-variant linear systems, as most of the more complex cases end up in that category for practical (numerical) treatment.

2.5 Causality

Dynamical system theory evolved with the progression of time. Just like number theory and algebra evolved, after the definition of positive integers, one was forced to define the number zero and then the negative integers. A natural system evolves in the positive time direction, that is, $x(t)$ depends only on what has happened before t, up to $t-$, or, for the discrete-time case, before k up to $k-1$. But why not consider, in addition, (1) systems that evolve in the opposite direction with decreasing time and (2) mixed systems where the present state and output are partially dependent on the past and on the future? We would call the first type *causal*, the second *anti-causal* and the third of *mixed causality*.

It would seem at first that the introduction of anti-causal systems presents an unnecessary additional mathematical difficulty. Luckily, that turns out not to be the case, but it surely needs additional mathematics. In the LTV case, causal systems correspond to (general) block lower triangular input–output transfer matrices, while anti-causal systems produce (general) block upper triangular input–output transfer maps. The two combined then lead to general (block) input–output matrices. Linear time-variant system theory is a theory of general block matrices, where by "general" we mean "without any a priori forced structure," like block Toeplitz.

The system's dynamics of an anti-causal system evolve with decreasing time – a new kind of system evolution dual to the original. Restricting ourselves to linear systems described by ordinary differential or difference equations, how would an anti-causal system be described? Looking first at the differential equation

$$\left\{\begin{array}{rcl} \dot{x}(t) & = & A(t)x(t) + B(t)u(t), \\ y(t) & = & C(t)x(t) + D(t)u(t), \end{array}\right. \tag{2.6}$$

we find that there is no other difference than that the integration of the state would run backward with decreasing times. The time reversal happens completely outside the system, it is in the eye of the beholder. (We know that Newtonian systems are time reversible!)

However, with discrete-time systems, the situation is different. According to our time indexing convention, a causal system moves from the index point k to the index point $k+1$, while an anti-causal system moves from the index point k to the index point $k-1$. A causal system cannot simply be turned into an anti-causal system and retain the same input–output relation. This is not really difficult, because the only difference between a causal and an anti-causal system is the direction of the time axis (increasing or decreasing), and we can simply define an anti-causal system as being a causal system for a time variable $-t$ or $-k$. There remains the issue of indexing the formal representation. The convention we adopt is to again take the state $x(k)$ as the input of the stage k, now in the opposite time direction, so that the description of an anti-causal discrete-time system becomes

$$\begin{cases} x(k-1) & = & f_k(x(k),u(k)), \\ y(k) & = & g_k(x(k),u(k)) \end{cases} \tag{2.7}$$

for some functions f_k and g_k.

Questions immediately arise whether the causal and the anti-causal descriptions can be mixed with each other and whether a causal description can be inverted to an anti-causal one with the same behavior, that is, the same overall relation between inputs and outputs. These questions are closely related to system inversion and treated in Chapters 7 and 8, so we do not discuss them further at this point.

2.6 The Behavior of a Recursively Defined Discrete-Time System

We defined behavior as *how a system is seen from the external world*, specifically what is called its "functional behavior," which defines how inputs and outputs are globally related to each other mathematically. Once a recursive definition of a system is obtained, the behavior is well defined, because the presumably invisible state can be eliminated, provided an initial state is given. So for a causal system, an initial time index K and an initial state $x(K)$, one computes in sequence $y(K)$, $x(K+1)$, $y(K+1)$, $\ldots$, while for an anti-causal system the sequence, starting from some $x(K)$, is $y(K)$, $x(K-1)$, $y(K-1)$, $\ldots$.

This raises the question what to do when the initial state is not given. Here we have to distinguish what we understand by *finitely indexed* and *infinitely indexed* systems, or, more accurately, systems with finite time support and systems that are allowed to start from $-\infty$ and/or run to $+\infty$. A finitely indexed system will not exist before some index K and after some other index $L \geq K$. Such a system starts with an *empty state* and is therefore essentially time-variant. We shall have to put some effort into representing such a state algebraically, but there are no major difficulties once one accepts the possibility of empty entries in matrices. A finitely indexed system will also end by producing an empty final state.

The situation is very different for time-invariant systems. For these systems, one has no choice but to assume that the system exists from time index $k = -\infty$ to $k = +\infty$, and a convention has to be made about initializing such a system and stopping it eventually. We do not discuss this point further here; this issue may come up occasionally (e.g., Chapter 10) and will be dealt with when that occurs.

An important point is how causal and anti-causal systems are combined into systems of mixed causality. Let us assume that we dispose of a causal system with the

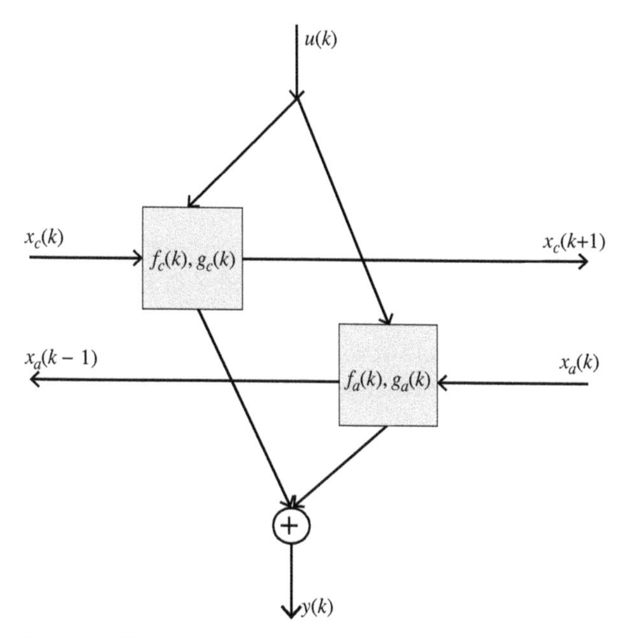

Figure 2.3 How a causal recursion combines with an anti-causal one.

state sequence $x_c(k)$, input $u(k)$ and output $y_c(k)$ given by

$$\begin{cases} x_c(k+1) & = & f_{c,k}(x_c(k),u(k)), \\ y_c(k) & = & g_{c,k}(x_c(k),u(k)) \end{cases} \tag{2.8}$$

and an anti-causal system sharing the input u_k with the causal system, state sequence $x_a(k)$ and output $y_a(k)$ given by

$$\begin{cases} x_a(k-1) & = & f_{a,k}(x_a(k),u(k)), \\ y_a(k) & = & g_{a,k}(x_a(k),u(k)), \end{cases} \tag{2.9}$$

where the total output $y(k) = y_c(k) + y_a(k)$ combines the causal with the anti-causal part in a linear way; see Fig. 2.3. Special attention has to be paid to how the indexing of the causal system combines with the indexing of the anti-causal system, given our indexing conventions. In particular, the state index $k + 1$ of the causal part seems to match time-wise with the index k of the anti-causal part. This may look strange in the diagram, but one should keep in mind that $x_c(k)$ and $x_a(k)$ appear at the beginning of the clock cycle k in both systems and produce their resulting $y_c(k)$ and $y_a(k)$ synchronously (to be used by the system in the time slot k)!

The Timing Assumption

The discussion in the previous section raises the question: What assumption underlies the discrete-time model used in our algorithms? Here it is:

All computations scheduled for time index k are completed during a time slot that is sufficiently large to allow for all the necessary intermediate transfers of data and sub-function executions.

This means that all data indexed by k are assumed stable before the end of the time slot, starting with the index k, and this is true for each index k. In particular, if an output "$y(k)$" is presented as an input "$u(k)$" to a subsequent computation, the assumption guarantees execution in the same time slot. (It is left to system designers to create new time slots or split operations as needed. Those are "design issues," and there are adequate methods available to deal with them.)

One can see that the assumption amounts to assuming the infinite speed of execution of all calculations, so that every variable labeled with k is available at a time infinitesimally later than the appearance of the clock tick k. One also assumes that the calculation order specified by the arrows in the data flow diagrams is respected and that the diagrams are well constructed in that there are no instantaneous, delay-free loops in the diagram.

2.7 The Modeling Issue

Often, the behavior of a system is all we can observe, that is, when we have no information about the system's internal description. Then the issue arises of what can be said about potential models of the system we are observing, or, more precisely, what models are compatible with the input–output observations. This is, of course, a very general problem, and it is difficult to solve in its full generality. So, one has to make some further general assumptions about the system one is considering. In this book, we assume linearity, but not time invariance. The hypothesis "the system is linear" is well verifiable, but we shall just simply assume that the system is indeed linear, and then try to figure out what models are compatible with the data. An important issue in this context is: What data do we need to be able to make effective identifications?

Identifying a time-variant system requires much more data than necessary under the hypothesis of time invariance. The global system response (transfer function) of a time-invariant system at each time point is the same. Hence, a shift in the input corresponds to a similar shift in the output in that case. This means that a single measurement is good for many, actually an infinite number of, time-shifted experiments. In contrast, a time-variant system requires separate information at each time point. However, and remarkably, there is a shortcut, and it is based on a further analysis of Nerode equivalence for the linear but time-variant case.

Nerode Equivalence in the Linear Case

Let us position ourselves at an index time point k of a discrete-time causal and linear system and consider two inputs u_1 and u_2 in the strict past of k that are Nerode equivalent. We write $\{u_1,u_2\} \in \mathrm{sp}_k$ to mean that u_1 and u_2 have their support on the "strict past with respect to k," that is, $-\infty : (k-1)$. Their Nerode equivalence means that given any input u_3 in the future of k marked as f_k, that is, $u_3 \in \mathrm{f}_k$, the system response $y_3 \in \mathrm{f}_k$ is equal for both strict past inputs, concatenated with u_3. In the formalism used in this book, the input is a column vector, and the concatenation of u_1 and u_3 is written as a stacked vector $\begin{bmatrix} u_1 \\ u_3 \end{bmatrix}$. The representation of the transfer operator T can be specialized to its operation centered at the time point k, and, being linear and causal, one can write, for $i = 1,2$,

$$T\begin{bmatrix} u_i \\ u_3 \end{bmatrix} = \begin{bmatrix} y_i \\ y_3 \end{bmatrix} \tag{2.10}$$

with y_i being the strict past response of the system to the strict past input u_i, which is independent of u_3 because of the assumed causality. Hence, and because of linearity, T applied on u_i, producing the strict past y_i, specializes to a linear operator K_k with $y_i = K_k u_i$, which acts on the strict past with respect to k. What about the future? y_3 depends on both $u_i, i = 1,2$, and u_3 in a linear fashion; hence, the response $y_3 = \left[T\begin{bmatrix} u_i \\ u_3 \end{bmatrix}\right]_{k:} = H_k u_i + E_k u_3$ for the linear operators H_k and E_k, which are partial operators of T. In summary, linearity specialized to the time point k decomposes T in three partial operators, K_k, H_k and E_k, such that

$$\begin{bmatrix} y_i \\ y_3 \end{bmatrix} = \begin{bmatrix} K_k & 0 \\ H_k & E_k \end{bmatrix}\begin{bmatrix} u_i \\ u_3 \end{bmatrix} = \begin{bmatrix} K_k u_i \\ H_k u_i + E_k u_3 \end{bmatrix}. \tag{2.11}$$

u_1 and u_2 are Nerode equivalent if and only if $H_k u_1 + E_k u_3 = H_k u_2 + E_k u_3$ independently of u_3, that is, if and only if $H_k u_1 = H_k u_2$, or, if and only if $H_k(u_1 - u_2) = 0$. (Thanks to linearity, we can cancel $E_k u_3$ in the equation: the specific $u_3 = 0$ is as good as any other u_3.)

This is the power of linearity, which we have used on the concatenation and on the decomposition of the transfer operator T. *Two inputs u_1 and u_2 are Nerode equivalent, if and only if their difference, $u_1 - u_2$, belongs to the kernel of H_k.* This observation defines *Nerode equivalence classes* in the input space. Each equivalence class corresponds to a specific, zero future input ($u_3 = 0$) response, and the corresponding single future output corresponds, also in a unique way, to the corresponding individual state.

H_k is the famous *Hankel operator* at the time point k. It is the partial operator of T that maps the strict past into the future. Its kernel is representative for the Nerode equivalence. The set of strict past inputs that map to the zero state is isomorphic to any other set of Nerode equivalent inputs (at k). Each state has a specific Nerode equivalence class attached to it in a unique way, *and* a specific future response (our y_3 for $u_3 = 0$), which is called the *natural response* for that state. So we find a representation of the state space at k in the strict past input space, *and* another one in the future output space. Figure 2.4 illustrates the correspondences.

Yet there is even more to be worked out further in the following chapters. The Nerode zero state equivalence set at k shares a lot of elements with the Nerode zero state equivalence set at $(k-1)$. If some input $u_{:(k-2)}$ generates the zero state at $k-1$, then the concatenation $\begin{bmatrix} u_{:(k-2)} \\ 0 \end{bmatrix}$ (with added zero at $k-1$) generates the zero state at k obviously. The Hankel operator H_k has, therefore, a lot in common with the Hankel operator H_{k-1}. Although H_k will contain new information on the system with respect to H_{k-1}, in many cases, this new information is limited. It is this fact that system

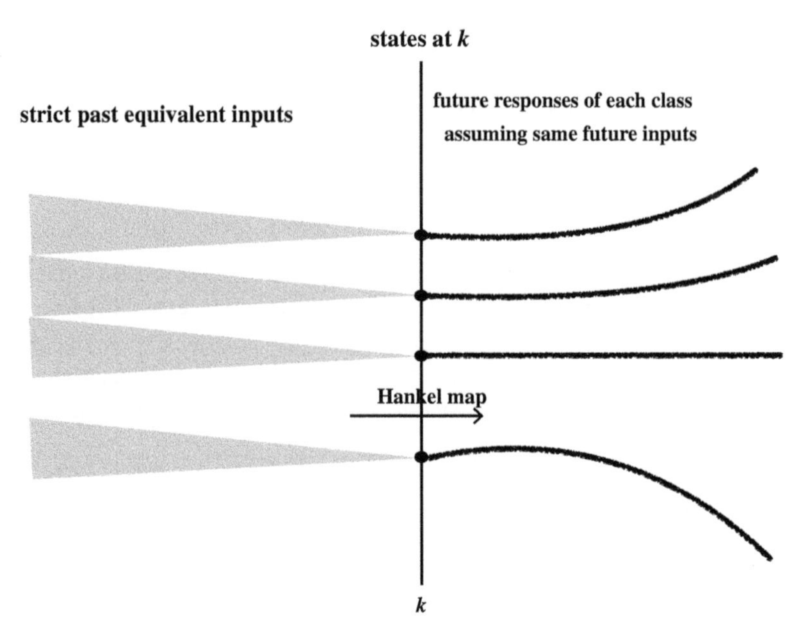

Figure 2.4 An illustration of Nerode equivalence classes.

identification or realization algorithms for time-variant systems exploit. The property can be expressed formally as follows: *let, for each k,* $\mathcal{N}_k$ *be the space of inputs u consisting of a strict past input part that produces the zero state at k concatenated with zero input from k to infinity; then* $\mathcal{N}_{k-1} \subset \mathcal{N}_k$. Exploiting this property mathematically is what most time-variant system theory boils down to! The collection of $\mathcal{N}_k$'s has been called a *nest algebra* in the mathematical literature.

2.8 Generalizations*

In this book, we restrict ourselves strictly to a monotonous time, either increasing or decreasing. Time forms a completely ordered (countable) lattice. Nothing prevents one from considering more general time lattices. One would be just an ordered lattice with origin, that is, a lattice in which two elements have at least one common ancestor (and hence also a minimal one). More general lattices have been considered as well, in particular regular 2D or 3D lattices as they occur in image- or video processing, that is, partial orders in space, or even systems defined on oriented graphs. Needless to say, system theory on such structures becomes much more complicated, because, in more general lattices, there are many ways in which evolution can take place, and hence descriptions inherit a large measure of arbitrariness. In this book, we take the view that time is a fully ordered lattice, and it can only be increasing or decreasing, while all other ordering principles (e.g., in space) are local at any given time point (and may even change between time points).

2.9 Concluding Remarks

The discussion in this chapter focuses on some very basic assumptions and considerations that are often overlooked or considered "trivial." However, the more basic the assumption, the more consequences it has, so it pays to take care in developing them. We did not do that fully as of yet; for that, we need a better vista on these very consequences, but they will soon appear when we proceed. Some of them are technical and determine how easy or complicated the resulting algebra will be. It has been thought for a long time that the continuous-time, time-invariant case is the simplest, but it will appear that this assessment does not really hold, although one can treat some of the estimation and control problems elegantly, that is, with simple expressions, in that case.

However, from the point of view of the simplest possible algebra, the discrete-time, time-variant case stands out, because it relates directly to matrix algebra. It allows for the simplest possible treatment of all the basic system-theoretic ideas that we have already met schematically in this chapter (e.g., state, reachability and observability), and this with an impressive generality, de facto capable of handling completely general matrices (while the time-invariant case is restricted to matrices with a peculiar structure: infinite-dimensional "block Toeplitz matrices" for transfer matrices and (infinite-dimensional) "block Hankel matrices" for what we have called Hankel maps).

Some assumptions appear to be fundamental in the sense that they have a great influence on the type of theory that follows. One example is whether one chooses to consider bounded systems on normed input and output spaces or whether one allows the system to start at some finite point in the past and then let it evolve unconstrained. These two cases lead to different types of system theory, which we shall have to deal with in parallel. More precisely, the important issue of system inversion leads to different results.

2.10 Discussion Items

We propose a set of topics for discussion, which may or may not have definite answers but are worth considering. We treat some of them in detail in further chapters.

1. Given a nonminimal state set of a system, how could one derive a minimal set?
2. Propose more examples of dynamical systems and logical state sets for them.
3. Given a system, one can create a system to observe it. What would be the state of the latter?
4. Systems can be built from other systems. How would that work? What would be the state of the overall system? How about its minimality?
5. An electrical circuit is a dynamical system. What is a reasonable state for it? The same question holds for a living cell.
6. State set equivalence: Given the representation of a state set for a system, could one define other equivalent state sets for it?

7. To analyze a new dynamical system, one would have to discover a state set for it. How could one do that?
8. The dynamical systems considered in this chapter have states that evolve either by a differential equation (for continuous time) or by a difference equation (for discrete time). What could be more general types of system?

2.11 Notes

1. A landmark book that more or less inaugurated the modern thinking on dynamical systems, and in particular "state-space methods" (in lieu of methods based on the Laplace or the z-transform), is the book of Kalman, Falb and Arbib, *Topics in Mathematical System Theory* [49]. This book highlights in particular the scope of the basic concepts introduced in this chapter and how they are active in a great variety of fields, to wit, robotics, information theory, control theory and artificial intelligence. The scope easily extends to computer science and digital system design. For the use of the notions "reachability" or "observability" in the latter field, see, for example, De Micheli's book on Digital Circuits [51, chapter 8].
2. A systematic treatment of optimal quadratic control is in the books of B. D. O. Anderson and J. B. Moore, *Optimal Filtering* [4] and *Optimal Control, Linear Quadratic Methods* [3]. The book treats the time-invariant continuous-time situation in detail. It is interesting to compare the pretty arduous continuous-time treatment with the simplicity of the discrete-time case.
3. Our basic concepts are not as innocuous as they may seem, but they address common computational or engineering situations. When one ventures into modern physics, one easily encounters cases in which the basic assumptions are not, or do not appear to be, valid. In Einstein's relativity theory, time depends on the reference frame, and events that are synchronous in one frame are not synchronous in another. Causality is not really impaired, but it looses universality: it is attached to a specific frame of reference and events that can influence each other in one frame but may not have that relationship in another (e.g., events whose simultaneity triggers a response in one frame may not do so in another). Also, quantum mechanics generates problems with our definition of state, in particular reachability and observability. In quantum mechanics, states are not directly observable nor controllable, and both observations and actions lead to probabilistic results. Nonetheless, also in quantum mechanics, it is possible to define a more refined notion of state, together with operations on it that are consistent with quantum mechanical principles. Even a less extreme example, like the brain system, works on principles that are substantially different from those formulated in this chapter. "Neuromorphic computing" would therefore need quite different basic assumptions or at least a specific narrative that translates what happens in the brain to, for example, a signal flow graph. This has been tried for what are called "artificial neural nets," but the connection between what happens in the brain and what happens in a neural net is tenuous, to say the least. These considerations make awareness of basic assumptions more than necessary!

3 LTV (Quasi-separable) Systems

In this chapter, we specialize the types of dynamical system to *linear, time-variant, discrete-time* and develop the formalism needed to accommodate for time variation in all its aspects, variable dimensions of inputs and outputs and variable state dimensions. This requires a substantial extension of matrix calculus, respecting its basic principles. We assume that the time is discrete and represented simply by an index that runs from $-\infty$ to $+\infty$. We develop straightforward, easy-to-use generic matrix representations for the state-space evolution and the input–output map, both for a causal and an anti-causal system. We introduce block diagonal representations, generic shifts and the transfer operator representation for the LTV case. Next, we discuss closely related notions: causality, anti-causality, stability and duality. Our aim is also didactical. We want to show how one can convert basic ideas into (numerical) algebra, in this case applied to the system concepts of the previous chapter.

Menu

Hors d'oeuvre
Why Linear Time-Variant (LTV) and Discrete-Time Systems?

First Course
The Formalism for Causal Systems

Second Course
Uniform Exponential Stability

Third Course
The Formalism for Anti-causal Systems
Duality

Fourth Course
The Diagonal Notation

Dessert
The LTI Case

3.1 The Formalism for Causal Systems

Let us first consider a system that runs forward in time (a causal system) and consider its local evolution at an index point k. At that point in time, the system will have reached a state x_k of dimension η_k and will receive an input u_k of dimension m_k. Stepping from k to $k+1$, it determines a new state x_{k+1} of (potentially different) dimension η_{k+1} and produces an output y_k of dimension n_k. In other words, there is a *local transition map* $(x_k, u_k) \mapsto (x_{k+1}, y_k)$ realized at the time index k, producing the next state x_{k+1} and a local output y_k, whereupon the system moves to the next index point $k+1$. The local transition map would, in general, be nonlinear, but we start out by assuming that it is linear, leaving the generalization to nonlinear for later.

For mathematical precision, one normally specifies to which type of vector space the various vectors involved belong. In this book, we compute with real or complex numbers ($\mathcal{R}$ or $\mathcal{C}$), even though many properties would also work in a finite field (like p-adic numbers), not considered here. So we write: $x_k \in \mathcal{R}^{\eta_k}$ or $x_k \in \mathcal{C}^{\eta_k}$. Actually, it does not matter much whether the numbers are real or complex for the theory to work, and, moreover, the real numbers can be considered a special case of complex numbers. We shall define our notation in such a way that things work out independently of which case is being considered (in many cases, the computations will just be real).

A linear map $a\colon \mathcal{R}^m \to \mathcal{R}^n : u \mapsto y = a(u)$ (often just written $y = au$) is represented by an $n \times m$ real matrix when the natural bases in the input and output spaces are used. Let us call A the matrix corresponding to the map or operator a, so that $y = Au$, or with indices $y_k = \sum_{i=1}^{m} A_{k,i} u_i$, we shall often write this as $y_k = \sum_{i=1:m} A_k^i u_i$, with the upper index in A_k^i indicating the column and the lower the row, although there might be confusion with exponentiation (the context will have to show what the upper index actually means). This is an example of matrix–column multiplication that assumes vectors to be represented by columns and the linear map by matrix–column multiplication. A row-based system would work equally well, and we shall sometimes use it, but most of the book uses the column representation for vectors. (In modern differential geometry, people use an "Einstein index notation" and do not really specify which type of matrix representation they use.)

It is often useful to subdivide vectors into subvectors or to assemble vectors into new, larger vectors. Such operations are very common in, for example, MATLAB, and we adopt MATLAB's notation conventions to construct new vectors. (We shall even extent the conventions; see further.) Column vectors can be concatenated: suppose, for example, that u_1 is a column vector of dimension n_1 and u_2 a column vector of dimension n_2; then one may define a new column vector u of dimension $n_1 + n_2$, which we denote as $\mathrm{col}(u_1, u_2)$, using the *constructor* "col." If $n_1 = n_2 = n$, then one could also set the two vectors next to each other and create an $n \times 2$ matrix $\begin{bmatrix} u_1 & u_2 \end{bmatrix}$. Many such constructs are possible. (We shall explain what is being done in detail when needed or else assume that it is clear from the context.) The distinctive characteristic of a matrix is that it is a rectangular block of data with precisely fitting dimensions. When assembling matrices, or doing operations with them, dimensions should always match properly. For example, a matrix A of dimensions $n \times m$ can only multiply a matrix B of dimensions $k \times \ell$ to the left of the latter when $m = k$. Similarly, a subdivision of the matrix $T = \begin{bmatrix} A & B \\ C & D \end{bmatrix}$ into blocks requires the number of rows of A and B as well

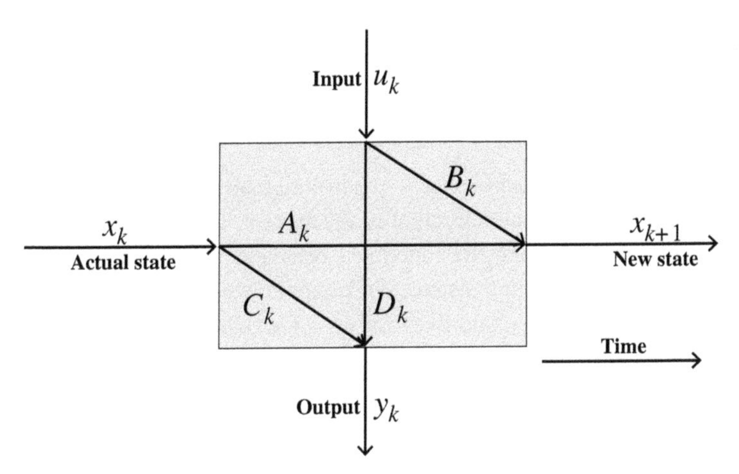

Figure 3.1 Signal flow diagram of a discrete-time, linear and time-variant causal system.

those of C and D to be equal and the number of columns of A and C as well as those of B and D.

With the conventions so far, our discrete-time, linear and time-variant system would then satisfy a transition equation at each index point k of the type

$$\begin{cases} x_{k+1} &= A_k x_k + B_k u_k, \\ y_k &= C_k x_k + D_k u_k, \end{cases} \tag{3.1}$$

in which A_k is a matrix of dimensions $\eta_{k+1} \times \eta_k$, B_k is a matrix of dimensions $\eta_{k+1} \times m_k$, C_k is a matrix of dimensions $n_k \times \eta_k$ and D_k is a matrix of dimensions $n_k \times m_k$. A_k is called the "state transition map," B_k is a map from the input at index k to the next state, C_k is a state-to-output map and D_k is a so-called *feed-through* that connects the input at index k directly to the output at index k. (From a computational point of view, there will be a slight delay, which is neglected, assuming a computer to calculate at infinite speed. In some system descriptions, the feed-through term is put zero. However, keeping the term simplifies many calculations and enhances generality. It also plays an essential role in the Kalman filter; see Chapter 9.)

Definition 3.1 The collection, for all k $\begin{bmatrix} A_k & B_k \\ C_k & D_k \end{bmatrix}$, of matrices of dimensions $(\eta_{k+1} + n_k) \times (\eta_k + m_k)$ is called a *realization*. The dimension η_k is called the (local) degree of the realization (see Fig. 3.1).

3.2 Compact Representations

Suppose we are doing a computation as follows: the calculation starts at index 1 with no input state (x_1 is empty). It takes in a first input u_1 of dimension m_1, computes $x_2 = B_1 u_1$ of dimension η_2 and $y_1 = D_1 u_1$ of dimension n_1, using some matrices B_1 and D_1 of appropriate dimensions. Then it moves to index 2, takes in u_2 of dimension

m_2, computes $x_3 = A_2x_2 + B_2u_2$ of dimension η_3 and $y_2 = C_2x_2 + D_2u_2$ of dimension n_2. Next, it moves to index 3, takes in u_3 of dimension n_3 and computes x_4 and y_3, etc.... up to a last index ℓ, whereby the last computation ends with an empty state $x_{\ell+1}$ and $y_\ell = C_\ell x_\ell + D_\ell u_\ell$.

What would be the *behavior*, that is, the overall map such a system has computed? Clearly, each state x_k remains internal to the system, for an external observer, only the input–output map, called the functional behavior, would matter. We obtain this behavior when we eliminate the state. For example, we have $y_2 = C_2x_2 + D_2u_2$, and hence $y_2 = C_2B_1u_1 + D_2u_2$, and likewise $y_3 = C_3A_2B_1u_1 + C_3B_2u_2 + D_3u_3$, etc...., expressed in a matrix form

$$\begin{bmatrix} y_1 \\ y_2 \\ y_3 \\ \vdots \end{bmatrix} = \begin{bmatrix} D_1 & 0 & 0 & \cdots \\ C_2B_1 & D_2 & 0 & \ddots \\ C_3A_2B_1 & C_3B_2 & D_3 & \ddots \\ \vdots & \ddots & \ddots & \ddots \end{bmatrix} \begin{bmatrix} u_1 \\ u_2 \\ u_3 \\ \vdots \end{bmatrix}. \tag{3.2}$$

The system transfer matrix in the middle, let us call it T, has dimensions $(n_1+n_2+n_3+\cdots)\times(m_1+m_2+m_3+\cdots)$. It is a matrix consisting of subblocks $T_{i,j}$ with dimensions $n_i \times m_j$. It is also a lower block triangular, matrix, reflecting the fact that an output y_i is only dependent on inputs u_j with $j \le i$. Its general element, assuming $j < i-1$, is $C_iA_{i-1}\cdots A_{j+1}B_j$, in the literature sometimes denoted by $C_iA^{>}_{i,j}B_j$, where $A^{>}_{i,j}$ is the *continuing product* $A_{i-1}\cdots A_{j+1}$, for $i > j$, and with the convention that $A^{>}_{i,i} = I$. The regularity in the construction of the entries is pretty apparent, so it will prove useful to introduce a more compact notation that summarizes the essentials.

To start with this, let us first look at vectors. If we do not want to see indices in detail, we put $u := \text{col}[u_j]_{j=1:\ell}$. But why restrict ourselves to starting times at index 1 or stopping times at ℓ? If we do not want to bother about the starting time, we can just pad u with empty entries for all index points less than 1 and larger than ℓ instead of introducing a stopping time. Let us put $u_j := -$ (i.e., *dash*, sometimes emphasized as $\boldsymbol{-}$) for all $j < 1$ and $j > \ell$, and we can write $u := \text{col}[u_j]$, where the index ranges from $-\infty$ to $+\infty$ is simply understood. The entry annotated by "–" (or $\boldsymbol{-}$) has dimensions 0×1, and it is compatible with the typical dimension of a vector, for example, $m \times 1$, now with $m = 0$. Multiple dashes are often abbreviated to a single dash (e.g., $- - -$ as $-$). One can always think of such dummies stacked on top and at the bottom of a vector. They play the role of "place holders," with the meaning "there is an index, but no entry." Such placeholders play an essential role in computer languages (in that context they are often denoted by a "perp" ($\perp$)), and they arise, for example, when a stack becomes empty after processing, or, as here, when at some indices in an indexed list, there is no entry. In our case, the definition should produce elements that are dimensionally compatible with other elements in the same vector.

Likewise, we could have a case where at some index points there are no input entries, while there is still a computation going on, or where the computation does not produce an output entry. So it makes sense to generalize our notion of "empty entry" further. We may have a matrix element of dimension 1×0, which we might indicate by a vertical dash ($|$) (or emphasized as $\mathbf{I}$), an element of dimensions 0×0 that we annotate with a simple dot ($\cdot$), and all these may be stacked as well in vectors, provided dimensions remain compatible. Clearly, an extension of matrix calculus is needed to allow us to do computations with such elements. Here are a few simple rules that keep everything compatible. Let us use the "$*$" to represent matrix multiplication explicitly and a any number:

Operation	Result
$- * \|$	$\cdot$
$\| * -$	$[0]$
$\cdot * -$	$-$
$\| * \cdot$	$\|$
$\cdot * \|$	illegal
$- * [a]$	$-$
$[a] * \|$	$\|$
$\begin{bmatrix} - & - \end{bmatrix} \begin{bmatrix} \| \\ \| \end{bmatrix}$	$\cdot$
etc....	

(Notice that we do not distinguish between $|$ and $[|]$, etc. . . . , for brevity – brackets are just "sugar coating.")

There might be a discussion about which of "$|$," "$-$" and "$\cdot$" has a right-inverse, left-inverse or any inverse at all. The following choices appear consistent: "$|$" *has the left inverse* "$-$" *because* $- * | = \cdot$, *while* "$-$" *has the right inverse* "$|$" *for the same reason.* "$\cdot$" *has* "$\cdot$" *as inverse, because* $\cdot * \cdot = \cdot$. "$\cdot$" *may therefore be considered to be the unit matrix of zero dimensions,* "$|$" *is a tall matrix of dimensions* 1×0, *while* "$-$" *is a flat matrix of dimensions* 0×1.

With such rules, we can still write for any input–output map $y = Tu$, in which all vectors and matrices are properly padded, and zero dimensions are allowed for the inputs, the states and the outputs, while all time series run from $-\infty$ to $+\infty$.

However, there is one issue with this approach that should already be mentioned here: as soon as vectors and matrices are allowed to have infinite indices, the matrix–vector product or matrix–matrix product may produce infinite sums that do not converge. We shall discuss in further chapters how one can deal with such a situation; at this point, let us assume that either it does not occur (because all summations are de facto finite) or that convergence is properly taken care of. The fact that systems evolve with unconstrained times has great technical consequences. So it makes sense to bring in that fact from the start. Another point of technical importance is that of orientation in an infinitely indexed vector or matrix. For this, we single out the element at index 0 and surround it with a box. Hence, we have $u = \mathrm{col}\left(\ldots, u_{-1}, \boxed{u_0}, u_1, \ldots\right)$, etc. . . . . (This element may, of course, be empty of dimensions 0×1.)

The next step is to represent the various actions in the system in a compact fashion. We start by remarking that each elementary computation takes place at a specific index point. Let us explore the global consequence of this simple fact. Concatenating the states, let's simply write

$$x := \text{col}\left(\ldots, x_{-1}, \boxed{x_0}, x_1, \ldots\right) \tag{3.3}$$

and observe that the operators A_k together not only map x to a backward shifted version of itself $\left(\text{namely col}\left(\ldots, x_0, \boxed{x_1}, x_2, \ldots\right)\right)$ but also each x_k maps only to its x_{k+1} and nothing else. To capture this globally, we define a *shift operator* Z: given any infinitely indexed vector (say x), let Zx be defined as the vector $Zx := \text{col}\left(\ldots, x_{-2}, \boxed{x_{-1}}, x_0, \ldots\right)$, that is, it just postpones x_k one index point. This operator admits, of course, an inverse Z^{-1} or equivalently Z'. In matrix terms, Z is a doubly infinitely indexed matrix, with varying dimensions and the following form (in which the 0 and I entries may be empty):

$$Z = \begin{bmatrix} \ddots & \ddots & & & \\ & I & 0 & & \\ & & I & \boxed{0} & \\ & & & I & 0 \\ & & & & \ddots & \ddots \end{bmatrix}, \tag{3.4}$$

in which the element of indices (0,0) is boxed. (As mentioned, it can be empty, but the unit elements must be on the first subdiagonal, which subsumes as a global unit matrix.) When the shift operator is applied to an input vector u of dimensions $\mathbf{m} = \text{col}\left(\ldots, m_{-1}, \boxed{m_0}, m_1, \ldots\right)$, it produces a numerically identical output of dimensions $\left(\ldots, \boxed{m_{-1}}, m_0, m_1, \ldots\right)$, hence the shift operator itself has dimensions

$$\left(\ldots, \boxed{m_{-1}}, m_0, m_1, \ldots\right) \times \left(\ldots, m_{-1}, \boxed{m_0}, m_1, \ldots\right) \tag{3.5}$$

and corresponds to a *downward shift* or causal shift on the input vector, but when applied to the right of a row vector (say u'), it corresponds to a horizontal *backward* shift.

More generally, and because of the definition of matrix multiplication, ZA has the same rows as A shifted one unit downward, while AZ has the same columns as A now shifted one unit to the left. Notice that the dimensions of Z are context dependent. Although Z can be represented by a matrix in a given context, it actually represents a collection of matrices, or, better still, it is a constructor. Notice also that Z (whether applied to the left or the right) shifts lower block triangular matrices to lower block triangular matrices and a block diagonal matrix to lower triangular with just one subdiagonal nonzero. So if A is causal, then both ZA and AZ are causal as well.

For the operators A_k, B_k, etc., which are just locally active at the index point k, we next define a (doubly infinitely indexed) block diagonal matrix, using a constructor that produces a block diagonal matrix from its arguments,

$A := \text{diag}\left(\ldots, A_{-1}, \boxed{A_0}, A_1, \ldots\right)$.
Similarly, $B := \text{diag}(B_k)_{k=-\infty:\infty}$,[1]
$C := \text{diag}(C_k)_{k=-\infty:\infty}$ and $D := \text{diag}(D_k)_{k=-\infty:\infty}$ will globally characterize the input-to-state, state-to-(next state), state-to-output and feedthrough maps that act locally at each index point, and the global state-space equations become

[1] Often written $[B_k]_{k=:}$ when no misunderstanding about the range is possible.

$$\begin{cases} x &= Z(Ax + Bu), \\ y &= Cx + Du. \end{cases} \tag{3.6}$$

This is probably as compact as one can get these time-variant equations. Whether such a compact representation is indeed useful has to be demonstrated, and we hope that the rest of the book will be convincing on this issue. It surely allows us to derive an equally compact representation for the state of the system, at least formally. Eliminating the state x, we obtain, in sequence,

$$x = (I - ZA)^{-1}ZBu \tag{3.7}$$

and $y = Tu$ with

$$T = D + C(I - ZA)^{-1}ZB, \tag{3.8}$$

provided a meaning can be given to $(I - ZA)^{-1}$.

3.3 Uniform Exponential Stability

Now the issue is: Is inverting $(I - ZA)^{-1}$ meaningful and what is the inverse then? Remarkably, there are several, totally different, answers possible to this question, each with its own significance (and history). First of all, suppose that the sequence x is known to begin somewhere (e.g., at index $k = 1$ in our previous example and is empty for indices $k < 1$); then one could write

$$w := (I - ZA)^{-1}x = (I + ZA + ZAZA + (ZA)^3 + \cdots)x, \tag{3.9}$$

remarking that the term $(ZA)^k x$ involves k subsequent forward shifts, since $[(ZA)^k x]_j = A_{j-1} \cdots A_{j-k+1}A_{j-k}x_{j-k}$ because $[(ZA)^k x]_j = [A(ZA)^{k-1}]_{j-1} = A_{j-1}[(ZA)^{k-1}]_{j-1}$, etc. Now, fix some j and look at the sum of all the terms in the jth component of the sum. When $j = 1$, only one term is nontrivial, that is, the first one: $w_1 = x_1$. When $j = 2$, we find $w_2 = x_2 + A_1x_1$, and then $w_3 = x_3 + A_2x_2 + A_2A_1x_1$, etc...., and when $j < 1$, w_j is empty. The sum contains less than j terms for every element in $w = (I - ZA)^{-1}x$ of the index less than j. When $j \leq k$, then x_{j-k} is empty $(= -)$, and hence the whole product is empty as well. This actually reflects the fact that the state-space equation is a forward recursion, whereby x_k is only dependent on previous values, which is just a finite set. However, the dependence grows with growing k, and so it may happen that for larger k's, these entries blow up. For example, suppose all $A_k = 2$ for $k \geq 1$; then one would have $[(ZA)^k x]_j = 2^{j-k}x_{j-k}$ for all $j > k$.

It is not difficult to write down the matrix for $(I - ZA)^{-1}$ for the case where A is a block diagonal matrix with entries A_k that are empty for $k < 1$ and $k > n$. It is

$$\begin{bmatrix} \boxed{\cdot} & - & - & - & \cdots & - & - \\ | & I & 0 & 0 & \cdots & 0 & 0 \\ | & A_2 & I & 0 & \cdots & 0 & 0 \\ | & A_3A_2 & A_3 & I & \cdots & 0 & 0 \\ \vdots & \vdots & \ddots & \ddots & \ddots & \vdots & \vdots \\ | & A_{n-1}A_{n-2}\cdots A_2 & A_{n-1}\cdots A_3 & A_{n-1}\cdots A_4 & \cdots & A_{n-1} & I \end{bmatrix}, \tag{3.10}$$

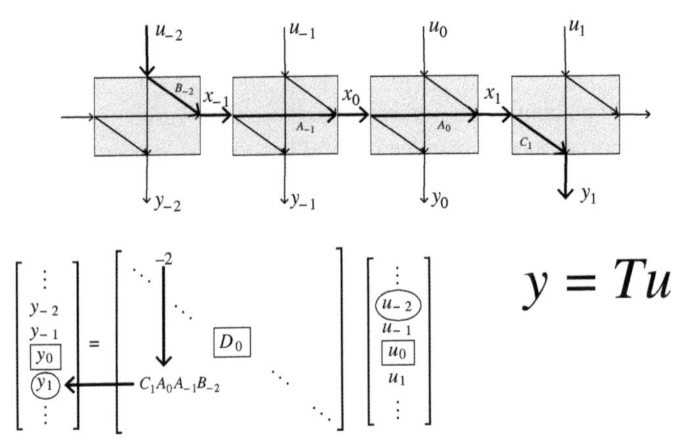

Figure 3.2 Example of an overall causal system build from local computations: only the (linear) contribution of u_{-2} to y_1 is shown.

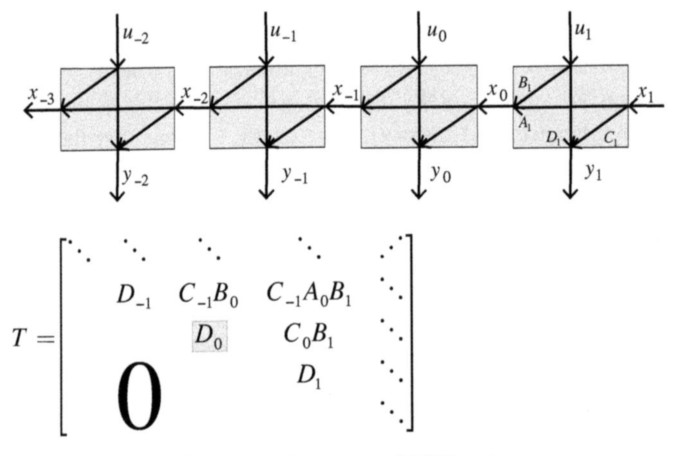

Figure 3.3 Schema for a general anti-causal LTV system.

it has dimensions $(\eta_1 + \eta_2 + \cdots + \eta_{n-1})^2$ and it is a square matrix, with unit matrices on the main diagonal. We see that if the A_k's are large, elements further away from the main diagonal might blow up even exponentially, making the resulting matrix numerically unstable. Such an instability must often be avoided. Although $(I-ZA)^{-1}$ remains well defined in the positive one-sided case just described and leads to finite arithmetic, it is often necessary to require boundedness.[2] This can be done as described in the following equation, where we consider the more general case that the system's indices

[2] Systems that need to be controlled can be unstable; hence, the controller is used to achieve stability. For such problems, a different system description is needed. On the other hand, instabilities in data-processing applications have to be avoided, making an adequate approach for the numerics necessary. We analyze this issue further in Chapter 7.

extend potentially to either $-\infty$, $+\infty$ or both, as would, for example, always be the case for time-invariant systems (Figs. 3.2 and 3.3).

Converging Neumann Series*

One way to impose some modus of stability, at least in the infinitely indexed case but also often needed for finite-dimensional matrices, is to assume that there exists a positive number $\sigma < 1$, such that there is an index M with the property that for all $\ell > M$ and all k (i.e., uniformly), $\|A^{>}_{k+\ell+1,k}\| \leq (\sigma)^{\ell}$, where $\|\cdot\|$ is by definition the Euclidean norm. The smallest σ is called the *spectral radius* of ZA. In the sequel, we shall denote the spectral radius of ZA by ℓ_A[3]:

$$\ell_A := \text{lim-inf}_{0\leq\sigma}(\exists M : \forall(\ell > M)\forall(k = -\infty{:}\infty)[\|A^{>}_{k+\ell+1,k}\| \leq (\sigma)^{\ell}]). \tag{3.11}$$

When $\ell_A < 1$ holds, A is said to be *uniformly exponentially stable* (u.e.s.), which amounts to say that the entries in the matrix $(I - AZ)^{-1}$ far from the main diagonal die out exponentially and uniformly with rate at most σ. This condition is, of course, automatically fulfilled when $(I - AZ)$ is a finite matrix, since a lower triangular block matrix with units on the main diagonal will always be invertible, but the result may still have an undesirably large condition number, a situation that then would have to be dealt with in each concrete case. We analyze this crucial situation in detail in Chapter 8.

3.4 Diagonal Shifts

Block diagonal matrices, responsible for local calculations, play a major role in the theory of LTV systems, much like constant matrices do in the theory of linear time-invariant systems. We introduce therefore a special type of shift for them. Suppose A is such a matrix; then consider ZAZ^{-1}. The left shift operator Z will shift the block rows of A one block down, while the right shift operator Z^{-1} shifts the columns one block to the right. The net result is a shift along the main diagonals in the southeast direction. We write $A^{\langle +1\rangle} := ZAZ^{-1}$. Similarly, a diagonal shift in the opposite northwest direction is $A^{\langle -1\rangle} := Z^{-1}AZ$. Equivalently, we have $ZA = A^{\langle +1\rangle}Z$ and $AZ = ZA^{\langle -1\rangle}$. Although Z does not commute with other operators, there is a kind of weak commutativity, involving a shifted version. As we shall see in further chapters, this is enough for most of LTV system theory to generalize what happens in the time-invariant theory. It also follows that $(ZA)^{\ell} = A^{\langle +1\rangle}A^{\langle +2\rangle}\cdots A^{\langle +\ell\rangle}Z^{\ell} = Z^{\ell}A^{\langle -(\ell-1)\rangle}\cdots A^{\langle -1\rangle}A$.

[3] The formula looks more complicated than it is! Just try to express the property that a block entry that is farther away from the main diagonal than M has norm less than σ^{ℓ}, where $\ell > M$ is the distance to the main diagonal.

3.5 Anti-causal Systems

Lower triangular block matrices represent *causal* calculations for the (block) matrix–column vector calculus since a state x_k or output y_k of a system whose transfer operator is (block) lower triangular depends only on inputs in the past up to index k (i.e., a causal system). Dually, we can consider systems that run backward in time, where y_k would only be dependent on inputs u_j, with $j \geq k$. Consider the transpose $Z' = Z^{-1}$ of Z: it defines a backward shift on vectors. Now, consider the backward running system

$$\begin{cases} x_{k-1} &= A_k x_k + B_k u_k, \\ y_k &= C_k x_k + D_k u_k \end{cases} \tag{3.12}$$

resulting, the same as before, in an *anti-causal* transfer operator

$$T = D + C(I - Z'A)^{-1}Z'B. \tag{3.13}$$

The corresponding matrix will then be upper triangular. In the case of a traditional (block) matrix with a state-space description starting at index n and running backward to 1, it looks as follows, when the support starts at 0 and ends at n (all entries with support outside $[0\!:\!n] \times [0\!:\!n]$ are empty):

$$\begin{bmatrix} \boxed{D_0} & C_0B_1 & C_0A_1B_2 & \cdots & C_0A_1\cdots A_{n-1}B_n \\ 0 & D_1 & C_1B_2 & \ddots & C_1A_2\cdots A_{n-1}B_n \\ 0 & 0 & D_2 & \ddots & \vdots \\ \ddots & \ddots & \ddots & \ddots & \vdots \\ \cdots & \cdots & \cdots & 0 & D_n \end{bmatrix} \tag{3.14}$$

with the general term $T_{i,j} = C_i A_{i+1} \cdots A_{j-1} B_j$ for $i + 1 < j$, which we sometimes write as $C_i A^{i<j} B_j$.

3.6 Duality

Given a causal state-space system $T = D + C(I - ZA)^{-1}ZB$, and using the regular matrix transposition ($'$), we may define the dual or transposed system $T' = D' + B'Z'(I - A'Z')^{-1}C' = D' + B'(I - Z'A')^{-1}Z'C'$. We see that the causal transfer operator $\begin{bmatrix} A & B \\ C & D \end{bmatrix}$ just changes into the anti-causal transfer operator $\begin{bmatrix} A' & C' \\ B' & D' \end{bmatrix}$. Time is reversed, the role of inputs and outputs are interchanged, the incoming state x_k comes out of the previous step, which is now located at the index $k + 1$, and the outgoing state will feed into stage $k - 1$. Although this convention may seem a bit strange at first (the kth state x_k is at a different location in the causal than in the anti-causal case), it turns out to be consistent with the normal indexing in the input–output matrices. Other conventions can be found in the literature; in particular, x_k can be taken as the output state in the causal filter, which then takes x_{k-1} as its input. This is, of course, immaterial; the convention adopted here seems the most convenient.

3.7 Example: A Simple Banded Matrix and Its Inverse

Let's find a simple realization for

$$T = \begin{bmatrix} d_1 & 0 & 0 & 0 \\ a_2 & d_2 & 0 & 0 \\ 0 & a_3 & d_3 & 0 \\ 0 & 0 & a_4 & d_4 \end{bmatrix}. \tag{3.15}$$

Let's retain the traditional matrix indexing scheme (i.e., starting at index 1). It follows that stage 0 is empty. At stage 1, we take in u_1 and immediately produce $y_1 = d_1 u_1$. Next, we have the choice to either immediately compute $a_2 u_1$ or take u_1 as a state and feed it to the next stage. Let's take that latter path and put $x_2 = u_1$. Hence $\begin{bmatrix} A_1 & B_1 \\ C_1 & D_1 \end{bmatrix} = \begin{bmatrix} | & 1 \\ | & d_1 \end{bmatrix}$. The next stage now takes in x_2 and u_2, and we compute immediately $y_2 = a_2 x_2 + d_2 u_2$. At this point, $u_1 = x_2$ is not needed further and can be discarded, but u_2 will be needed in the next computation. Hence, we put $x_3 = u_2$ and have $\begin{bmatrix} A_2 & B_2 \\ C_2 & D_2 \end{bmatrix} = \begin{bmatrix} 0 & 1 \\ a_2 & d_2 \end{bmatrix}$. The next stage is not much different, and we find $\begin{bmatrix} A_3 & B_3 \\ C_3 & D_3 \end{bmatrix} = \begin{bmatrix} 0 & 1 \\ a_3 & d_3 \end{bmatrix}$. Finally, the fourth stage takes in $x_4 = u_3$ and u_4, and we compute $y_4 = a_4 x_4 + d_4 u_4$ so that $\begin{bmatrix} A_4 & B_4 \\ C_4 & D_4 \end{bmatrix} = \begin{bmatrix} - & - \\ a_4 & d_4 \end{bmatrix}$. The following steps are empty again (see Fig. 3.4).

Anticipating the section on matrix inversion, let us observe that if all the D_i's are invertible, one obtains an easy realization of the inverse of a system (forgetting about stability), by reversing the feedthrough arrow from the input to the output (see Fig. 3.5); see the appendix for the definition of the data flow graphs we use. This procedure can be applied directly to the state-space description:

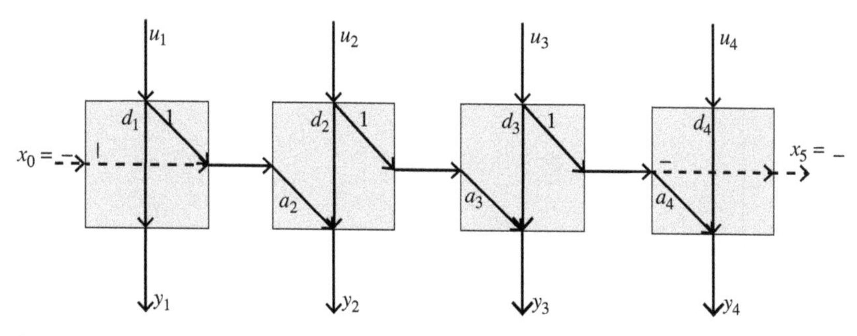

Figure 3.4 A straight realization for transfer operator (3.15).

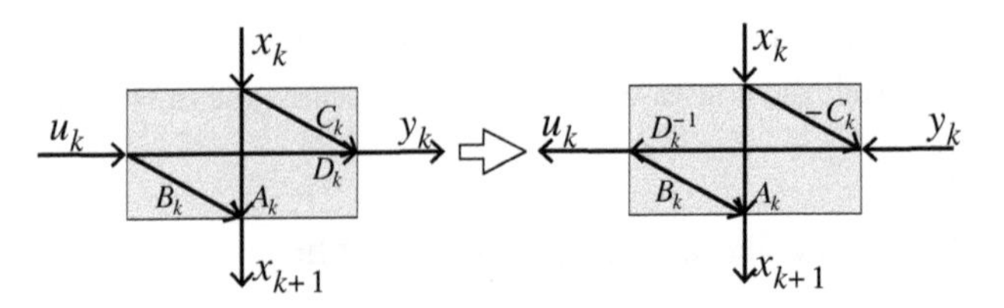

Figure 3.5 Finding a formal (potentially unstable) inverse by arrow reversal.

$$\begin{bmatrix} A & B \\ C & D \end{bmatrix} \mapsto \begin{bmatrix} A - BD^{-1}C & BD^{-1} \\ -D^{-1}C & D^{-1} \end{bmatrix} \tag{3.16}$$

from which a realization for the inverse is easily found in sequence:

$$\begin{bmatrix} | & d_1^{-1} \\ | & d_1^{-1} \end{bmatrix}, \begin{bmatrix} -d_2^{-1}a_2 & d_2^{-1} \\ -d_2^{-1}a_2 & d_2^{-1} \end{bmatrix}, \begin{bmatrix} -d_3^{-1}a_3 & d_3^{-1} \\ -d_3^{-1}a_3 & d_3^{-1} \end{bmatrix}, \begin{bmatrix} - & - \\ -d_4^{-1}a_4 & d_4^{-1} \end{bmatrix}. \tag{3.17}$$

Not only does this allow for a quick and easy calculation of the inverse matrix (which would actually not be too hard by back-substitution), but also it shows that the inverse system has a realization whose state dimension equals the original; hence, it can perform computations (solving equations) with the same computational complexity as the original, although the matrix now has a full triangular lower part.

3.8 Block Matrices and the Diagonal Notation

Suppose A, B, C, and D are the (block) diagonal matrices representing a realization; then we could write down a (hierarchical) block matrix $\begin{bmatrix} A & B \\ C & D \end{bmatrix}$ having these block diagonal matrices as components, provided the dimensions match, of course. More graphically, this representation looks as follows:

$$\left[\begin{array}{cccc|cccc} \ddots & & & & \ddots & & & \\ & A_{-1} & & & & B_{-1} & & \\ & & \boxed{A_0} & & & & \boxed{B_0} & \\ & & & A_1 & & & & B_1 \\ & & & \ddots & & & & \ddots \\ \hline \ddots & & & & \ddots & & & \\ & C_{-1} & & & & D_{-1} & & \\ & & \boxed{C_0} & & & & \boxed{D_0} & \\ & & & C_1 & & & & D_1 \\ & & & \ddots & & & & \ddots \end{array}\right] \tag{3.18}$$

Alternatively, one could write the same realization as a single block diagonal consisting of local block matrices as follows:

$$\begin{bmatrix} \ddots & & & & \\ & \begin{bmatrix} A_{-1} & B_{-1} \\ C_{-1} & D_{-1} \end{bmatrix} & & & \\ & & \boxed{\begin{bmatrix} A_0 & B_0 \\ C_0 & D_0 \end{bmatrix}} & & \\ & & & \begin{bmatrix} A_1 & B_1 \\ C_1 & D_1 \end{bmatrix} & \\ & & & & \ddots \end{bmatrix}. \tag{3.19}$$

Both representations are, of course, fully equivalent: via an obvious permutation of appropriate rows and columns. One could develop a formal equivalence theory for such structures, but we shall not do so, assuming that the type of representation used will always be clear from the context, and thereby avoiding an unnecessary cluttering of symbols. How this type of liberty functions is seen in the following, hopefully obvious, example; when, for example, $A = \begin{bmatrix} A_{11} & A_{12} \\ A_{21} & A_{22} \end{bmatrix}$ is a matrix of block diagonals, then a correct expression for ZA is

$$ZA = \begin{bmatrix} ZA_{11} & ZA_{12} \\ ZA_{21} & ZA_{22} \end{bmatrix} \tag{3.20}$$

because "Z" is nothing else but a generic shift operator, which applies equally on the (block) indices of all rows (when applied left) or on the (block) indices of all the columns (when applied right). One could actually state the formal equivalence

$$\begin{bmatrix} \operatorname{diag}(A_k) & \operatorname{diag}(B_k) \\ \operatorname{diag}(C_k) & \operatorname{diag}(D_k) \end{bmatrix} \equiv \operatorname{diag} \begin{bmatrix} A_k & B_k \\ C_k & D_k \end{bmatrix}, \tag{3.21}$$

showing that the matrix constructor commutes in a sense with the diagonal constructor (provided dimensions of the arguments agree, of course). However, the matrices on both sides of this expression have a different order.

3.9 Example: The Realization of a Series

With the ideas just expounded, it is easy to generalize the previous example and produce realizations in a "companion form" as in the classical case. Consider the global realization

$$\begin{bmatrix} A & B \\ C & D \end{bmatrix} = \left[\begin{array}{ccc|c} 0 & I & 0 & 0 \\ 0 & 0 & I & 0 \\ 0 & 0 & 0 & I \\ \hline T_3 & T_2 & T_1 & T_0 \end{array}\right], \tag{3.22}$$

in which the I's are (finite or infinite) unit matrices, the 0's (finite or infinite) are zero matrices and the T_i's are block diagonals, all with matching dimensions as needed by the matrix, of course. Then this realizes $T = T_0 + T_1 Z + T_2 Z^2 + T_3 Z^3$ as is easily verified from

$$(I - ZA)^{-1} ZB = \begin{bmatrix} I & -Z & 0 \\ 0 & I & -Z \\ 0 & 0 & I \end{bmatrix}^{-1} ZB = \begin{bmatrix} Z & Z^2 & Z^3 \\ 0 & Z & Z^2 \\ 0 & 0 & Z \end{bmatrix} \begin{bmatrix} 0 \\ 0 \\ I \end{bmatrix}. \tag{3.23}$$

The method generalizes easily further to larger expansions. In the case of the smaller Example 3.7, we have

$$T = \begin{bmatrix} d_1 & & & \\ & d_2 & & \\ & & d_3 & \\ & & & d_4 \end{bmatrix} + \begin{bmatrix} | & & & \\ & a_2 & & \\ & & a_3 & \\ & & & a_4 \end{bmatrix} \begin{bmatrix} - & & & \\ 1 & 0 & & \\ & 1 & 0 & \\ & & 1 & 0 \end{bmatrix} \tag{3.24}$$

and the realization

$$\left[\begin{array}{cccc|cccc} | & & & & 1 & & & \\ & 0 & & & & 1 & & \\ & & 0 & & & & 1 & \\ & & & - & & & & - \\ \hline | & & & & d_1 & & & \\ & a_2 & & & & d_2 & & \\ & & a_3 & & & & d_3 & \\ & & & a_4 & & & & d_4 \end{array}\right] \tag{3.25}$$

(it is worthwhile checking this out carefully!)

3.10 Discrete-Time and Linear Time-Invariant Systems*

A time-invariant system runs from arbitrarily small negative indices to arbitrary large positive ones; it is by definition an infinitely indexed system. This mere fact makes a number of issues related to such systems rather peculiar. Historically, such systems were studied first, and it was soon thought that time-variant systems would be more complicated, which, in light of modern time-variant theories, appears not to be true at all. Some attempts to generalize results for time-invariant systems to the time-variant case failed miserably (we shall discuss some in further chapters), and it took a long time to understand that the algebraic principles underlying general, time-variant dynamical systems were at the same time simpler, more general and more intuitive than in the LTI case. Much is somewhat similar to the move from continuous-time to discrete-time systems, and in particular from the Laplace transform to the z-transform. Discrete-time systems are basically much simpler than continuous-time systems. The modern approach that we advocate turns things around: one discusses the discrete-time, time-variant case first and then moves on to the continuous-time, time-invariant case.

Concerning the notion of time invariance, one may adopt an internal and an external or behavioral view. The internal view is that none of the constituents of the system change with time. If the system has a "natural" state-space description, time invariance means that this description is the same at all times. In the case of discrete-time, linear systems with finite-dimensional states, this means that the same transition map $\begin{bmatrix} A & B \\ C & D \end{bmatrix}$ is valid for all k running from $-\infty$ to $+\infty$. In that case, the corresponding global block diagonal matrices A, B, C, and D are diagonal *block Toeplitz*, meaning that their diagonal entries are all the same matrix. In that case, the shift Z commutes with the operators and can simply be written as a "scalar" z, since we have $zA = Az$. Also, the diagonal calculus can be replaced by z-transform calculus, in which the constant local transition map is used. The time-invariant transfer function can then be written with a lower case z as is common with the z-transform:

$$T(z) = D + C(I - zA)^{-1}zB, \tag{3.26}$$

in which z is a forward shift. (*Remark:* in much of the engineering literature, the forward shift is written as z^{-1}, with z being the backward shift. We do not follow that convention, mainly to avoid formulas with lots of z^{-1}'s.)[4] This is useful, because then $(I - zA)^{-1}$ can be interpreted as a rational matrix:

$$(I - zA)^{-1} = \frac{1}{\zeta_A(z)} M_A(z), \tag{3.27}$$

in which $\zeta_A(z) = \det(I - zA)$ and $M_A(z)$ is the matrix of minors of $(I - zA)$. (*Notice:* if $\chi_A(\lambda) = \det(\lambda - A)$ is the characteristic polynomial of A of dimension $\eta \times \eta$, then $\zeta_A(z) = z^{\eta}\chi_A(z^{-1})$.) $\zeta_A(z)$ is a polynomial of degree η, and $M_A(z)$ is a matrix of polynomials of degree at most $\eta - 1$. The transfer function can then be expressed as

$$T(z) = \frac{1}{\zeta_A(z)}[D\zeta_A(z) + zCM_A(z)B]. \tag{3.28}$$

Much of discrete-time LTI system theory consists in studying this expression, using various properties from linear algebra and analysis. In particular, the zeros of $\zeta_A(z)$ are the poles of $T(z)$, and $T(z)$ will be "stable," if all these lie strictly outside the unit disc $\mathbf{D}$ of the complex plane. It may be seen that this condition is equivalent to the more general condition of "uniform exponential stability" that we discussed earlier. Suppose that $a \in \mathbf{D}$, with $a \neq 0$ being an eigenvalue of A; then $1/a$ is a pole of $T(z)$, and it lies outside the unit disc. It would be called a "stable pole." Suppose that $a \neq 0$ is an eigenvalue of order ℓ of A, then one can show that the corresponding time series decays as $\ell k|a|^k$ for indices k running to $+\infty$ when this pole is excited, which will always be majorized eventually by $(|a| + \epsilon)^k$ for any arbitrarily small $\epsilon > 0$. One way to study these phenomena is to convert the state transition matrix A to a Jordan canonically form; in further chapters, we shall hint at some of those. In particular, when all the eigenvalues of A are zero, we shall have $\zeta_A(z) = I$, with A being nilpotent and $T(z)$ purely polynomial, called a "moving average filter."

[4] Although we shall not pursue this path formally, there is an "isomorphism" involved here, where a calculus of (doubly infinitely indexed) Toeplitz matrices is made to correspond with infinite matrix series in the variable z. For example, the input–output causal Toeplitz operator T corresponds to the *transfer function* $T(z) = T_0 + zT_1 + z^2T_2 + \cdots$.

Conversely, suppose that a is an eigenvalue of A, with $|a| > 1$, and that the corresponding pole $1/a$ is not canceled out by all entries in $M_A(z)$; then the system will definitely be unstable. A somewhat dubious case arises when the eigenvalue a has the modulus $|a| = 1$, resulting in a kind of borderline stability when the eigenvalue is single. (When the eigenvalue is multiple, it is unstable.) However, there are good reasons to even call the case of a single-boundary eigenvalue unstable, which is what is usually done in the literature.

3.11 Discussion Issues

- Given two LTV systems, one could combine them in various ways. Since each represents an input–output operator, one could, for example, add them when they have the same inputs and outputs ($T = T_1 + T_2$) or multiply them, in case the output of one can be taken as input to the next ($T = T_2 T_1$). Suppose each has an appropriate realization, what would be a realization of the sum or the product? One could, of course, consider more general, networked cases: How could a more general theory look formally?
- The various construction mechanisms for block- and block diagonal matrices seem to be only partially compatible with each other. Would there be a more logical system, that is, at the same time simple, MATLAB-like and allowing for arbitrary, matrix-compatible constructs?
- With respect to the computation of $(I - ZA)^{-1}$ in Section 3.3 and Example 3.7: an easy way to see how signals are put together (in particular, the state at a certain index) is by drawing the data flow diagram, much in the taste of Figs. 3.2 and 3.3. You can easily check the correctness of the formulas given that way. Which properties should such a diagram have so that it corresponds to an executable computation?
- Stability as we defined it in Section 3.3, whether u.e.s. in the time-variant case or depending on the location of poles of the transfer function, is based on the resulting evolution of the state or the response of the system when time increases. Or, to put it differently, it describes whether the effect of a single disturbance dies down when the system progresses in time. Another notion that is often used is "boundedness of the input–output map, when a square norm is used on the input and the output spaces." Is there a relation between stability and this type of boundedness? (This is a complex issue that may require quite a bit of research.)

3.12 Notes

- The significant interest in linear time-variant systems started most likely with the groundbreaking work of R. E. Kalman on state estimation theory, in the early

1960s [48]. For a nice overview of much of the early work, see [46]. The gist of the new movement in system theory was the introduction of state-space descriptions, so that system properties could be studied intimately via the properties of the evolution of the state rather than purely from an input–output point of view, as was the standard until then. Notions such as reachability, controllability and observability became the central concepts. They allowed the development of new approaches not only to estimation theory but also to control theory and network theory. Even though many of these were very successful, a big gap remained between the properties of time-variant and time-invariant systems, because in the latter, full use could be made of transform theory via the properties of poles and zeros of the transfer function or the eigenvalues of the state transition matrix. Only in the beginning of the 1990s, a bridging concept was discovered that allowed many, if not most, of the properties of time-invariant systems to be generalized. The present chapter introduces that concept for the case of discrete-time systems, namely the globalization of the system description via instantaneous diagonal operators and the shift operator Z. In the remainder of this book, we study the emerging basic concepts in detail. We just mention at this point that "inner–outer" factorization will turn out to be the central method. From a mathematical point of view, inner–outer factorization is exemplary of what happens in what has been termed "nest algebras" by Ringrose [58] and Arveson [7], which provide for the basic theoretical framework common to both time-invariant and time-variant systems. The case of discrete-time, time-variant systems using diagonal calculus is central to the nest algebra approach that makes the treatment of time-variant and time-invariant look very similar and was first proposed in [31]; see [29] for a full account.

- One of the salient features of using diagonals as basic entities is that (block) diagonals act as the "scalars" or basic building blocks of the theory, very much like real or complex numbers do in elementary one-input one-output system theory, or $n \times m$ real or complex matrices in multiport theory, with m being the dimension of the input vectors and n that of the output vectors. In other words: only the character of the basic state operators $\{A, B, C, D\}$ changes, but not, in a large part, their algebra. They no longer commute with shifts, but the structure remains rich enough to allow for most, if not all, basic notions and operations. This will become apparent in the following chapters and is also the main motivation why the theory is developed in this way.
- There has been quite some interest for semiseparable and, later, quasi-separable systems in the numerical literature; see, for example, the reference book of Raf Vandebril, Marc Van Barel and Nicola Mastronardi [55]. Originally, a matrix was said to be semiseparable if both its lower part and upper part are restrictions (projections) of full, rank-one matrices (i.e., matrices of the type uv'). The notion was soon generalized to what has been called "quasi-separability," meaning that both upper and lower parts have Hankel operators of low rank. Quasi-separability has presently acquired the latter general meaning and is therefore equivalent to the time variant of

mixed causal–anti-causal type. The interest of part of the numerical community has focused on efficiently solving eigenvalue problems for semiseparable systems (as in the book cited). The emphasis in this book is on problems related to system theory, such as identification, estimation and control, interpolation, model reduction and embedding (see references in the relevant chapters). All these efforts are, of course, complementary.

4 System Identification

System identification tries to answer the question: Given input–output measurements of a system, what could be a system model that reproduces the measurements? This question does not have a simple answer in general without at least some information concerning which kind of system one is dealing with: it is the problem of identifying what is inside a black box without knowledge of what it could possibly be, using only information from accessible inputs and outputs. Our goal then can only be modest. The first question we shall be capable of answering is: Given the knowledge that the system under consideration is a discrete-time and linear system, and given its transfer operator (i.e., its impulse response for each time point), what are compatible system models? Next, we show how to satisfy the requirement of having the complete transfer function available and answer the question for "given partial information on the transfer function" under some conditions.

Menu

Hors d'oeuvre
The Matrix Case

Intermezzo
Specializing to Time Invariance*

Main Course
Realizations with Partial Data

Dessert
The Case with Infinite Indices

4.1 Introduction

In Chapter 3, we determined the matrix representation for the input–output behavior of a causal dynamical systems with a state-space realization – at least in the discrete-time, linear case. System identification takes the opposite path: it starts from an input–output description and then tries to figure out a state-space realization that produces the specified behavior.

Characterizing the relation between input–output behavior on the one hand and the internal structure of the system on the other provides the basis for system synthesis.

This consists in transforming a desired behavioral description (an input–output description) into a feasible architecture.

In this chapter, we work primarily on causal systems – the anti-causal case being dual and equivalent algebraically. Hence, we assume the transfer matrices describing the input–output behavior to be lower block triangular. It soon turns out that for systems with relatively small state-space description, the full input–output description is highly redundant, and the identification can be done with a limited amount of well-chosen data. In another direction, we could try to match limited system descriptions to a limited number of experiments, a point that we shall only touch. The procedures are applicable to mixed causal–anti-causal systems as well.

4.2 The Finite Matrix Case

We assume that we are given a lower block triangular transfer matrix T, and we want to find a *realization* (i.e., a state-space description) that reproduces it. From Chapter 3, we already know that just finding a potentially highly redundant, nonminimal realization is relatively easy: just decompose T in its block diagonals and a standard "companion form" realization follows; see Section 3.9. This is not really what is desired, because the dimensions of such a realization are as large as the original data, and the realization does not provide any structural information on the system. The whole attraction of state-space descriptions comes from the fact that many systems possess a low-order system description, which can be used to execute efficient computations, for example, matrix–vector multiplication or system inversion – to be treated in following chapters – state estimation or system control. Hence, what is needed is a realization that is as small as possible. We shall soon see that there are indeed realizations with the smallest possible dimension at each index point and that these dimensions are uniquely determined by the input–output matrix – a very strong result.

At this point, our basic insights from Chapter 2 come to the rescue. The state of the system at a given index k is *what the system has to remember from its past so as to allow the determination of its future evolution, given future inputs* – see Fig. 2.2. To put it differently: past inputs produce the same state when the observation of any future system evolution cannot distinguish between them, whichever future test inputs are used. (We then say that these inputs are Nerode equivalent.) We also saw that Nerode equivalence in a linear system can be determined using the value zero as the test input for the future, that is, purely from the Hankel map.

The *Hankel map* was defined as the map of strict past system inputs to future outputs, with zero as the future input. Let us recall the procedure. At every index point k, we subdivide any input sequence into a strict past part and a future part, that is, $u = u_{p_k} + u_{f_k}$, with u_{p_k} being the input up to and including index $k - 1$ and u_{f_k} the input from k to infinity (the subdivision w.r.t. the index k puts the "present" k in the future part), and we do likewise for y. With this, the (causal) input–output operator T decomposes as follows:

$$T = \begin{bmatrix} T_{k-} & 0 \\ H_k & T_{k+} \end{bmatrix} = \left[\begin{array}{cccc|cccc} \ddots & & & & & & & \\ \ddots & T_{k-3,k-3} & & & & & & \\ \ddots & T_{k-2,k-3} & T_{k-2,k-2} & & & \mathbf{0} & & \\ \cdots & T_{k-1,k-3} & T_{k-1,k-2} & T_{k-1,k-1} & & & & \\ \hline \cdots & T_{k,k-3} & T_{k,k-2} & T_{k,k-1} & T_{k,k} & & & \\ \ddots & T_{k+1,k-3} & T_{k+1,k-2} & T_{k+1,k-1} & T_{k+1,k} & T_{k+1,k+1} & & \\ \ddots & T_{k+2,k-3} & T_{k+2,k-2} & T_{k+2,k-1} & T_{k+2,k} & T_{k+2,k+1} & T_{k+2,k+2} & \\ \iddots & \ddots & \ddots & \vdots & \vdots & \ddots & \ddots & \ddots \end{array}\right]$$

Figure 4.1 The decomposition of T at the index point k into strict past and future.

$$y = Tu \mapsto \begin{bmatrix} y_{p_k} \\ y_{f_k} \end{bmatrix} = \begin{bmatrix} T_{k-} & 0 \\ H_k & T_{k+} \end{bmatrix} \begin{bmatrix} u_{p_k} \\ u_{f_k} \end{bmatrix}. \tag{4.1}$$

Hence, in MATLAB notation,[1] we have $T_{k-} := T_{:(k-1),:(k-1)}$, $H_k := T_{k:,:(k-1)}$ and $T_{k+} := T_{k:,k:}$; see Fig. 4.1. T_{k-} maps strict past to strict past, T_{k+} maps future (including present) to future (including present), while the crucial Hankel operator H_k maps strict past to future at the index point k.

From the relation $y_{f_k} = H_k u_{p_k} + T_{k+} u_{f_k}$, we see that two inputs in the (strict) past will produce the same output in the future when their difference belongs to the kernel of H_k, no matter what the future input u_{f_k} is. Conversely, if, for some future input, two past inputs produce different future outputs, then the difference between these two past inputs cannot be in the kernel of H_k. It follows that *the minimal state dimension of the system at the index point k has to be equal to or larger than the rank of* H_k. (We shall constructively show that it may be chosen equal.)

It will turn out that finding an adequate realization consists in determining matrices $\{A_k, B_k, C_k, D_k\}$ at every index k, which reproduce H_k and thereby also T. Let us therefore first derive conditions these matrices have to satisfy so that they reproduce each H_k. Next, we use the knowledge so gained to turn the tables and derive an algorithm that produces matrices that satisfy the necessary properties. (This is a common strategy in algebra: assume a solution, find a set of properties and if these turn out to be necessary and sufficient, use them to determine the solution.) Finally, we have to prove that the so derived realization matches the original data (showing that the properties used are indeed sufficient). From Eq. (3.2), we find, assuming the model,

$$H_k = \begin{bmatrix} \cdots & C_k A_{k-1} A_{k-2} B_{k-3} & C_k A_{k-1} B_{k-2} & C_k B_{k-1} \\ \ddots & C_{k+1} A^{k \geq k-2} B_{k-3} & C_{k+1} A_k A_{k-1} B_{k-2} & C_{k+1} A_k B_{k-1} \\ \ddots & C_{k+2} A^{k+1 \geq k-2} B_{k-3} & C_{k+2} A^{k+1 \geq k-1} B_{k-2} & C_{k+2} A_{k+1} A_k B_{k-1} \\ \iddots & \ddots & \ddots & \vdots \end{bmatrix}, \tag{4.2}$$

[1] In MATLAB, ranges of indices are indicated as follows: 1:5 means from 1 to 5 inclusive, 1: runs from 1 to infinity, and multiple ranges are separated by comma's. Hence, a doubly infinite matrix with rows running from $-\infty$ to -1 and columns from 0 to ∞ has ranges $: -1, 0 :$.

and we see that H_k factorizes into two operators:

$$H_k = \begin{bmatrix} C_k \\ C_{k+1}A_k \\ C_{k+2}A_{k+1}A_k \\ \vdots \end{bmatrix} \begin{bmatrix} \cdots & A_{k-1}A_{k-2}B_{k-3} & A_{k-1}B_{k-2} & B_{k-1} \end{bmatrix}$$
$$:= \mathbf{O}_k \mathbf{R}_k. \quad (4.3)$$

Interpreting this decomposition by applying u_{p_k} on the right, we see

$$\mathbf{R}_k u_{p_k} = \begin{bmatrix} \cdots & A_{k-1}A_{k-2}B_{k-3} & A_{k-1}B_{k-2} & B_{k-1} \end{bmatrix} u_{p_k} = x_k, \quad (4.4)$$

the state for this realization, and, dually, $\mathbf{O}_k x_k = y_{f_k}$, when u_{f_k} is zero. Referring back to Chapter 2, we realize that we have discovered the *reachability operator* $\mathbf{R}_k$ at index k, and, dually, the *observability operator* $\mathbf{O}_k$ at the same index for this particular realization.

Before "turning the tables," we have to derive important (necessary and sufficient) properties of the decomposition of the H_k, given a realization. First, we see that $C_k = [\mathbf{O}_k]_k$ and $B_{k-1} = [\mathbf{R}_k]_{k-1}$.

Remark: $\mathbf{O}_k$ is a block matrix with one block column running from k to $+\infty$, whose kth element is C_k (the top element of a column matrix starting at k). A dual remark applies to $\mathbf{R}_k$, but the indexing convention makes B_{k-1} its rightmost element, because the strict past runs up to $k-1$.

Next, considering $[\mathbf{O}_k]_{(k+1):}$ (the beheaded $\mathbf{O}_k$, which we also write as $\mathbf{O}_k^{\uparrow}$), we see that

$$[\mathbf{O}_k]_{(k+1):} = \mathbf{O}_{k+1} A_k. \quad (4.5)$$

So, if $\mathbf{O}_{k+1}$ is left invertible, that is, if there exists a *pseudo-inverse* $\mathbf{O}_{k+1}^{+}$ such that $\mathbf{O}_{k+1}^{+}\mathbf{O}_{k+1} = I$, then $A_k = \mathbf{O}_{k+1}^{+}[\mathbf{O}_k]_{(k+1):}$.[2] Dually,

$$A_k[\mathbf{R}_k] = [\mathbf{R}_{k+1}]_{:(k-1)}, \quad (4.6)$$

and hence $A_k = [\mathbf{R}_{k+1}]_{:(k-1)}\mathbf{R}_k^{+}$, provided $\mathbf{R}_k$ is right invertible. These observations give rise to a few definitions.

Definition 4.1 We say that a realization is reachable if and only if all $\mathbf{R}_k$ are right invertible. Equivalently, the rows of each $\mathbf{R}_k$ are linearly independent.

Definition 4.2 We say that a realization is observable if and only if all $\mathbf{O}_k$ are left invertible. Equivalently, the columns of $\mathbf{O}_k$ are linearly independent.

Definition 4.3 We say that a realization is minimal if and only if it is both reachable and observable.

[2] Often the Moore–Penrose pseudo-inverse $X^{\dagger}$ is used. However, one may also allow more general pseudo-inverses X^{+} here. Those are often computationally less demanding.

When a realization is minimal, then the corresponding factorizations of the H_k are minimal as well, and conversely, when all factorizations of H_k are minimal, then the resulting realization is minimal. This means that, in the minimal case, the columns of $\mathbf{O}_k$ form a basis for the range of H_k, while by the same token, the columns of $\mathbf{R}_k'$ form a basis for the co-range of H_k (i.e., the range of H_k'), as happens with any minimal factorization of a matrix.

The gist of realization theory is that, conversely, minimal factorizations of H_k produce minimal realizations. We formulate this as a theorem. (This may be the most important theorem in system theory, which we could call the "generalized Kronecker theorem"; see the notes at the end of the chapter on this.)

Theorem 4.4 *The rank of the Hankel operator H_k at each index k is the minimal local degree of the system. Let, for each k, $H_k := \mathbf{O}_k\mathbf{R}_k$ be a minimal factorization of H_k; then a corresponding minimal realization is given by*

$$\begin{cases} A_k = \mathbf{O}_{k+1}^+[\mathbf{O}_k]_{(k+1):} \\ B_k = [\mathbf{R}_{k+1}]_k \\ C_k = [\mathbf{O}_k]_k \\ D_k = T_{k,k}. \end{cases} \tag{4.7}$$

In addition, $A_k = [\mathbf{R}_{k+1}]_{:(k-1)}\mathbf{R}_k^+$.

Proof

For a given minimal factorization of each H_k, we first show that the realization defined by (4.7) reproduces all the entries in the transfer operator T. This we do by first showing that the given realization reproduces all the left factors $\mathbf{O}_k$. This is clear by definition for the first entry of each $\mathbf{O}_k$, with the given definition, $C_k = [\mathbf{O}_k]_k$. For the remaining entries $[\mathbf{O}_k]_{(k+1):}$, we observe that, with the given definitions, $\mathbf{O}_{k+1}A_k = \mathbf{O}_{k+1}\mathbf{O}_{k+1}^+[\mathbf{O}_k]_{(k+1):} = \Pi_{k+1}[\mathbf{O}_k]_{(k+1):}$ with $\Pi_{k+1} := \mathbf{O}_{k+1}\mathbf{O}_{k+1}^+$, which is a (perhaps skew)[3] projection operator onto the range of H_{k+1}. But *the columns of* $[\mathbf{O}_k]_{(k+1):}$ *belong to the range of H_{k+1} since we assumed a minimal factorization*, so Π_{k+1} projects them on themselves, and hence $[\mathbf{O}_k]_{(k+1):} = \mathbf{O}_{k+1}A_k$. It follows that the $\{A_k, C_k\}$ define all the subsequent $\mathbf{O}_k$, starting from the highest relevant value of k. A dual reasoning now works for the $\mathbf{R}_k$, with the definition $A_k = [\mathbf{R}_{k+1}]_{:(k-1)}\mathbf{R}_k^+$. Hence, it remains to be shown that the two definitions for A_k produce the same result.

From the way the Hankel operators are intertwined, we have $[H_k]_{(k+1):,:} = [H_{k+1}]_{:,:(k+1)}$, and hence $[\mathbf{O}_k]_{(k+1):}\mathbf{R}_k = \mathbf{O}_{k+1}[\mathbf{R}_{k+1}]_{:(k-1)}]$. Pre- and post-multiplication of these respectively by $\mathbf{O}_{k+1}^+$ and $\mathbf{R}_k^+$ produces the equality of the definitions of A_k.

This defines all the (block) entries below the main diagonal or T in terms of the $\{A_k, B_k, C_k\}$. The entries on the main diagonal are directly given as $D_k = T_{k,k}$, and all the entries in the strict upper block triangle of T are zero because of the assumed

[3] Depending on the pseudo-inverse used. In the case where the Moore–Penrose inverse is used, then the projection is orthogonal.

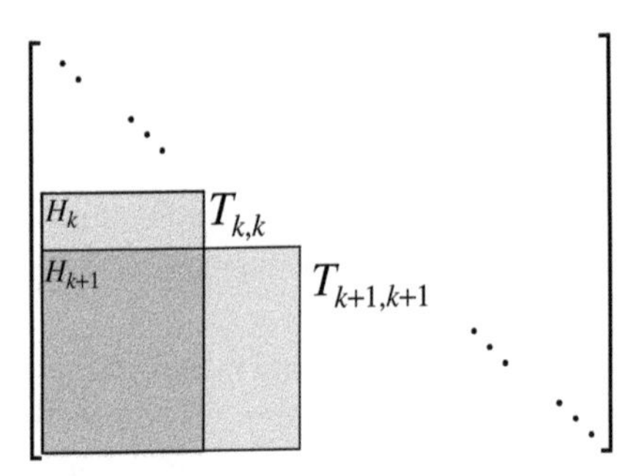

Figure 4.2 How the Hankel operators at k and $k+1$ are intertwined.

causality. Conversely, given the transfer operator T, the H_ks are uniquely defined, and hence their rank, which equals the local degree. □

Figure 4.2 illustrates the intertwining property of the Hankel operators. Concerning the pseudo-inverses, up to this point we have not supposed them to be Moore–Penrose inverses, so actually there is a large collection of pseudo-inverses possible, which, however, would all produce a valid realization given any minimal factorization. Also, with the choice of bases for the columns or rows of each H_k, there is a lot of freedom possible, which we shall soon exploit.

Example Consider, for example,

$$T = \begin{bmatrix} 1 & & & \\ 0 & 1 & & \\ 0 & 0 & 1 & \\ 1 & 0 & 0 & 1 \end{bmatrix} \tag{4.8}$$

with the normal matrix indexing schema. We have $H_1 = \begin{bmatrix} | \\ | \\ | \\ | \end{bmatrix}$ empty. Next,

$$H_2 = \begin{bmatrix} 0 \\ 0 \\ 1 \end{bmatrix}, H_3 = \begin{bmatrix} 0 & 0 \\ 1 & 0 \end{bmatrix} = \begin{bmatrix} 0 \\ 1 \end{bmatrix} \begin{bmatrix} 1 & 0 \end{bmatrix}, H_4 = \begin{bmatrix} 1 & 0 & 0 \end{bmatrix}, \tag{4.9}$$

and hence $C_1 = |$ is empty, and $C_2 = 0$, $C_3 = 0$ and $C_4 = 1$. Likewise, $B_1 = 1$, $B_2 = 0$, $B_3 = 0$, and B_4 is empty. We can take $\mathbf{O}_2^+ = \begin{bmatrix} 0 & 0 & 1 \end{bmatrix}$, $\mathbf{O}_3^+ = \begin{bmatrix} 0 & 1 \end{bmatrix}$ and $\mathbf{O}_4^+ = [1]$, so that A_4 is empty, $A_3 = 1 * 1 = 1$, $A_2 = \begin{bmatrix} 0 & 1 \end{bmatrix} \begin{bmatrix} 0 \\ 1 \end{bmatrix} = 1$ and A_1 again is empty. The result is shown in Fig. 4.3.

Remark that in this example, other values for, for example, $\mathbf{O}_2^+$ would be $\begin{bmatrix} x & y & 1 \end{bmatrix}$, with arbitrary values for x and y, and that this would not influence the result. Other minimal realizations would be found by using different minimal

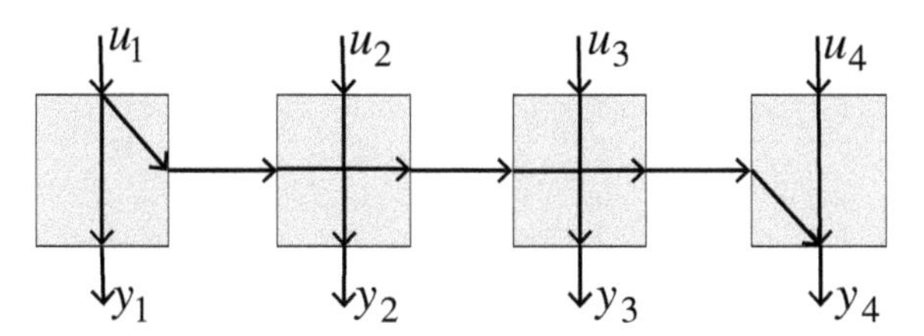

Figure 4.3 A realization example.

factorizations of the subsequent Hankel operators H_k; for example, one could write $H_3 = \begin{bmatrix} 0 \\ 2 \end{bmatrix} \begin{bmatrix} 1/2 & 0 \end{bmatrix}$, and one would have obtained a somewhat different realization. (In this simple example, the differences are not that big.)

4.3 Realizations Using Reduced Past to Future Matrices

The realization theory of Section 4.2 may be satisfying in its completeness; it suffers from overkill in practical situations. Under certain conditions that are often satisfied, the matrices used (see Fig. 4.2) can be drastically reduced. Here is how this works. Suppose that for some reason we know upper limits to the sizes $a_k - k$ with $a_k > k$ and $k - b_k$ with $b_k < k$ needed for each partial $[H_k]_{k:a_k,b_k:k}$ to reach the full rank δ_k, with, in addition, $a_{k+1} \geq a_k$ and $b_{k+1} \geq b_k$. This would, for example, be the case when the transfer matrix is a block-banded matrix or the inverse of a block-banded matrix; or else, and assuming that it is known that elements decay rapidly at some distance away from the diagonal, one can assume them to contribute little. (Note that this can be a hazardous assumption in the case of very large matrices, in particular in the case of matrices representing higher dimension finite element discretizations.) It turns out that minimal factorizations of somewhat enlarged submatrices of H_k are then sufficient to find a realization.

To show this, consider minimal factorizations of $[H_k]_{k:a_k,b_k:k} := \widehat{\mathbf{O}}_k \widehat{\mathbf{R}}_k$, a partial Hankel operator as shown in Fig. 4.4. Then, assuming a realization and following the same strategy as in Section 4.2, it would follow

$$\begin{cases} A_k = \widehat{\mathbf{O}}_{k+1}^{+} H_{k+1:a_{k+1},b_k:k} \widehat{\mathbf{R}}_k^{+} \\ B_k = [\widehat{\mathbf{R}}_{k+1}]_k \\ C_k = [\widehat{\mathbf{O}}_k]_k \\ D_k = T_{k,k} \end{cases} \tag{4.10}$$

very much as before. The reason of this is that factorizations of the somewhat larger partial Hankels produce factorizations of smaller partial Hankels, and they can be extended to factorizations of larger ones when full rank has been achieved. For a full proof, see [29], but it is not hard to see that

$$H_{k+1:a_{k+1},b_k:k} = \widehat{\mathbf{O}}_{k+1} A_k \widehat{\mathbf{R}}_k, \tag{4.11}$$

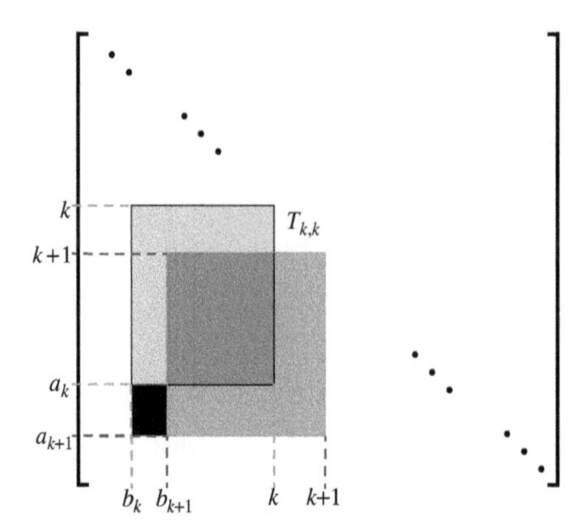

Figure 4.4 Realization with partial Hankel operators.

which is the main relation used in the proof. Besides the partial Hankel operators, knowledge of the darkly marked submatrix at the bottom left is needed. We leave it to the reader to investigate how the ideas of Theorem 4.4 can be refined to handle the present case.

4.4 The Time-Invariant Case*

The time-invariant case is in essence not much different from the matrix case just treated, except that the matrices involved are now infinite dimensional. A fully responsible treatment necessitates the introduction of a mathematical framework that allows for such matrices, because matrix–vector products are no longer guaranteed to converge. Although the theory can straightforwardly be extended to that case given certain restrictions, we suffice here with an informal discussion. Because of time invariance, all Hankel operators are now equal; let's look at H_0, and let $T(z) = T_0 + zT_1 + z^2T_2 + \cdots$; then

$$H_0 = \begin{bmatrix} \cdots & T_3 & T_2 & T_1 \\ \ddots & T_4 & T_3 & T_2 \\ \ddots & T_5 & T_4 & T_3 \\ \cdot^{\cdot^{\cdot}} & \ddots & \ddots & \vdots \end{bmatrix}. \tag{4.12}$$

This is a famous "block Hankel matrix," which is normally written as

$$H := \begin{bmatrix} T_1 & T_2 & T_3 & \cdots \\ T_2 & T_3 & T_4 & \cdot^{\cdot^{\cdot}} \\ T_3 & T_4 & T_5 & \cdot^{\cdot^{\cdot}} \\ \vdots & \cdot^{\cdot^{\cdot}} & \cdot^{\cdot^{\cdot}} & \ddots \end{bmatrix} \tag{4.13}$$

after reordering of the columns (corresponding to reversing the input order). The Hankel matrix is highly regular (it has identical blocks on anti-diagonals), and it is handy to introduce a constructor that produces a Hankel matrix from a series of blocks (notice a different definition of H_k in this section, now referring to the block dimension, while before, it was a Hankel operator pertaining to the kth index point):

$$H_k := \mathrm{Han}(T_1,T_2,\cdots,T_k) := \begin{bmatrix} T_1 & T_2 & \cdots & T_k \\ T_2 & T_3 & \iddots & T_{k+1} \\ T_3 & T_4 & \iddots & T_{k+2} \\ \vdots & \iddots & \iddots & \vdots \\ T_k & T_{k+1} & \cdots & T_{2k-1} \end{bmatrix}. \tag{4.14}$$

The famous result (originally due to Kronecker in the scalar case) is now that $T(z)$ is a rational transfer function, if and only if $H := H_\infty$ is finite dimensional (i.e., its rows and columns span finite-dimensional spaces). This somewhat delicate question (because the T_k might not be bounded when k increases, producing unbounded H_k's) can luckily be dealt with without recourse to infinite matrices. Suppose that for a given Hankel matrix, we can ascertain that there is an index k so that $\mathrm{Han}(T_1 \cdots T_\ell)$ has the same rank δ for all $\ell \geq k$; then this can be used as a criterion for finite dimensionality of H, even when H turns out not to be bounded in any reasonable metric. Actually, one can even take k as the minimal such value, in which case, one would call it the *order* of the system, with the rank η being its *degree*. Taking a finite H_ℓ with $\ell > k$ and assuming the H_ℓ to be block indexed starting at index 1, let us factorize it minimally as $H_\ell = \mathbf{O}_\ell \mathbf{R}_\ell$, where $\mathbf{O}_\ell$ has δ columns and $\mathbf{R}_\ell$ has δ rows. One can now easily see that $H_{\ell-1}$ has a minimal factorization derived from the previous factorization, as $H_{\ell-1} = [\mathbf{O}_\ell]_{1:(\ell-1)}[\mathbf{R}_\ell]_{1:(\ell-1)}$, also of rank δ. The construction of the previous section applies here as well:

$$\begin{cases} A := [\mathbf{O}_\ell]^+_{1:(\ell-1)}[\mathbf{O}_\ell]_{2:} (= [\mathbf{R}_\ell]_{2:}[\mathbf{R}_\ell]^+_{1:(\ell-1)}) \\ B := \mathbf{R}_1 \\ C := \mathbf{O}_1 \\ D := T_0 \end{cases} \tag{4.15}$$

and the proof runs precisely like in Theorem 4.4.

Once the full rank δ is reached at the block index k, all the subsequent finite-dimensional Hankel matrices can actually be constructed from H_k because the additional rows and columns are all linearly dependent on those in H_k, and contain sufficient blocks that are already known, which determine the further coefficients in the factorization. This *partial realization* method generalizes actually to the general LTV case; see Section 4.5.

4.5 Identifying a Running System*

Up to this point, we assumed the given input–output matrix, and the problem was to find a minimal realization. However, direct access to all the information needed, that is, all the entries of the matrix (namely all the impulse responses at each index point), may not be available, and the realization has to be made from indirect measurements, or it may be that only partial information is available. We already observed that Section 4.4, where we were able to construct a realization from smaller submatrices of the Hankel operators, provided some additional information on the system, that is, the maximum delay needed to reach the dimension of the state at each index point. A further question would be: Can one identify the system just based on observing inputs and outputs?

Let us first, for exploration's sake, look at systems with one input and one output. Suppose that the input is u and the output is y, and we know already that the system is causal. Can we specify the matrix from that information? Clearly, one has to bring in all the extra information one has and find input–output matrices that are compatible with it, for example, time invariance. This can be fully accounted for by observing that if $u \mapsto y$, then also $Zu \mapsto Zy$, etc. ... Thus, a full set of input–output maps (I/O maps) becomes available, merely by shifting (assuming u starts at some point)

$$\begin{bmatrix} u_0 & 0 & 0 & \cdots \\ u_1 & u_0 & 0 & \ddots \\ u_2 & u_1 & u_0 & \ddots \\ \vdots & \ddots & \ddots & \ddots \end{bmatrix} \mapsto \begin{bmatrix} y_0 & 0 & 0 & \cdots \\ y_1 & y_0 & 0 & \ddots \\ y_2 & y_1 & y_0 & \ddots \\ \vdots & \ddots & \ddots & \ddots \end{bmatrix}. \tag{4.16}$$

Suppose now that u_0 is neither unreasonably small nor large; then this map, when restricted to dimension n, specifies the first n entires of the I/O map:

$$\begin{bmatrix} T_0 & 0 & 0 & \cdots \\ T_1 & T_0 & 0 & \ddots \\ T_2 & T_1 & T_0 & \ddots \\ \vdots & \ddots & \ddots & \ddots \end{bmatrix} = \begin{bmatrix} y_0 & 0 & 0 & \cdots \\ y_1 & y_0 & 0 & \ddots \\ y_2 & y_1 & y_0 & \ddots \\ \vdots & \ddots & \ddots & \ddots \end{bmatrix} \begin{bmatrix} u_0 & 0 & 0 & \cdots \\ u_1 & u_0 & 0 & \ddots \\ u_2 & u_1 & u_0 & \ddots \\ \vdots & \ddots & \ddots & \ddots \end{bmatrix}^{-1}. \tag{4.17}$$

(Observe that the result will automatically be finite Toeplitz.) Hence, a partial representation of the I/O map in the sense of Section 4.4 has been obtained.

In the case of a time-variant system, this does not work, because we lose the shift property and hence cannot construct a full, invertible input matrix that will determine the transfer function, not even partially. Not enough information is available to construct the full I/O map. Suppose, to begin with, that a single input–output pair (u, y) is the only information available; then what is a system with minimal state dimension that will reproduce the pair? First, the system we are identifying is known to be linear, so we should surely have $ua \mapsto ya$ for any number a. Next, to be a bit more specific, let us start at $k = 0$, and suppose we input just u_0, leaving all the next entries 0; then we should certainly have, because of assumed causality, the first output to be y_0, all the following entries being unknown. (Since we do not have the impulse response for a single non-zero input at $k = 0$ available, we end up with quite a bit of freedom.) Next, a similar argument is valid for the inputs $\mathrm{col}[u_0, u_1]$ followed by zeros, for which we only know that it must produce $\mathrm{col}[y_0, y_1]$ and arbitrary further entries. It follows that the following map will hold (with unspecified and hence potentially free to choose entries marked "?"):

$$\begin{bmatrix} T_{0,0} & 0 & 0 & \cdots \\ T_{1,0} & T_{1,1} & 0 & \ddots \\ T_{2,0} & T_{2,1} & T_{2,2} & \ddots \\ \vdots & \ddots & \ddots & \ddots \end{bmatrix} \begin{bmatrix} u_0 & u_0 & u_0 & \cdots \\ 0 & u_1 & u_1 & \ddots \\ 0 & 0 & u_2 & \ddots \\ \vdots & \ddots & \ddots & \ddots \end{bmatrix} \mapsto \begin{bmatrix} y_0 & y_0 & y_0 & \cdots \\ ? & y_1 & y_1 & \ddots \\ ? & ? & y_2 & \ddots \\ \vdots & \ddots & \ddots & \ddots \end{bmatrix}. \tag{4.18}$$

Let us now, for discussion's sake, make the rough assumption that all the entries u_k up to $k = n$ (supposing we have to go that far) are nicely invertible, then the u matrix above is invertible, and we may see that the ranks of the Hankel operators build on the strictly lower part of the T matrix will be the same as for the corresponding lower Hankel operators in the output matrix, all consisting of ?'s, because the inverse of the u matrix is again upper, and the strictly lower parts of the product of the y matrix with the inverse of the u matrix does not depend on the upper part of the y matrix. Hence, if we want to find a realization that is minimal in the state dimensions, we should choose all these ranks to be as small as possible. The simplest would be to put all the ?'s equal to zero. That would then simply yield a diagonal T with $T_{k,k} = y_k/u_k$ obviously the simplest possible solution; it is *algebraically minimal* as it contains exactly the same number of free parameters as the problem has.

What now when one entry $u_k = 0$? We see that in this case u_k has no influence on y_k (multiplying u_k with whatever constant produces zero): the value of y_k has to be generated, because of causality, from previous values of u, and relevant "?" can no longer be zero. (For example, if one wishes $\begin{bmatrix} T_{0,0} & 0 \\ T_{1,0} & T_{1,1} \end{bmatrix} \begin{bmatrix} u_0 & u_0 \\ 0 & 0 \end{bmatrix} = \begin{bmatrix} y_0 & y_0 \\ ? & y_1 \end{bmatrix}$, one has no alternative but to choose $T_{1,0} = y_1/u_0$ and hence $? = y_1$.) In the more general case, one can fill in the ? entries recursively so as to minimize the ranks of the subsequent Hankel operators, taking care of correct ranges as given by the right-hand side (the y matrix); we leave the details for the "Discussion items." So we see that in the LTV case, we end up with realizations that have much smaller state dimensions than what would be the case for LTI systems, although they are subjected to some constraints forced by linearity and causality. The Hankel theory gives us a good grasp on the situation.

Let us now briefly consider generalizations to multi-input multi-output systems (so-called MIMO systems). In the LTI case, a single input–output pair will not suffice to characterize the full transfer function: one would need a sufficient number of independent inputs, at least the same number as the number of input ports. This, together with shifted versions, would then provide a full characterization of the transfer function, by inversion of the input data as done in the one input case. For LTV systems, the situation is even more complex: not only is information needed about inputs applied to all the input ports, but also at every time point. Again as before, such situations can be analyzed using, in particular, the Hankel theory that we have developed, but the situation quickly becomes complex and specific for the example at hand. See the notes for some literature references on the issues.

4.6 Working with Infinite Indices*

With the exploration done so far, it is not too hard to imagine what to do when we deal with LTV systems running for all times (from $-\infty$ to $+\infty$). Let us assume that we dispose of a full set of impulse responses at each index point or, equivalently, of the matrix input/output operator T, which we assume to be causal, and that we are asked to compute a realization, assuming the system does have one. The most practicable way then is to assume that we can *compute* any $T_{i,k}$ $(i \geq k)$ or that it can be given when asked for (because we cannot assume an infinite amount of data to be given numerically).

From our knowledge so far, we know that a realization with finite states will exist, if and only if each Hankel H_k (which is now an infinitely indexed matrix) has finite rank, by which it is meant that the rank of each sub-Hankel $[H_k]_{k:,\ell:(k-1)}$ of H_k will be bounded, uniformly in $\ell < k$. Indeed, assuming a minimal realization to exist, one has

$$
\begin{aligned}
[H_k]_{k:,\ell:(k-1)} &= \begin{bmatrix} C_k A_{k+1} \cdots A_{\ell-1} B_\ell & \cdots & C_k B_{k-1} \\ C_{k+1} A_k \cdots A_{\ell-1} B_\ell & \cdots & C_{k+1} A_k B_{k-1} \\ \vdots & \vdots & \vdots \end{bmatrix} \\
&= \begin{bmatrix} C_k \\ C_{k+1} A_k \\ \vdots \end{bmatrix} \begin{bmatrix} A_{k+1} \cdots A_{\ell-1} B_\ell & \cdots & B_{k-1} \end{bmatrix},
\end{aligned}
\tag{4.19}
$$

so the maximal rank of all these Hankel operators over ℓ will be equivalent to the dimension of the state space at the index point k. If δ_k is this dimension, then there will be an index ℓ_k, such that for all $\ell \leq \ell_k$, the rank of $[H_k]_{k:,\ell:(k-1)}$ will be δ_k, and the factorization shown will be a minimal factorization for all $\ell \leq \ell_k$. Conversely, adopting the strategy of Section 4.3, and finding ℓ_k and ℓ_{k+1} that satisfy the full rank condition, a realization at point k can be found by choosing compatible bases as explained in that section; we leave it to the interested reader to figure out the details.

While the previous paragraph does produce a solution to the identification problem, it is somewhat unsatisfactory, in that quantities ℓ_k and ℓ_{k+1} have to be found that are dependent on unknown quantities, namely the degrees δ_k and δ_{k+1} of the state at the index point k. One can, of course, determine the ranks of Hankel submatrices progressively, but that may amount to a lot of work without guarantee that one has gone out far enough to reach the maximal rank.

The situation is, of course, inherent to the identification problem: in some systems and especially time-variant systems, things may happen at unforeseen points in time, but they can be detected by the rank conditions developed so far. On the one hand, that is satisfying because it is a precise condition; on the other, it may require a lot of computation, which in many systems would be unnecessary, as one can put a bound on potential time delays in the system's operation. If the latter is the case, then one can often state that (1) there is a finite upper limit to the state dimension δ_k over all k and (2) there is an upper limit d to the distance $d_k := k - \ell_k$, uniformly over k, for $[H]_{k:k+d_k,k-d_k,k-1}$ to reach the state dimension, also uniformly over k, so that partial realization theory can be used on submatrices $[H_k]_{k:k+d_k,k-d_k:k-1}$. Systems with this property are sometimes called *locally finite systems*.

4.7 Discussion Items

1. Find (minimal) realizations for interesting "elementary" systems given by their transfer operator, in particular,

   (a) $T = \begin{bmatrix} \boxed{1} & & & \\ \frac{1}{2} & 1 & & \\ & \ddots & \ddots & \end{bmatrix}$ (both finitely indexed and infinitely indexed);

   (b) $T = \begin{bmatrix} \boxed{1} & 2 & & \\ & 1 & 2 & \\ & & \ddots & \ddots \end{bmatrix}$ (finitely and infinitely indexed).

   (c) What about the inversion of the two T's just defined and the realization of the inverses? (Please explore this issue. We shall have to discuss system inversion intensely in several of the following chapters.)

   (d) A Jacobi matrix with shifted main diagonal

$$J = \begin{bmatrix} a_0 & \boxed{b_0} & & & \\ b_0 & a_1 & b_1 & & \\ & b_1 & \ddots & \ddots & \\ & & \ddots & \ddots & \ddots \end{bmatrix} \tag{4.20}$$

   (also finitely and infinitely indexed). Standard Jacobi matrices have coefficients a_i arbitrary real and b_i strict positive, also arbitrary. Suppose a b_k is "accidentally" zero; then what happens? Why is it useful to shift the main diagonal forward?

2. An important issue, not covered in this chapter, is what happens when "partial" identification is done: one uses data of a certain order, does the identification and obtains a realization. How does the realization obtained relate to the original system? (Some questions, e.g., what gets interpolated, may be easy to answer; others, e.g., how well does the partially identified system approximate the original, seem much harder.)
3. There are other ways to approximate or identify a system (or an approximation), given data or information about it. One is to use various kinds of "interpolations" and even specific properties the system may be known to have. For example, it may be known that the system is "bounded" with respect to some norm, and hence the approximation should satisfy the same constraint. How this can be done is discussed in the advanced Chapter 16.
4. In the time-invariant case, it is known that approximating or identifying a system on the basis of either the impulse response or the resulting transfer function may

not be the best thing to do. In particular, the frequency response of the system may be dependent on the given coefficients in a highly sensitive way (meaning that tiny variations in the value of the coefficients entail big variations in frequency response).

4.8 Notes

Needless to say, system identification is a central problem in system theory, and a lot of literature has been devoted to it. Far from giving a full account, we want to mention a few salient contributions.

- The first results of what later became identification theory are due to the German mathematician Leopold Kronecker, who showed that a Maclaurin series belongs to a rational function if and only if the Hankel matrix build on its coefficients has a finite rank. Kronecker's work was easily extended to rational matrix functions and has become a standard fixture of, among other theories, matrix function and Hardy space theory [59].
- In the 1960s and in the wake of the development of state-space theory, a search was on to evaluate methods of identifying the state-space description (the so-called realization) of a linear, discrete-time and time-invariant system and to derive both conditions and algorithms to ascertain that the system given by its (multiport) impulse response has a finite-dimensional state space. The best known is probably the contribution of B. L. Ho and R. E. Kalman [44], but also L. M. Silverman and H. E. Meadows [62] solved the problem independently in the same period. The method presented in this chapter, based on the global Hankel operator, was first presented by A. J. Van der Veen in his thesis [69] and subsumes much of the previous work.
- Identification using real-life measurements is a more involved issue that we could only touch in this chapter, although it uses the same basic principles (how the system maps strict past inputs to the future). Original work on this topic and further references can be found in [70–72].

A nice observation is that system identification based on the knowledge of the input–output transfer data (the transfer matrix or transfer operator) amounts to the determination of the range and the co-range of all the relevant Hankel operators derived from the transfer data, which in turn connects to their subsequent minimal factorizations! This observation greatly simplifies both the methods for identification and the proofs, as compared to the original approaches.

Very often, the determination of a realization of one or the other transfer operator is part of a more general problem such as solving systems of equations, matrix factorization, spectral factorization or the approximation of a matrix with a low degree realization. The methods presented in this chapter often reappear as building blocks to solve those more elaborate issues.

5 State Equivalence, State Reduction

The state of a system is part of its interior and is often hidden and inaccessible. Consequently, a state is by no means uniquely characterized by the input–output behavior of a system. There may be redundancy in the state (think of the memory of a computer: it may contain many items that are not relevant to a given problem); different state sets may produce the same input–output behavior: they are equivalent from the point of view of the input–output map. In this chapter, we deal with these two issues in the context of discrete-time, finitely indexed LTV systems (the finite matrix case), ending it with some considerations concerning infinitely indexed LTV and LTI systems. Changing the state description without changing the behavior can produce desirable effects; in particular, it can produce interesting canonical forms: the input or output normal forms and balanced realizations, all of which play important roles in further developments (approximation, inversion, estimation, control and synthesis).

Menu

Hors d'oeuvre
Introspection

First Course
Equivalences of Minimal States

Second Course
Reduction to Minimal
A First Square-Root Algorithm

Third Course
The LTI Case

Dessert
The Case of Infinite Indices

5.1 Equivalences of Minimal LTV System Realizations

In Chapter 4, we discovered that a minimal realization of a discrete-time LTV system can be obtained through factorization of the Hankel operator H_k at each index point k of a system with the input–output map T: $H_k = \mathbf{O}_k\mathbf{R}_k$. The reachability operator $\mathbf{R}_k$ maps the "strict past" of the system to a (minimal) state, while the observability operator $\mathbf{O}_k$ maps the contribution of the state at the index point k to the future. (Because of linearity, that contribution is a linear component of the output from k on.)

Factorizations of a matrix are by no means unique, even when they are minimal: choose for each k an arbitrary invertible matrix S_k, and $H_k = \widehat{\mathbf{O}}_k \widehat{\mathbf{R}}_k$, with $\widehat{\mathbf{R}}_k := S_k \mathbf{R}_k$ and $\widehat{\mathbf{O}}_k := \mathbf{O}_k S_k^{-1}$ will likewise be a good minimal factorization. The new state becomes $\widehat{x}_k := S_k x_k$, and the new realization

$$\begin{cases} \widehat{x}_{k+1} = \widehat{A}_k \widehat{x}_k + \widehat{B}_k u_k, \\ y_k = \widehat{C}_k \widehat{x}_k + D_k u_k, \end{cases} \tag{5.1}$$

with

$$\begin{bmatrix} \widehat{A}_k & \widehat{B}_k \\ \widehat{C}_k & D_k \end{bmatrix} := \begin{bmatrix} S_{k+1} A_k S_k^{-1} & S_{k+1} B_k \\ C_k S_k^{-1} & D_k \end{bmatrix}. \tag{5.2}$$

Such state transformations are commonly used to produce interesting "canonical" forms for the realization; canonical means characterizing, often uniquely in a certain sense. The first idea is to choose an orthonormal basis for either reachability spaces or observability spaces.

Definition 5.1 A minimal realization of a system is in an *input normal form*, when all its $\begin{bmatrix} A_k & B_k \end{bmatrix}$ are co-isometric (i.e., $A_k A_k' + B_k B_k' = I$). Dually, the realization of a minimal system is in an *output normal form*, when all its $\begin{bmatrix} A_k \\ C_k \end{bmatrix}$ are isometric (i.e., $A_k' A_k + C_k' C_k = I$).

These canonical choices amount to finding an orthonormal basis for the co-range and the range of the Hankel operator H_k at each index point k. Determining such bases from the transfer data is not easy: we are dealing with the full range of indices. So it makes sense to study how the result can be obtained with less information and preferably recursive calculations.

Suppose, therefore, we are given a *minimal*, but otherwise arbitrary realization $\{A, B, C, D\}$; then what does it take to put it in one or the other normal form? To start with the input normal form, we have to find a transformation S_k at each index point, which is so that the resulting $\begin{bmatrix} \widehat{A}_k & \widehat{B}_k \end{bmatrix} := \begin{bmatrix} S_{k+1} A_k S_k^{-1} & S_{k+1} B_k \end{bmatrix}$ is co-isometric, or

$$S_{k+1} A_k S_k^{-1} (S_k^{-1})' A_k' S_{k+1}' + S_{k+1} B_k B_k' S_{k+1}' = I \tag{5.3}$$

for each index k. Let $M_k := (S_k' S_k)^{-1}$; then this reduces to the recursive equation

$$M_{k+1} = B_k B_k' + A_k M_k A_k'. \tag{5.4}$$

This latter equation is called a *Lyapunov–Stein equation*, and it can be determined recursively going forward from k to $k+1$, provided one knows an initial, starting matrix. M_k should be strictly positive definite, so that its inverse can be factorized subsequently into $S_k' S_k$, with S_k square nonsingular. Let us investigate further whether such a solution indeed exists, and how it can be obtained.

To start, suppose that the system is at first empty, and the first index point at which it becomes active is k_0. The state x_{k_0} is assumed empty, and $x_{k_0+1} = B_{k_0} u_{k_0}$. Hence,

M_{k_0} will be empty, and $M_{k_0+1} = B_{k_0}B'_{k_0}$. Is the latter nonsingular? Yes, indeed. We assumed the realization to be minimal, and the first non-empty Hankel operator factorizes as $H_{k_0+1} = \mathbf{O}_{k_0+1}B_{k_0}$, which we assumed to be a minimal factorization. Hence, B_{k_0} has independent rows, and M_{k_0+1} is therefore nonsingular. *M_{k_0+1} is actually the Gramian of the chosen row basis of $\mathbf{R}_{k_0+1}$* in the given realization.[1]

This last property generalizes recursively. The next $M_{k_0+2} = B_{k_0+1}B'_{k_0+1} + A_{k_0+1}B_{k_0}B'_{k_0}A'_{k_0+1}$ will again be nonsingular, because in the realization, $\mathbf{R}_{k_0+1} = \begin{bmatrix} A_{k_0+1}B_{k_0} & B_{k_0+1} \end{bmatrix}$ had been chosen nonsingular and positive definite (automatically). The recursion keeps on progressing until it terminates at the last index k_ℓ. All this will go well if the subsequent transition matrices A_k are properly bounded; actually their continuous product $A_k A_{k-1} \cdots A_{k_0+1}$ comes into play, which may blow up exponentially. This eventuality is especially of concern in the infinitely indexed case and will be discussed to some extent later in Section 5.6, but numerical stability problems may already occur in the finitely indexed case (this issue is potentially inherent in most recursions).

For the output normal form, something similar happens dually (duality also reverses time!). Requiring $\begin{bmatrix} \widehat{A}_k \\ \widehat{C}_k \end{bmatrix} = \begin{bmatrix} S_{k+1}A_k S_k^{-1} \\ C_k S_k^{-1} \end{bmatrix}$ to be isometric now leads, with $N_k := S'_k S_k$, to the *backward* Lyapunov–Stein equation

$$N_k = A'_k N_{k+1} A_k + C'_k C_k, \tag{5.5}$$

and N_k can be interpreted as the Gramian of the local observability basis at the index point k, which because of minimality will be nonsingular.

5.2 A Square-Root Algorithm to Compute Normal Forms

A pertinent question is whether the S_k in the previous calculations of a normal form could not be determined directly (e.g., not via the product $N_k = S'_k S_k$). The answer is that they can be, with great numerical benefits. Let us look first at the output normal form (the backward recursion); the case of the input normal form is just dual (and with a forward recursion). At stage k, one may assume that S_{k+1} is known, and the goal is now to find an orthonormal (partial) realization such that

$$\begin{bmatrix} S_{k+1}A_k S_k^{-1} \\ C_k S_k^{-1} \end{bmatrix} = \begin{bmatrix} S_{k+1}A_k \\ C_k \end{bmatrix} S_k^{-1}, \tag{5.6}$$

hence, equivalently, an orthonormalization of $\begin{bmatrix} S_{k+1}A_k \\ C_k \end{bmatrix}$. For this, one performs a QR factorization of

[1] If we also assume that the input u_{k_0} maps one-to-one to the first non-empty state x_{k_0+1}, then B_{k_0} will have to be square nonsingular as well. This is an assumption that is usually made: inputs to be nonredundant. If they are not, then their dimension can be reduced to the co-range of B_{k_0}.

$$\begin{bmatrix} S_{k+1}A_k \\ C_k \end{bmatrix} =: \begin{bmatrix} Q_{1,1} & Q_{1,2} \\ Q_{2,1} & Q_{2,2} \end{bmatrix} \begin{bmatrix} R_k \\ 0 \end{bmatrix} = \begin{bmatrix} Q_{1,1} \\ Q_{2,1} \end{bmatrix} R_k. \tag{5.7}$$

Then, premultiplying both sides with their complex conjugates, one gets the relevant Lyapunov–Stein equation back for this case, because of the unitarity of the Q matrix. However, one gets more. Since the left-hand side has to be nonsingular, R_k is also square nonsingular, and from the Lyapunov–Stein equation, we have that actually $R_k = S_k$, which is the desired state transformation at the index k. It then follows directly that

$$\begin{bmatrix} \widehat{A}_k \\ \widehat{C}_k \end{bmatrix} = \begin{bmatrix} Q_{1,1} \\ Q_{2,1} \end{bmatrix}, \tag{5.8}$$

and the QR factorization simply becomes

$$\begin{bmatrix} S_{k+1}A_k \\ C_k \end{bmatrix} =: \begin{bmatrix} \widehat{A}_k & Q_{1,2} \\ \widehat{C}_k & Q_{2,2} \end{bmatrix} \begin{bmatrix} S_k \\ 0 \end{bmatrix} = \begin{bmatrix} \widehat{A}_k \\ \widehat{C}_k \end{bmatrix} S_k. \tag{5.9}$$

Thus, a simple recursive QR factorization solves the Lyapunov–Stein equation in the square-root form. (S_k can be taken as an upper triangular square root of $N_k = S_k' S_k$.[2]) In Chapter 6 (on canonical factorizations), the components $Q_{1,2}$ and $Q_{2,2}$ will also acquire significance.

Why is solving a square-root equation preferable to solving the Lyapunov–Stein equation directly? First of all, finding S_k directly guarantees the positivity of $N_k = S_k' S_k$. But there is more. One can show that the condition number of the linear equations one has to solve to find N_k is the square of the condition number for finding S_k (see the mathematical literature for the definition of the condition number). This means that one loses half of the significant bits in the direct computation of N_k as compared to the computation via S_k. The result is that only the square-root equation will give results that are accurate within the range allowed by the precision offered by the computer. The square-root calculation must hence be preferred, except perhaps in the most elementary cases.

5.3 Balanced Realizations

Balanced realization theory plays an important role in model reduction theory, that is, when one wants to represent a system with a reduced but still accurate model. A balanced realization treats the input and output sides of the realization equally in some sense, to be defined next.

We start out with a minimal realization $\{A, B, C, D\}$ in the input normal form (for example). The local reachability Gramian $\mathbf{R}_k \mathbf{R}_k' = I$ is then unitary. Let the

[2] Strictly speaking, the square root $N_k^{1/2}$ of N_k is a symmetric or Hermitian matrix and is unique. However, for most applications, any minimal size factorization $N_k = S_k' S_k$ works, in which case, S_k is viewed as a kind of generalized square root.

corresponding observability Gramian be $N_k := \mathbf{O}_k'\mathbf{O}_k$, and let us find its eigendecomposition: $N_k = U_k'\Sigma_k^2 U_k$, with U_k unitary and Σ_k diagonal strictly positive. (That Σ_k is nonsingular is, of course, a consequence of the minimality assumption.) Let us now define a new realization $\{\widehat{A}_k, \widehat{B}_k, \widehat{C}_k, D_k\} := \{S_{k+1}A_k S_k^{-1}, S_{k+1}B_k, C_k S_k^{-1}, D_k\}$, with $S_k := \Sigma_k^{1/2}U_k$. Then we put $\widehat{\mathbf{R}}_k = \Sigma_k^{1/2}U_k\mathbf{R}_k$ and $\widehat{\mathbf{O}}_k = \mathbf{O}_k U_k'\Sigma_k^{-1/2}$ (leaving H_k unchanged), and hence the resulting modified $\widehat{M}_k = \Sigma_k$ and $\widehat{N}_k = \Sigma_k$. Both the reachability and observability Gramians have become diagonal strictly positive definite and equal to each other.

One shows fairly easily that the realizations so obtained correspond to a singular value decomposition (SVD) of each Hankel operator H_k – actually, this would be a way to derive them as well. Hankel matrices can be updated recursively and combined with the generation of a realization, as indicated in Section 5.2. To obtain the balanced form, one computes the SVD of S_k in the square-root version given by Eq. (5.9), which is an additional operation, but it can be done very efficiently (and it is a local operation at each index point k).

5.4 Reduction to a Minimal Realization

The next point we have to discuss is how to reduce a nonminimal realization to a minimal one. Let $\{A, B, C, D\}$ be a presumably nonminimal realization; how can it be reduced, without going back to the factorization of the Hankel operators? This question often occurs when systems are combined with each other, for example, when their transfer operators are added, put in cascade or, more generally, combined into a connected network; "cancellations" may occur between the systems, for example, when a system is cascaded with its inverse, just a direct feedthrough results.

What is needed is a method to discover superfluous parts of the state space. There are two main ways the dimension of the state space may be too large: (1) some states (at some index k) cannot be generated from the input (they are called "unreachable") and (2) some states cannot be discriminated on the basis of future observations (their difference is "unobservable"). These notions can be made more precise by looking at relevant Gramians, which will be singular when the state dimension is too large. (We already encountered Gramians in Section 5.3.)

Let us start with reachability. The given realization will have reachability operators for each k,

$$\mathbf{R}_k = \begin{bmatrix} \cdots & A_{k-1}A_{k-2}B_{k-3} & A_{k-1}B_{k-2} & B_{k-1} \end{bmatrix}, \tag{5.10}$$

and related Gramian $M_k := \mathbf{R}_k\mathbf{R}_k'$. When the rows of $\mathbf{R}_k$ form a basis, and hence M_k is nonsingular, then all states x_k are reachable with appropriate inputs running from the beginning point of the system to $k-1$. However, if M_k is singular, then there exists a (actually many) nonsingular matrix S_k such that

$$M_k = \begin{bmatrix} S_{k,1} & S_{k,2} \end{bmatrix} \begin{bmatrix} \widehat{M}_k & 0 \\ 0 & 0 \end{bmatrix} \begin{bmatrix} S_{k,1}' \\ S_{k,2}' \end{bmatrix}, \tag{5.11}$$

with $\widehat{M}_k$ being a square nonsingular matrix. For later use, let us be more precise. The columns of $S_{k,1}$ must form a basis for the range of M_k. It really does not matter what $S_{k,2}$ is so long as it complements the basis generated by the columns of $S_{k,1}$ (please check this statement!). It may be useful, but not strictly necessary, to choose the columns of $S_{k,2}$ to form a basis for the kernel of M_k'. In that case, one will have

$$\begin{bmatrix} S_{k,1} & S_{k,2} \end{bmatrix}^{-1} = \begin{bmatrix} S_{k,1}^{\dagger} \\ S_{k,2}^{\dagger} \end{bmatrix}, \tag{5.12}$$

in which the daggers stand for the respective Moore–Penrose inverses: $S_{k,1}^{\dagger} = (S_{k,1}' S_{k,1})^{-1} S_{k,1}'$. (However, whatever the choice of an appropriate $S_{k.2}$, we can always have $\widehat{M}_k = (S_{k,1}^{+})' M_k S_{k,1}^{+}$ for any pseudo-inverse $S_{k,1}^{+}$, if so desired.)

Consider now, at each k, the transformation $\widehat{x}_k = S_k^{-1} x_k$. Then we have $x_k = \begin{bmatrix} S_{k,1} & S_{k,2} \end{bmatrix} \begin{bmatrix} \widehat{x}_{k,1} \\ \widehat{x}_{k,2} \end{bmatrix}$, and subsequently with

$$\begin{bmatrix} \widehat{A}_k & \widehat{B}_k \\ \widehat{C}_k & D_k \end{bmatrix} = \begin{bmatrix} S_{k+1}^{-1} A_k S_k & S_{k+1}^{-1} B_k \\ C_k S_k & D_k \end{bmatrix} \tag{5.13}$$

and using the induced partitioning at each index point k:

$$\begin{cases} \begin{bmatrix} \widehat{x}_{k+1,1} \\ \widehat{x}_{k+1,2} \end{bmatrix} = \begin{bmatrix} \widehat{A}_{k;1,1} & \widehat{A}_{k;1,2} \\ \widehat{A}_{k;2,1} & \widehat{A}_{k;2,2} \end{bmatrix} \begin{bmatrix} \widehat{x}_{k,1} \\ \widehat{x}_{k,2} \end{bmatrix} + \begin{bmatrix} \widehat{B}_{k;1} \\ \widehat{B}_{k;2} \end{bmatrix} u_k, \\ y_k = \begin{bmatrix} \widehat{C}_{k;1} & \widehat{C}_{k;2} \end{bmatrix} \begin{bmatrix} \widehat{x}_{k,1} \\ \widehat{x}_{k,2} \end{bmatrix} + D_k u_k, \end{cases} \tag{5.14}$$

and, as the forward recursion on M_k is $M_{k+1} = B_k B_k' + A_k M_k A_k'$, we also have, because of Eq. (5.11),

$$\begin{bmatrix} \widehat{M}_{k+1} & 0 \\ 0 & 0 \end{bmatrix} = \begin{bmatrix} \widehat{B}_{k;1} \\ \widehat{B}_{k;2} \end{bmatrix} \begin{bmatrix} \widehat{B}_{k;1}' & \widehat{B}_{k;2}' \end{bmatrix} + \begin{bmatrix} \widehat{A}_{k;1,1} & \widehat{A}_{k;1,2} \\ \widehat{A}_{k;2,1} & \widehat{A}_{k;2,2} \end{bmatrix} \begin{bmatrix} \widehat{M}_k & 0 \\ 0 & 0 \end{bmatrix} \begin{bmatrix} \widehat{A}_{k;1,1}' & \widehat{A}_{k;2,1}' \\ \widehat{A}_{k;1,2}' & \widehat{A}_{k;2,2}' \end{bmatrix}. \tag{5.15}$$

The $(2,2)$ entry in this expression gives

$$0 = \widehat{B}_{k;2} \widehat{B}_{k;2}' + \widehat{A}_{k:2,1} \widehat{M}_k \widehat{A}_{k:2,1}' \tag{5.16}$$

From this, the positivity and non-singularity of $\widehat{M}_k$ as well as the full right-hand side, it follows that $\widehat{B}_{k;2} = 0$ and $\widehat{A}_{k;2,1} = 0$ so that the hatted realization has the form

$$\left[\begin{array}{cc|c} \widehat{A}_{k;1,1} & \widehat{A}_{k;1,2} & \widehat{B}_{k;1} \\ 0 & \widehat{A}_{k;2,2} & 0 \\ \hline \widehat{C}_{k;1} & \widehat{C}_{k;2} & D_k \end{array}\right]. \tag{5.17}$$

The state $\widehat{x}_{k;2}$ (at each k) will always remain zero, because there is no input that can change it. (At least when it started at zero or was empty; it could lead a life all by itself, of course, but then some "deus ex machina" would have been necessary to introduce a value different from naught.) Hence, $\widehat{x}_{k;2}$ can be canceled out, and we have found a reduced realization for the system, which is reachable:

$$\left[\begin{array}{c|c} \widehat{A}_{k;1,1} & \widehat{B}_{k;1} \\ \hline \widehat{C}_{k;1} & D_k \end{array}\right]. \tag{5.18}$$

As before, one would execute the reduction to a reachable system using a square-root algorithm directly on the state transformations S_k:

$$\begin{bmatrix} A_k S_k & B_k \end{bmatrix} = \begin{bmatrix} S_{k+1;1} \mid 0 \end{bmatrix} \begin{bmatrix} Q_{1,1} & Q_{1,2} \\ Q_{2,1} & Q_{2,2} \end{bmatrix} \tag{5.19}$$

for some Q still to be identified. To avoid ambiguities with pseudo-inverses, let us now work exclusively with Moore–Penrose inverses. For example, as $S_{k:1}$ has the full column rank (for each k), we have $S_{k;1}^{\dagger} = (S_{k;1}' S_{k;1})^{-1} S_{k;1}'$. To complement $S_{k;1}$, one then chooses $S_{k;2}$ column-wise orthogonal to $S_{k;1}$, that is, $S_{k;1}' S_{k;2} = 0$, and then one has

$$\begin{bmatrix} S_{k;1} & S_{k;2} \end{bmatrix}^{\dagger} = \begin{bmatrix} S_{k;1} & S_{k;2} \end{bmatrix}^{-1} = \begin{bmatrix} S_{k;1}^{\dagger} \\ S_{k;2}^{\dagger} \end{bmatrix}. \tag{5.20}$$

(This provides coherence between pseudo-inverses.) Actually, we do not have to compute the $S_{k;2}$ at all since

$$\widehat{A}_{k;1,1} = S_{k+1;1}^{\dagger} A_k S_{k;1}; \quad \widehat{B}_k = S_{k+1;1}^{\dagger} S_{k;1}; \quad \widehat{C}_k = C_k S_{k;1} \tag{5.21}$$

are solely dependent on $S_{k,1}$ and $S_{k+1;1}$.

In summary: The reduction algorithm hence boils down to a *minimal factorization* for each index k

$$\begin{bmatrix} A_k S_k & B_k \end{bmatrix} = S_{k+1} \begin{bmatrix} \widehat{A}_k & \widehat{B}_k \end{bmatrix}. \tag{5.22}$$

This is a forward recursion in which S_k is given initially at step k, as well as the original A_k and C_k, and the new, now reachable realization $\widehat{A}_k, \widehat{C}_k$ together with the new, left invertible partial transformation matrix S_{k+1} to be used in the next step $k+1$ are determined (observe: *any* minimal factorization will do!). The resulting $\widehat{C}_k$ has to be updated as well and is simply given by $\widehat{C}_k = C_k S_k$. At the beginning of the recursion, starting with the smallest non-trivial index, the initial data is empty.

A dual operation will reduce the nonminimal state-space realization to one that is observable. This is immediate because observability is reachability in the dual system and vice versa, of course with time reversal. Again, the reduction can be computed using a square-root algorithm, now progressing backward in time.

This boils down also to *a recursive minimal factorization* now of the form (at stage k)

$$\begin{bmatrix} R_{k+1} A_k \\ C_k \end{bmatrix} = \begin{bmatrix} \widehat{A}_k \\ \widehat{C}_k \end{bmatrix} R_k \tag{5.23}$$

in which R_{k+1}, A_k and C_k are known and a right invertible partial R_k, $\widehat{A}_k$ and $\widehat{C_k}$ are determined via a minimal factorization (agein: any minimal factorization will do.) Also $\widehat{B}_k = R_{k+1} B_k$ has to be properly updated.

When a minimal realization is desired, one must proceed in two steps: first, a reduction to a reachable system with a forward recursion; and second, a reduction to an observable system on the already reduced reachable system. It turns out that together this will result in a minimal system. (The second reduction to the observable system does not destroy the reachability.)

5.5 LTI Systems*

The main difference between an LTV and an LTI system is that for the latter one needs fixed-point solutions when solving recursive equations. Otherwise, algebraic manipulations on the state space are the same, for example, when figuring out the reachability Gramian, instead of recursing $M_{k+1} = B_k B_k' + A_k M_k A_k'$, where, say, $M_k = S_k S_k'$, and one has to find $M = SS'$ or, better, S so that

$$M = BB' + AMA'. \tag{5.24}$$

This is a set of linear equations in the entries of M, which can be solved directly, but at the cost of numerical accuracy, as we know already. Because of the importance of this (Lyapunov–Stein) equation, let us consider further. When A is strictly stable, that is, when all its eigenvalues are strictly inside the unit complex disk, then the unique positive definite solution of this equation is the reachability Gramian:

$$M = \sum_{k=0}^{\infty} A^k BB'(A')^k = \mathbf{RR}'. \tag{5.25}$$

Thanks to the strict stability, $A^k \to 0$ exponentially, and the sum converges.

There are two main problems with this equation:

1. Some eigenvalues of A may be close to the unit circle, so that the condition number of the system of linear equations is high; this gets to be problematic when the linear system is solved directly, resulting in an M that is not positive definite.
2. If, on the contrary, a recursive solution is used based on the equation $M = \sum_{k=0:\infty} A^k BB'(A')^k$ or, better, the recursive square-root version (see further), then positive definiteness is ensured, but not only can conversion be very slow, also the computation of A^k may become problematic.

Example To illustrate some of the issues, let us consider the case $A = \begin{bmatrix} a & 0 \\ b & a \end{bmatrix}$, $B = \begin{bmatrix} 1 \\ 0 \end{bmatrix}$, in which $|a| < 1$, but close to the unit circle, b is arbitrary, but perhaps pretty large (although that does not matter too much). We have $A^k = \begin{bmatrix} a^k & 0 \\ ka^{k-1}b & a^k \end{bmatrix}$. When $|a|$ is close to 1, then $k|a^{k-1}b|$ grows much larger than $|b|$. The function $t|a|^{t-1} = te^{(\ln|a|)(t-1)}$ has a maximum for $t_m = -1/(\ln|a|)$, for example, when $|a| = 0{,}99$, $t_m \approx 100$, with the maximum being ≈ 100. So A^{100} is still very large, and it would take a much larger k to have $A^k \approx$ machine precision! The square-root recursion takes ages to converge, even in the doubling variety – a highly efficient way to get the result faster is discussed further in this section. Given the accumulation of errors with such long calculations, the result would be far from accurate. In this case, the direct method easily leads to a result, namely (assuming a real)

$$M = \frac{1}{1-a^2}\begin{bmatrix} 1 & \frac{ab}{1-a^2} \\ \frac{ab}{1-a^2} & \frac{b^2(1+a^2)}{(1-a^2)^2} \end{bmatrix}, \tag{5.26}$$

which is obviously a matrix that is close to singular, thanks to the various factors $\frac{1}{1-a^2}$ in the expression (think floating point!).

Discussion of Various Solution Methods

The source of the problem is the explicit or implicit occurrence of all positive powers of A. Papers have been written about this issue with titles like "9 Bad Methods to Compute the Power of a Matrix." The various alternative methods for solving the Lyapunov–Stein equation deal with this issue in different ways. There is no ultimate method known that works well in all cases. It is also true that most problems are either small, or not that badly conditioned. Nonetheless, much occurring cases in selective filtering are notoriously difficult because eigenvalues are very close to the boundary. (Their computation often requires floating point calculations with a large number of significant digits.) We review the main methods to end with the most reliable one – the use of the Schur eigenvalue form – at the cost of computing eigenvalues, *even for a linear problem.*

Direct Linear Solver

The Lyapunov–Stein equation is an equation defining entries in a matrix. When the matrix is of size $\eta \times \eta$, this means η^2 entries. Due to symmetry, the number is reduced to $(\eta + 1)\eta/2$. Most matrix equations can be formalized using the so-called vec calculus, credited to Kronecker. When A is a matrix, then $\text{vec}(A)$ is a vector formed by stacking its columns in order: $\text{vec}A_{:,k:m} = \text{col}[A_{:,k}, \ldots, A_{:,m}]$. Furthermore, one uses Kronecker products of matrices: $[A \otimes B]_{i,j} := [a_{i,j}B]$ is a new block matrix obtained by multiplying each element of A with the full B matrix. Assuming that the original matrices consist of scalar real elements, one shows the crucial property

$$\text{vec}(AMB) = (B' \otimes A)\text{vec}M. \tag{5.27}$$

The Lyapunov–Stein equation $M - AMA' = BB'$ then becomes

$$(I - (A \otimes A))\,\text{vec}M = \text{vec}(BB'), \tag{5.28}$$

with the additional symmetry constraint $M_{i,j} = M_{j,i}$. It is not hard to see that the system (5.28) is consistently redundant and can easily be reduced to a nonredundant system with $(\eta + 1)\eta/2$ equations. Details of this reduction should be obvious (and can easily be formalized!). Less obvious is that a nonsingular system of equations is obtained, provided all eigenvalues of A are located in the open unit disk. (Hint: use the canonical Schur form to show this – see also the further section on using the Schur form.) Let us evaluate advantages/disadvantages:

+ medium simplicity, closed form expression, very large complexity (most direct solvers would show a computational complexity of $\eta^6/2$);
– poor conditioning, resulting M not guaranteed positive definite.

Power Methods

One could, of course, recurse directly. Define $M_0 := BB'$, $B_0 := B$, and then compute recursively $B_{k+1} := AB_k$ so that $M_{k+1} = M_k + B_{k+1}B'_{k+1}$ – very pedestrian, but reasonably well conditioned when convergence is fast, because it is de facto a square-root method. Notice that there is an implicit calculation of A^k until convergence. Convergence can therefore be very slow, resulting in a high computational complexity and affected with an unwieldy accumulation of errors. Convergence can be sped up greatly by using a *doubling* or *telescoping* method: one puts $M_0 := BB'$, $A_0 := I$, and then $A_{k+1} := (A_k)^2 (= A^{2k})$, $M_{k+1} = M_k + A_{k+1}M_k A'_{k+1}$. The doubling procedure must typically be refined further using a square-root version, just like in the direct recursion: write $M_k = B_k B'_k$ and recur on B_k rather than on M_k (in many applications, the minimal square root B with $M = BB'$ is the desired quantity, and it can be obtained in a much more stable numerical way than directly):

$$\begin{bmatrix} B_k & A_k B_k \end{bmatrix} = \begin{bmatrix} B_{k+1} & 0 \end{bmatrix} Q_{k+1}, \tag{5.29}$$

which will eventually converge when all eigenvalues of A are in the open unit disk. Needless to say, the convergence of the doubling procedure is much faster. (Quadratic in the error after getting close: when ν correct digits have been obtained, then in the next step, one will have 2ν digits, so it goes very fast once one is close; the problem is, of course, the convergence of A^k.) Advantages/disadvantages:

+ high simplicity, positivity guaranteed;
– often high numerical complexity and low accuracy when convergence is slow.

Direct Evaluation in a Square-Root Form

An attractive way to stabilize the system of Lyapunov–Stein Eqs. (5.24) is by converting A to its Schur form. (There are numerically stable and reasonably efficient algorithms to do so.) Let $A = USU'$, with U being a unitary matrix and S a lower block triangular Schur form, with either real eigenvalues on the main diagonal or 2×2 blocks of the type $\begin{bmatrix} \alpha & \beta \\ -\beta & \alpha \end{bmatrix}$ representing eigenvalues of the type $\alpha \pm j\beta$ ($j := \sqrt{-1}$); then Eq. (5.24) transforms to

$$\widehat{M} = \widehat{B}\widehat{B}' + S\widehat{M}S', \tag{5.30}$$

with $\widehat{M} = U'MU$ and $\widehat{B} = U'B$. Suppose S, $\widehat{B}$ and hence $\widehat{M}$ are all scalars; then the equation reduces to $(1 - |S|^2)\widehat{M}^2 = |\widehat{B}|^2$ and $\widehat{M} = (1 - |S|^2)^{-1}|\widehat{B}|$. In the matrix case, a recursion can be set up on the "square-root version" of this equation. Because of its independent interest, we give an algorithm in the appendix of this chapter.

The procedure has been proven to be numerically stable; see the bibliographical notes for further references. The important point is that it is a square-root procedure, inheriting its numerical stability properties. The disadvantage of this direct method is the necessity to compute the Schur form of the state transition matrix A, which, in the case of even moderately large systems, can easily be prohibitive and would be hard to streamline on a parallel processor. Advantages and disadvantages:

+ most stable method numerically;
– computationally intensive.

5.6 Infinitely Indexed LTV Systems*

One issue still to be settled about solving the Lyapunov–Stein equation recursively relates to the question of a starting value. (So far, we assumed a cold, empty start.) Although this question cannot be settled in general (it is system dependent), we can state some important properties of the initial condition. Concentrating on the reachability Gramian and the forward equation: $M_{k+1} = B_k B_k' + A_k M_k A_k'$, and assuming some starting value M_{k_0} is known, we see that its contribution in the solution for M_k with $k > k_0$ is given by $A^{k_0-1<k} M_{k_0} (A^{k_0-1<k})'$, which goes exponentially to zero for $k \to \infty$ when the system is u.e.s. (remember: $A^{k_0-1<k} := A_{k_0} \cdots A_{k-1}$). So, for a u.e.s. system, the initial condition does not matter if one starts early enough.

For systems whose behavior at $-\infty$ is known in the case of the reachability Gramian, or $+\infty$ in the case of the observability Gramian, much more can be said. In that case, the initial value can be computed by solving a fixed-point equation, and the recursion starts as soon as the system becomes time variant – of course, also in this case, the initial value will not matter much if the system remains stable (u.e.s.).

The observations just made for the respective square-root algorithms as well; however, we skip the technical proof.

5.7 Discussion Topics

- *The Lyapunov–Stein recursion is not invertible:* The forward Lyapunov–Stein recursion for the reachability Gramian will be unconditionally numerically stable when the state transition operator A is u.e.s. (considering the LTV case). This means that any error made early on in the recursion will die down when progressing with it. The Lyapunov–Stein recursion cannot be inverted in that case. In fact, because of the u.e.s. property, the eventual values of the Gramian become fully independent of the initial values, sometimes even after a few steps. Hence, the initial values *cannot* be determined from later values.
- *Balanced model reduction:* Based on the balanced form, and at each index point, an approximate (local) system with a smaller amount of states can be derived by neglecting small singular values in the joint diagonal reachability/observability Gramian. Would it be possible to derive error bounds that describe how well the behavior of the reduced system approximates the original?

5.8 Notes

- The determination of the rank of a collection of vectors of the same dimension and the calculation of a basis (often an orthonormal basis) is a key step in many computations for digital signal processing. Important and often used algorithms have been developed for that purpose. We mention in order of precision: QR, rank-revealing QR and the SVD, with the latter being the method of choice when the highest possible accuracy is desired because of its inherent numerical stability when executed well. However, in many practical cases, QR may suffice (but it remains up to the design engineer to decide whether accuracy is sufficient, of course). For references to these techniques, see basic books on numerical analysis, for example, [40, 64, 65, 74].
- In the case of an LTV system, identification and state-space reduction concern not just the estimation of a single basis but a recursive set of bases: one typically needs a basis for the actual reachability or observability space at each index point. The Lyapunov–Stein equation provides the link between the subsequent stages, and it can be solved with a recursive QR- or SVD algorithm. In the LTI case, the situation is more involved, because then a fixed-point solution has to be found. Many methods for this have been proposed, and we have treated a few salient ones in this chapter. For a survey of what the numerical community proposes in this respect, see [60].
- In subsequent chapters, we shall see that the Lyapunov–Stein equation plays a central role in dynamical system theory. Actually, this distinction should be awarded to the square-root algorithm that solves the inner–outer factorization problem, which we shall treat at length in Chapter 8, but there also, Lyapunov–Stein plays a role in the background. This is not surprising in view of the important role reachability and observability play in the derivation of state-space realizations. This importance will, of course, extend to control, optimization and synthesis problems-some of them to be treated in further chapters.
- *Using the sign matrix:* In the control literature, another method to solve the Lyapunov–Stein equation has received some attention, the sign method. More information on the use of this method in our context can be found in [56]. The method is also an iterative method and may have an advantage when the A matrix is sparse, although this statement has to be clarified, as the direct recursive method is also well suited to sparse cases and has the added advantage that it guarantees positivity.

5.9 Appendix*: Fixed-Point Lyapunov–Stein Equation

We give a brief summary on how to find the square-root fixed-point solution of the LTI Lyapunov–Stein equation based on the Schur eigenvalue form of the transition operator A, because of the importance of the issue also for LTV theory and the interest of the algorithm itself (which may not be so well known). The equation to be solved is

$$M = BB' + SMS', \tag{5.31}$$

and we assume that S is in the lower Schur eigenvalue form, that is, S lower triangular with, of course, its eigenvalues on the main diagonal. (In case the original A does not have that form, it can be converted to it via a unitary state transformation: $AU = US$, with U unitary, and B must then be adapted as well. U and S are normally determined with a well known numerically stable algorithm.) We want to find a "square-root" L of M, that is, a lower triangular matrix such that $M = LL'$. The recursion is possible, thanks to the following observation. Suppose a block subdivision of S and L with $S = \begin{bmatrix} S_{1,1} & 0 \\ S_{2,1} & S_{2,2} \end{bmatrix}$ and $L = \begin{bmatrix} L_{1,1} & 0 \\ L_{2,1} & L_{2,2} \end{bmatrix}$, where both $S_{1,1}$ and $L_{1,1}$ square of dimensions $k \times k$, and conformally $B = \begin{bmatrix} B_1 \\ B_2 \end{bmatrix}$ given at stage k of the recursion; then we have

$$L_{1,1}L'_{1,1} = B_1B'_1 + S_{1,1}L_{1,1}L'_{1,1}S'_{1,1}, \tag{5.32}$$

so that one can move from stage k in the recursion to stage $k+1$ by adding a new row to B_1, $S_{1,1}$ and $L_{1,1}$ and solving just for the last row of the latter. This can be done in the square-root form, as we show now. We have generally

$$\begin{bmatrix} B & SL \end{bmatrix} \begin{bmatrix} B' \\ L'S' \end{bmatrix} = \begin{bmatrix} L & 0 \end{bmatrix} \begin{bmatrix} L' \\ 0 \end{bmatrix}, \tag{5.33}$$

where L has been augmented with zeros to match the dimensions of $\begin{bmatrix} B & SL \end{bmatrix}$, and hence there exists (at least one) unitary matrix (A may have complex eigenvalues) such that

$$\begin{bmatrix} B & SL \end{bmatrix} = \begin{bmatrix} L & 0 \end{bmatrix} \begin{bmatrix} Q_{1,1} & Q_{1,2} \\ Q_{2,1} & Q_{2,2} \end{bmatrix}, \tag{5.34}$$

in which the columns of $\begin{bmatrix} Q_{1,1} & Q_{1,2} \end{bmatrix}'$ form an orthonormal basis for the co-range (corresponding to the row space) of $\begin{bmatrix} B & SL \end{bmatrix}$ and the columns of $\begin{bmatrix} Q_{2,1} & Q_{2,2} \end{bmatrix}'$ for its kernel (which the array that has generated Q so far produces automatically).

To keep the notation simple, let us now just add a row (and adjoining column where needed) to the last equation, indicated with lowercase symbols; the situation being typical for the move from k to $k+1$, with the new S written as $\begin{bmatrix} S & 0 \\ s_1 & s_2 \end{bmatrix}$, the new L as $\begin{bmatrix} L & 0 \\ \ell_1 & \ell_2 \end{bmatrix}$ and the new Q as $\widehat{Q} = \begin{bmatrix} Q_{1,1} & Q_{1,2} & 0 \\ \widehat{Q}_{2,1} & \widehat{Q}_{2,2} & \widehat{Q}_{3,1} \\ \widehat{Q}_{3,1} & \widehat{Q}_{3,2} & \widehat{Q}_{3,3} \end{bmatrix}$ with fitting dimensions:

$$\begin{bmatrix} B & SL & 0 \\ b & s_1L + s_2\ell_1 & s_2\ell_2 \end{bmatrix} = \begin{bmatrix} L & 0 & 0 \\ \ell_1 & \ell_2 & 0 \end{bmatrix} \begin{bmatrix} Q_{1,1} & Q_{1,2} & 0 \\ \widehat{Q}_{2,1} & \widehat{Q}_{2,2} & \widehat{Q}_{3,1} \\ \widehat{Q}_{3,1} & \widehat{Q}_{3,2} & \widehat{Q}_{3,3} \end{bmatrix}. \tag{5.35}$$

Notice that in this new expression the original $\begin{bmatrix} Q_{1,1} & Q_{1,2} \end{bmatrix}$ can be retained, because an orthonormal basis for $\begin{bmatrix} B & SL \end{bmatrix}$ is still needed, but now augmented with a scalar 0, while a new row is added to accommodate the new data and the kernel is accordingly modified as well. Notice also that the entries s_2 and ℓ_2 are just scalar, and ℓ_2 in particular has to be nonzero if the pair $\{A, B\}$ is reachable. (For the more general nonminimal case, additional considerations must be made, which we skip here for brevity.) In this expression b, s_1 and s_2 are known as well as $Q_{1,1}$ and $Q_{1,2}$, while ℓ and the $\widehat{Q}$-entries have to be determined. (As before, they will follow automatically from the LQ array, that we now proceed to update.)

Postmultiplying with the Hermitian conjugate of the $\widehat{Q}$-array (assuming we know it), we find that Q is to be determined so that

$$\begin{bmatrix} B & SL & 0 \\ b & s_1 L + s_2 \ell_1 & s_2 \ell_2 \end{bmatrix} \begin{bmatrix} Q'_{1,1} & \widehat{Q}'_{2,1} & \widehat{Q}'_{3,1} \\ Q'_{1,2} & \widehat{Q}'_{2,2} & \widehat{Q}'_{3,2} \\ 0 & \widehat{Q}'_{2,3} & \widehat{Q}'_{3,3} \end{bmatrix} \text{ becomes equal to } \begin{bmatrix} L & 0 & 0 \\ \ell_1 & \ell_2 & 0 \end{bmatrix}. \tag{5.36}$$

From the first block column (which has k columns), we obtain

$$bQ'_{1,1} + s_1 L Q'_{1,2} + s_2 \ell_1 Q'_{1,2} = \ell_1, \tag{5.37}$$

in which the only unknown is the (row) ℓ_1. Since s_2 is scalar (it is an eigenvalue of S), we find

$$\ell_1 = (bQ'_{1,1} + s_1 L Q'_{1,2})(I - s_2 Q'_{1,2})^{-1}, \tag{5.38}$$

and good arguments can be made for the existence of the inverse, in the nonsingular case. (Normally, both s_2 and $Q_{1,2}$ are strictly contractive, but it is enough that s_2 is in the open unit disk!)

From this point on, the normal LQ factorization algorithm applies: one passes the newly added row $\begin{bmatrix} b & s_1 L + s_2 \ell_1 & s_2 \ell_2 \end{bmatrix}$, through the orthogonal array build so far (still leaving ℓ_2 to be determined), and the result then has to be followed by a compression of the columns $k + 1$ to the end – the standard procedure for an LQ factorization. More precisely, when inputting the partially known $\begin{bmatrix} b & s_1 L + s_2 \ell_1 & s_2 \ell_2 \end{bmatrix}$ to the already existing array $\begin{bmatrix} Q'_{1,1} & Q'_{2,1} & \\ Q'_{1,2} & Q'_{2,2} & \\ & & 1 \end{bmatrix}$, out comes $\begin{bmatrix} \ell_1 & bQ'_{2,1} + (s_1 L + s_2 \ell_1) Q'_{2,2} & s_2 \ell_2 \end{bmatrix}$, and the new array layer has to orthogonalize the part $\begin{bmatrix} x & s_2 \ell_2 \end{bmatrix}$ with $x = bQ'_{2,1} + (s_1 L + s_2 \ell_1) Q'_{2,2}$ known. The compression boils down to produce $\begin{bmatrix} x & s_2 \ell_2 \end{bmatrix} = \ell_2 \begin{bmatrix} q_{1,1} & q_{1,2} \end{bmatrix}$, with $\begin{bmatrix} q_{1,1} & q_{1,2} \end{bmatrix}$ co-isometric. Using a positive choice for ℓ_2, this requires $\ell_2 = xx'/\sqrt{1 - |s_2|^2}$, which determines ℓ_2 as well as the next layer in the array. The new contribution to the array forms a unitary matrix $\begin{bmatrix} q_{1,2} & q_{1,2} \\ q_{2,1} & q_{2,2} \end{bmatrix}$, and the updated transformation matrix becomes

$$\widehat{Q} = \begin{bmatrix} I & & \\ & q_{1,1} & q_{1,2} \\ & q_{2,1} & q_{2,2} \end{bmatrix} \begin{bmatrix} Q_{1,1} & Q_{1,2} & \\ Q_{2,1} & Q_{2,2} & \\ & & 1 \end{bmatrix} = \begin{bmatrix} Q_{1,1} & Q_{1,2} & 0 \\ q_{1,1} Q_{2,1} & q_{1,1} Q_{2,2} & q_{1,2} \\ q_{2,1} Q_{2,1} & q_{2,1} Q_{2,2} & q_{2,2} \end{bmatrix}. \tag{5.39}$$

All recursive quantities have now been updated. Although the implicit determination of the missing data for a straight LQ factorization may seem a bit cumbersome, only L and $Q_{1,2}$ are $k \times k$ matrices; all the other quantities (matrices) have typically small size (depending on the size of B, of course). So the order of computation at each step is a bit more than $\mathcal{O}(k^2)$.

6 Elementary Operations

One of the main reasons for considering systems described by state-space realizations, or, equivalently quasi-separable systems, is the derivation of efficient algorithms for various matrix operations, that is, algorithms that are linear rather than quadratic (or even worse) in the overall dimensions of the matrices involved. In this chapter, we show how elementary matrix operations, namely matrix–vector multiplication, addition of matrices and multiplication of matrices, are performed efficiently using state-space representations. More complex operations such as matrix inversion or matrix approximation will be discussed in a further chapter, but they also have the same property: linear numerical complexity in the overall size of the transfer operator. This chapter ends with a characterization of invertible systems based on their realization, called *outer* systems.

Menu

Hors d'oeuvre
Algebraic Minimality

First Course
Matrix–Vector Multiplication

Second Course
Reduction to Minimal
Addition of Quasi-separable Systems

Third Course
Multiplication

Cheese Dish
Elementary Inversion

Dessert
Outer Systems

6.1 Algebraic Minimality

Before starting on the efficiency path, one very important issue has to be dealt with first, namely, *algebraic minimality*. The state-space realizations derived in the previous chapters often have more variables than algebraically specified by the system's behavior, or, to put it differently, the variables used in the state-space representation have algebraic dependencies that make some or even many of them redundant, given the required behavior. To illustrate the point, suppose you have a scalar LTI system with an (irreducible) transfer function

$$T(z) := d + \frac{b_n z^n + \cdots + b_1 z}{a_n z^n + \cdots + a_1 z + 1}. \tag{6.1}$$

Then a minimal realization in a companion form is

$$\begin{bmatrix} A & B \\ C & D \end{bmatrix} = \left[\begin{array}{ccccc|c} 0 & 1 & 0 & \cdots & 0 & 0 \\ 0 & 0 & 1 & \cdots & 0 & 0 \\ \ddots & \ddots & \ddots & \cdots & \vdots & \\ 0 & 0 & 0 & \cdots & 1 & 0 \\ -a_n & -a_{n-1} & -a_{n-1} & \cdots & -a_1 & 1 \\ \hline b_n & b_{n-1} & b_{n-1} & \cdots & b_1 & d \end{array}\right] \tag{6.2}$$

(it is a good exercise to show that this is a correct realization for the given transfer function given – hint: use the LU factorization to find $(I - zA)^{-1}$), and we see that the number of free parameters (coefficients) in the transfer operator corresponds exactly to the number of free parameters in the realization, namely $2n + 1$, which would not be the case when the matrix A is a full matrix with n^2 entries. From the point of view of computational complexity, $2n+1$ is optimal. However, the companion form realization is likely far from optimal from an accuracy or sensitivity point of view. In particular, pole locations are very sensitively dependent on the companion form, especially when the poles are close to each other, as is often the case in practice. (Here sensitivity means small variations in the coefficients cause large variations in the pole locations.) So it may very well be that the system is theoretically stable (poles inside the unit disk of the complex plane), but the realization leads to unstable computations (at least one pole has moved outside due to rounding errors). It turns out that better forms exist that are also algebraically minimal. To design them is an interesting and important synthesis issue that is not treated in this book: it is one of the main topics in a high-performance system and circuit design. If the results of the desired operations have to be algebraically minimal and at the same time must fulfill stringent numerical properties, then the transformation of the realization to an adequate algebraically minimal form will most likely be necessary. Nonetheless, we can state that the decomposition of the realization into a cascade of small systems (the so-called sections) very often gives a desirable solution. Although we do not treat the synthesis issue in this book, we do build the necessary groundwork for it in this chapter, namely by showing how larger transfer operators originate from compositions of smaller ones.

6.2 Matrix–Vector Multiplication

Let an LTV causal realization $T = D + C(I - ZA)^{-1}ZB$ be a realization of a causal LTV system with transfer operator T, and assume that one wants to compute $y = Tu$, with u starting at some index point k_0 (i.e., empty before k_0) and assume that the initial state x_{k_0} is empty, k_0 being the "creation time." Then one can simply put the recursion $x_{k+1} = A_k x_k + B_k u_k$ and $y_k = C_k x_k + D_k u_k$ to work starting at $k = k_0$ with $x_{k_0} = -$ (i.e., empty) of dimension $0{\times}1$,[1] and keep on recurring. If the dimensions of A_k at any point k are $\eta_{k+1} \times \eta_k$, m_k of u_k and n_k of y_k, then the number of multiplications (and additions) at the index k will roughly be $(\eta_{k+1} + n_k)(\eta_k + m_k)$. Taking averages for the various quantities denoted as η, m and n, assuming $m = n$, and assuming that N steps have to be executed, this results in an overall complexity of $N(\eta + m)^2$, instead of N^2. The complexity is hence linear in the number of steps instead of quadratic, but quadratic in the dimension of the state. This will be computationally advantageous when $N \gg \eta + m$.

In the case of a doubly sided matrix with a causal and an anti-causal part, $T = D + C_c(I - ZA_c)^{-1}ZB_c + C_a(I - Z'A_a)^{-1}Z'B_a$, the multiplication splits in a forward recursion (for the causal part) and a backward recursion (for the anti-causal part), totally independently from each other except for multiplication with the constant term, which may be assigned to either the causal or the anti-causal part.

6.3 Adding Two Systems

Let $T_1 \sim_c \begin{bmatrix} A_1 & B_1 \\ C_1 & D_1 \end{bmatrix}$ and $T_2 \sim_c \begin{bmatrix} A_2 & B_2 \\ C_2 & D_2 \end{bmatrix}$; then a potentially nonminimal realization for $T := T_1 + T_2$, assuming the matching input and output dimensions, is

$$T \sim_c \left[\begin{array}{cc|c} A_1 & 0 & B_1 \\ 0 & A_2 & B_2 \\ \hline C_1 & C_2 & D_1 + D_2 \end{array}\right], \tag{6.3}$$

that is, the global state is the concatenation of the two individual states. However, this new state may not have minimal dimensions, and might hence be reduced. As we saw in the previous chapter, the reduction is based on reducing the reachability and observability Gramians to nonsingular. Suppose that we already know the individual Gramians of T_1 and T_2; then the new Gramians can be computed, with less effort, as follows.

Starting with the reachability Gramian ($M_k = B_{k-1}B'_{k-1} + A_{k-1}M_{k-1}A'_{k-1}$), suppose $M_{1,k}$ and $M_{2,k}$ are the respective local reachability Gramians of T_1 and T_2 and $M_{t,k}$ is the local reachability Gramian of $T_1 + T_2$; then we find

$$M_{t,k} = \begin{bmatrix} M_{1,k} & M_{t,k;1,2} \\ M'_{t,k;1,2} & M_{2,k} \end{bmatrix}, \tag{6.4}$$

in which the entry $M_{t,k;1,2}$ satisfies the Lyapunov–Stein equation $M_{t,k+1;1,2} = B_1 B'_2 + A_1 M_{t,k;1,2} A'_2$. Hence, solving just one more Lyapunov–Stein equation produces the

[1] The overall x is a column stack of the individual x_k: $x = \mathrm{col}(x_k)_{k=-\infty}^{+\infty}$; hence, an empty x_k has a $0{\times}1$ entry.

reachability Gramian of the sum. (This can, of course, be done with a square-root algorithm; we leave details for the discussion – see also Chapter 5.) For example, in case $T_1 = T_2$, we obtain the Gramian $\begin{bmatrix} M_1 & M_1 \\ M_1 & M_1 \end{bmatrix}$ for the reachability matrix, which is, of course, singular with the same rank as the original and can be reduced immediately.

As to the observability Gramian ($N_k = C_k'C_k + A_k'N_{k+1}A_k$), it gives rise to a backward Lyapunov–Stein equation, but, again, the individual $N_{i,k}'$s may be borrowed, and only the mixed term has to be computed. The situation may seem awkward, because obtaining a minimal total realization requires both a forward and a backward computation, but that is how it is: it is the price one has to pay for computational reduction. The two opposite recursions *cannot in general* be avoided, unless the system has a special form that makes one or the other recursion stably invertible in time: reachability depends intrinsically on the past and observability on the future; one cannot, in general, be deduced from the other.

Nonetheless, in most cases, reduction to minimality is not necessary. Not only are cancellations rare when one system is connected with another, but when priority to numerical stability is given through the use of orthogonal transformations or filtering, minimality is not necessary, provided some elementary measures are taken. For example, orthogonal transformations can be made slightly contractive in all cases (e.g., by adequate rounding), so that the "energy" in the system (represented by the square norm of the evolving state) remains bounded.

The causal theory generalizes easily to double-sided matrices as the causal and anti-causal parts can be treated independently from each other.

6.4 Matrix–Matrix Multiplication

The next case would be $T = T_2T_1$, assuming that the output dimensions of T_1 match the input dimensions of T_2, and let us first look at the causal case, that is, assuming both T_1 and T_2 causal. It is easy to see that a realization for T is again obtained by concatenating the states and is given by

$$\begin{bmatrix} A_t & B_t \\ C_t & D_t \end{bmatrix} := \left[\begin{array}{cc|c} I & 0 & 0 \\ 0 & A_2 & B_2 \\ \hline 0 & C_2 & D_2 \end{array}\right] \left[\begin{array}{cc|c} A_1 & 0 & B_1 \\ 0 & I & 0 \\ \hline C_1 & 0 & D_1 \end{array}\right] = \left[\begin{array}{cc|c} A_1 & 0 & B_1 \\ B_2C_1 & A_2 & B_2D_1 \\ \hline D_2C_1 & C_2 & D_2D_1 \end{array}\right], \quad (6.5)$$

and, again, this realization may not be minimal. (That happens when there are cancellations between the numerators and the denominators in the LTI case, but these notions are no longer useful in the LTV context.)

Again, the overall Gramians will have to be reduced if one wants to obtain a minimal realization for the product. Here also, some additional efficiency may be obtained from the original Gramians, but not as much as in the previous case (we leave details for the discussion).

For general systems with both causal and anti-causal parts, the situation is more complex. We get four terms in the product: causal × causal, causal × anti-causal, anti-causal × causal and anti-causal × anti-causal. As we know already how to deal with causal × causal, and the anti-case is just similar, let us now look at, for example, anti-causal × causal: $T = T_2T_1$, where $T_1 = C_1(I - ZA_1)^{-1}ZB_1$ and $T_2 = C_2(I - Z'A_2)^{-1}Z'B_2$. (The constants D_1 and D_2 do not contribute to the mixed form; they can best be taken as belonging to either the causal or the anti-causal part of the originals.) The result is worth a lemma.

Lemma 6.1 Decomposition by parts lemma. *(Let, in the infinitely indexed case, A_1 and A_2 be uniformly exponentially stable and C_1 and B_2 be bounded.) We have*

$$(I - Z'A_2)^{-1}Z'B_2C_1(I - ZA_1)^{-1}Z = (I - Z'A_2)^{-1}Z'A_2M + M + MA_1(I - ZA_1)^{-1}Z, \tag{6.6}$$

where M is the unique diagonal operator $M = \mathrm{diag}[M_k]$, which satisfies the Lyapunov–Stein equation

$$M_{k+1} = B_{2,k}C_{1,k} + A_{2,k}M_kA_{1,k}. \tag{6.7}$$

Proof

By premultiplication with $(I - Z'A_2)$, postmultiplication with $(I - A_1Z)$ and using $Z(I - A_1Z)^{-1} = (I - ZA_1)^{-1}Z$, one checks that the decomposition reduces to

$$ZMZ' = B_2C_1 + A_2MA_1, \tag{6.8}$$

which is exactly the given Lyapunov–Stein recursion. All these operations are legal because, even in the case of infinitely indexed systems, the inverses exist as bounded operators,[2] and the Lyapunov–Stein equation has a unique solution, thanks to the u.e.s. assumption. □

It follows that the mixed term then decomposes additively in a strictly causal, instantaneous and strictly anti-causal term:

$$C_2(I - Z'A_2)^{-1}Z'A_2MB_1 + C_2MB_1 + C_2MA_1(I - ZA_1)^{-1}ZB_1. \tag{6.9}$$

Similarly, a product of a causal with an anti-causal term will decompose additively, but now a backward Lyapunov–Stein equation will have to be solved.

6.5 Outer Systems

Preliminary remark: definitions and properties in this section are only valid for finitely indexed systems. The definition for infinitely indexed systems is considerably more involved and requires the introduction of non-bounded causal systems.

Definition 6.2 A (finitely indexed) causal system T is said to be left outer if there exists a causal system T_ℓ such that $T_\ell T = I$; similarly, T is right outer if there exists a causal system T_r such that $TT_r = I$; T is said to be outer if there is a causal system T^{-1} such that $T^{-1}T = TT^{-1} = I$.

[2] See [29].

Outer systems play an important role wherever a system has to be inverted. Luckily, there is a very simple way to characterize the various types of outerness, which we conveniently formalize in the following theorem. It turns out that the outerness properties of a causal or anti-causal system T are exclusively dependent on T's instantaneous diagonal term $D_T = T_0$, and not any further on the dynamics. Nonetheless, it will be interesting to derive state-space realizations for outer inverses, given original state-space realizations.

Theorem 6.3 *A causal and finitely indexed T is left outer if and only if D_T has a left inverse (is a diagonal of nonsingular "tall" matrices); it is right outer if and only if D_T has a right inverse (is a diagonal of nonsingular "flat" matrices), and it is outer if and only if D_T is a diagonal of square invertible matrices.*

Proof

We start by giving proof for the left-outer case and shall derive the other cases from it. The proof depends on the unicity of diagonal expansions of a causal operator, corresponding to the fact that two matrices of equal dimensions are equal if and only if all their block diagonals are equal, which is just a fancy way to say that all entries are equal.

Necessity: Suppose T has a causal left inverse T_ℓ; then we must have

$$\begin{aligned} T_\ell T &= [D_{T_\ell} + Z \cdot (\text{lower})][D_T + Z \cdot (\text{lower})] \\ &= D_{T_\ell} D_T + Z \cdot (\text{lower}) = I; \end{aligned} \tag{6.10}$$

hence $D_{T_\ell} D_T = I$.

Sufficiency: by arrow reversal. We make a constructive argument, assuming that D_T has a left inverse $D_T^+ = D_{T_\ell}$ (any left inverse will do). Let T have the realization $\{A_T, B_T, C_T, D_T\}$, and expressing u in terms of y, we obtain

$$\begin{cases} Z^{-1}x &= A_T x + B_T u, \\ u &= -D_T^+ C_T x + D_T^+ y; \end{cases} \tag{6.11}$$

see Fig. 3.5. Eliminating u in the first equation gives the state-space realization for T_ℓ:

$$T_\ell \sim_c \begin{bmatrix} A_T - B_T D_T^+ C_T & B_T D_T^+ \\ -D_T^+ C_T & D_T^+ \end{bmatrix}, \tag{6.12}$$

which produces a causal realization for the inverse T_ℓ. The computational schema evidently exhibits a computation (and realization) for the (causal) inverse, but a direct evaluation of the product $T_\ell T$ also gives the result.

This shows left invertibility. Right invertibility is equivalent to left invertibility on the dual system $T' = D_T' + B_T'(I - Z' A_T')^{-1} Z' C_T'$, now using Z' as a shift instead of Z. Finally, a system is fully invertible if and only if it is both left and right causally invertible. (We leave it as an exercise to write out a full proof for right invertibility, paralleling the proof for left invertibility.) □

Remark: Empty entries in the transfer operator have to be handled with care! For example, the empty operator $\boxed{-} : \mathbf{R} \to \emptyset$ (as in $- \cdot 1 = -$) has no left inverse. It does have a right inverse: $|$, since $- \cdot | = \cdot$. Therefore, the operator $\begin{bmatrix} 1 & \boxed{1} \\ 0 & 1 \end{bmatrix}$ is neither left nor right outer; its diagonal operator is $D = \begin{bmatrix} - & & \\ & \boxed{1} & \\ & & | \end{bmatrix}$, which has no inverse, neither right nor left!

The inversion of outer systems (of any kind) may at first seem very simple and obvious, but in the case of LTI systems or infinitely indexed LTV systems, there is a serious problem: the resulting inverse system may not be stable, and will indeed not be stable in many cases, leading to erroneous results. To see what can happen, let us look at the case where D is invertible. In the finitely indexed case, we obtained a realization of the inverse system as

$$T^{-1} \sim_c \begin{bmatrix} A - BD^{-1}C & BD^{-1} \\ -D^{-1}C & D^{-1} \end{bmatrix}, \tag{6.13}$$

and we see the Schur complement of D appearing as a state transition matrix of the inverse system. There is no reason why this matrix should be u.e.s. In the LTI case (where this theory is equally valid), it turns out that the eigenvalues of $\Delta := A - BD^{-1}C$ correspond to the *zeros* of the system, which may have arbitrary locations in the complex plane. (To be precise, if a is an eigenvalue of Δ, then $1/a$ is a zero of T.) So this means that the elementary realization of the inverse is only computationally valid when Δ is in some sense stable, for example, when Δ is u.e.s. or, perhaps, borderline stable (to be defined later). *Only in such a case, T is said to be outer*. Therefore, the notion of "outer" is substantially different for finitely indexed systems and general systems.

In Chapter 7, we shall meet a totally different kind of system that we call "inner," and whose inverse is intrinsically anti-causal. Inner systems play an equally important and complementary role determining the properties of the inverses or pseudo-inverses of systems.

6.6 Topics for Discussion

- Suppose you add two identical causal systems. How can the direct realization be reduced? Suppose you multiply them. Can the direct realization still be reduced? Under what conditions would a product of two systems be reducible?
- Let us look at the difference between "arrow inversion" on the input–output line and the inversion of the whole system matrix $\begin{bmatrix} A & B \\ C & D \end{bmatrix}$. Consider a few cases:
  - A finite-dimensional causal shift $Z = \begin{bmatrix} \boxed{-} & & & \\ 1 & 0 & & \\ & \ddots & \ddots & \\ & & 1 & | \end{bmatrix}$.

  - The case $D = I$. What happens when $\begin{bmatrix} A & B \\ C & I \end{bmatrix}$ is square invertible? Suppose only arrow reversal on the input/output line is done; then what happens? When the system is finitely indexed, can $\begin{bmatrix} A & B \\ C & D \end{bmatrix}$ be square invertible? (Consider both finitely indexed and infinitely indexed systems one side toward $+\infty$ and both sides.)
  - More general cases?
- A question which one may already raise at this point is: Does a system that has a contractive transfer operator have a contractive realization? (Can you show the converse: a contractive realization produces a contractive system?) This somewhat delicate issue is considered in what is called embedding theory, where one shows that this is indeed the case, and methods to determine such a realization are proposed; see, for example, [29] or Chapter 17.

6.7 Notes

- Although the material in this chapter may seem pretty standard, the generality of the results may be surprising, mostly thanks to the definition of "dummies." First, there is the LTV "decomposition by parts lemma": this is the key technical property that makes much of the whole theory work. It allows, in this precise setting, for "dichotomy": the splitting of a mixed causal–anti-causal term in a causal and an anti-causal part – an operation that is bread-and-butter for system theory.
- Next, on "elementary inversion": one should resist the temptation to think that D_k nonsingular is a necessary condition for system inversion. The shift operator Z has the simple inverse Z' and illustrates the fact that a causal (bounded) operator may very well have a bounded anti-causal inverse. When a bounded, causal system has a bounded *causal* inverse, then it will belong to the class of "outer" systems. In the case of finitely indexed systems, the existence of the inverse D_k (and hence also that D_k is square) is necessary and sufficient for the system to be outer. In the infinitely indexed case, the situation is much more complicated.
- What we call "outer" is the terminology originally adopted in what is known as "Hardy space theory" in complex analysis (see e.g., Rudin, *Real and Complex Analysis* [59]). Note, in particular, Szegö's theorem that gives a characterization of an outer function in terms of its Fourier transform. However, in traditional systems and control engineering for single-input single-output systems (SISO systems), an outer transfer function is called a *minimum phase* transfer function. This terminology is sometimes used, by extension, also for non-scalar transfer functions (so-called MIMO systems: multiple input multiple output.)

7 Inner Operators and External Factorizations

Chapters 7 and 8 are devoted to classical factorization theory in the nonclassical setting of time-variant systems. Factorizations have played a major role in the development of system theory. They are an essential ingredient in solving many problems in control theory and numerical algebra, as shall be demonstrated in the subsequent chapters. In this chapter, we treat what is traditionally called "coprime factorization," that is, the representation of a causal system as the ratio of two anti-causal matrix factors, one of which (the denominator) characterizes the dynamic behavior of the system. One type of denominator is a causal, unitary transfer function and is called an *inner function.* Another is polynomial in the shift Z and is presented in Chapter 10. This chapter starts with the definition and properties of inner functions, followed by the coprime factorization theory and its relation to other important system characterizations, such as the system realization itself, the Hankel operator and the characterization of what may be termed *natural state spaces* in the strict past input space and, isomorphically, the future output space.

Menu

Hors d'oeuvre
Inner Matrices

First Course
Various Types of Inner Operator

Second Course
Fractional System Representations Using Inners in the Denominator

*Third Course**
The LTI Case

Dessert
Historical Notes

7.1 Inner Systems

State-Space Representation of Inners

As in almost all numerical analysis, isometric and unitary matrices play an important role, think about QR factorization or the SVD. That is also the case with quasi-separable or LTV systems, not to talk about Circuit Theory, where the notion of "losslessness" plays a similar and related role. The historical term *inner* to name such operators when they are also *causal* has to do with the combination poles and zeros they exhibit in Complex Function Analysis. The motivation for the term in the LTV setting will be discussed in Chapter 8 on inner–outer factorization. In engineering practice, the term *modulus one*, *pure phase function* or *lossless transfer function* is often used. We define and adopt "inner" here to stay in line with the original historical use, pioneered by Hardy space theory in the early twentieth century. Also in the LTV theory, inner operators get special and remarkable appearances. They play an essential role toward the solution of many problems in numerical analysis, filtering, optimization and control.

A Discussion on Inners and Causality

The most obvious and elementary inner is the identity. In the LTV case, an identity system is subdivided in blocks of variable dimensions. Somewhat more generally, a block diagonal matrix with unitary blocks would, of course, also be unitary, causal and also nondynamical. One step further would be a block diagonal matrix D whose individual blocks are either all isometric, resulting in a global causal isometry ($D'D = I$), or co-isometric ($DD' = I$), resulting in a global causal co-isometric operator.

There are many more interesting unitary, isometric or co-isometric causal matrices. To start with, let us consider a causal shift operator Z. We defined the causal shift as a generic causal operation on sequences represented by a (finitely indexed or infinitely indexed) column vector. To restrict ourselves to the simplest possible case, let the indexed vector x consist of scalar entires indexed from 0 to n: $x = \text{col}\left[\boxed{x_0} \cdots x_n\right]$; then we have

$$Z \begin{bmatrix} \boxed{x_0} \\ \vdots \\ x_n \end{bmatrix} = \begin{bmatrix} \boxed{-} \\ x_0 \\ \vdots \\ x_n \end{bmatrix}. \tag{7.1}$$

The downward causal shift makes an empty entry to appear at index 0 and shifts x_0 one index unit "forward" to the future. Which operator or matrix achieves this? The trick to constructing the matrix for Z is to figure out the column and row dimensions. The dimensions of x determine the column dimensions of Z, while the dimensions of the shifted version Zx determine the relevant row dimensions of Z, with all other dimensions zero and suppressed in the representation: $\left[\boxed{0}, 1, \ldots, 1\right] \times \left[\boxed{1}, 1, \ldots, 1, 0\right]$; hence,

$$Z = \begin{bmatrix} \boxed{-} & & & \\ 1 & 0 & & \\ & \ddots & \ddots & \\ & & 1 & | \end{bmatrix}, \tag{7.2}$$

or Z is a unit matrix with a main diagonal of zeros (and dummies)!

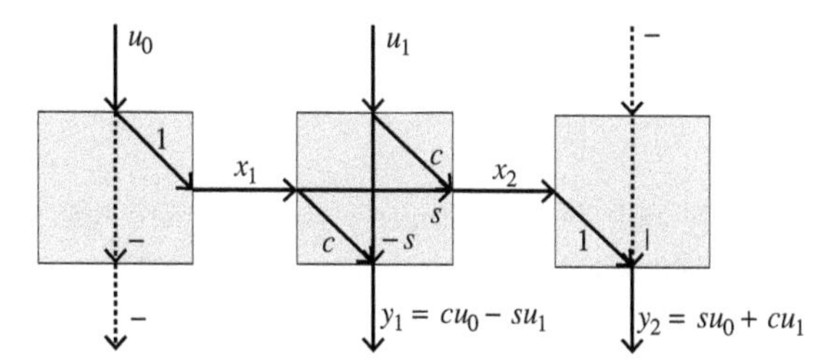

Figure 7.1 Flowgraph of a single rotor with scalar inputs and outputs.

More generally, what can be an inner system with scalar entries? One shows that a general form of a causal unitary finitely indexed operator with scalar entries is

$$\begin{bmatrix} \boxed{-} & & & \\ * & * & & \\ \vdots & \ddots & \ddots & \\ * & * & \cdots & | \end{bmatrix}. \tag{7.3}$$

This will follow from Theorem 7.2 when considering finitely indexed realizations. For example, with c being the cosine and s the sine of some given angle,

$$U = \begin{bmatrix} \boxed{-} & & \\ c & -s & \\ s & c & | \end{bmatrix} \tag{7.4}$$

is a causal unitary transfer operator that rotates a two-dimensional input over the given angle. Its matrix representation is necessarily a square skeleton with a shifted main diagonal. See Fig. 7.1 for a flow graph realizing U. The realization corresponding to the flow graph is

$$U \sim_c \operatorname{diag}\left(\boxed{\begin{bmatrix} | & 1 \\ \cdot & - \end{bmatrix}}, \begin{bmatrix} s & c \\ c & -s \end{bmatrix}, \begin{bmatrix} - & \cdot \\ 1 & | \end{bmatrix} \right). \tag{7.5}$$

Causality and the condition of scalar entries force a shift in the main diagonal. In the literature, much has been made of the fact that a lower triangular unitary matrix is necessarily diagonal with diagonal entries having unit magnitude, but we see here that this is only the case because an "unnatural" order is traditionally imposed on the inputs and outputs: causality requires a delay in the production of the result of a rotation, when the entries are forced to be scalar!

The situation is, of course, a bit different for isometric or co-isometric matrices or, in more colloquial parlance, for tall or flat matrices with orthonormal columns or rows respectively. We shall see that in the finitely indexed case, these can be considered sub-operators of unitary operators. (This property does not hold in general for the infinitely indexed case.) It is easy to construct isometric causal systems with a "normal" main diagonal, for example (with $c^2 + s^2 = 1$),

$$\begin{bmatrix} \boxed{c} & \\ c^2 & s \\ -cs & c \end{bmatrix}. \tag{7.6}$$

Such matrices are necessarily tall. In the infinitely indexed case, the distinction between "tall" and "square" disappears, and entirely new cases appear.

Definitions

To avoid confusion between the various types of inner, here are the definitions.

Definition 7.1 The transfer operator of a system is *left-inner* when it is causal and isometric, *right-inner* when it is causal and co-isometric, and *inner (or bi-inner)* when it is causal and unitary. Similar definitions hold for anti-causal systems, in which case the transfer operator of a system is called *left co-inner* (or left conjugate inner), *right co-inner*, or *co-inner* respectively.

Our interest will, of course, be focused on systems that have state-space realizations (or, equivalently, are quasi-separable). The following theorem summarizes the important properties relating the transfer operator of an inner system to its possible realizations.

Theorem 7.2 *A left-inner operator has a minimal realization that is isometric (and u.e.s. in the infinite indexed case). Dually, a right-inner operator has a minimal realization that is co-isometric (and u.e.s. in the infinite indexed case). An operator that is inner has a minimal realization that is unitary and u.e.s. Conversely, a finitely indexed causal system that has an isometric realization is left-inner. It is right-inner if it has a co-isometric realization and bi-inner if it has a unitary realization.*

Remark: In the infinitely indexed case, it is not true that an isometric or unitary realization necessarily produces an isometric or unitary transfer operator, although the converse holds true. Here we give a proof for finitely indexed systems. The general case is treated in [29].

Proof

Suppose T isometric ($T'T = I$), and consider a minimal realization for T in the output normal form (i.e., with all the observability operators $\mathbf{O}_k$ isometric or $\mathbf{O}_k'\mathbf{O}_k = I$ for all k). This choice already makes the pair $\begin{bmatrix} A_k \\ C_k \end{bmatrix}$ isometric, since $\mathbf{O}_k = \begin{bmatrix} C_k \\ \mathbf{O}_{k+1}A_k \end{bmatrix}$, and hence

$$\mathbf{O}_k'\mathbf{O}_k = I = C_k'C_k + A_k'\mathbf{O}_{k+1}'\mathbf{O}_{k+1}A_k = C_k'C_k + A_k'A_k. \tag{7.7}$$

Next, consider the southwest (or left-bottom) corner of the transfer operator T, which contains D_k as its upper-right element:

$$\begin{bmatrix} C_k\mathbf{R}_k & D_k \\ \mathbf{O}_{k+1}A_k\mathbf{R}_k & \mathbf{O}_{k+1}B_k \end{bmatrix}. \tag{7.8}$$

The last column has to be isometric and orthogonal to all the previous columns, due to the isometry of T. It then follows:

1. $D_k'D_k + B_k'\mathbf{O}_{k+1}'\mathbf{O}_{k+1}B_k = D_k'D_k + B_k'B_k = I$, and
2. $(D_k'C_k + B_k'\mathbf{O}_{k+1}'\mathbf{O}_{k+1}A_k)\mathbf{R}_k = 0$,

and hence also $B_k'A_k + D_k'C_k = 0$, making the realization fully isometric, because we assumed the realization to be minimal, and hence $\mathbf{R}_k$ right invertible since it forms a row basis. The converse property, namely a finitely indexed isometric realization results in an overall isometric map, boils down to showing that a chain of orthogonal transformations is itself an orthogonal transformation and is left as an exercise. All the other properties claimed are *mutatis mutandis* proved in a similar way. □

A consequence of this is that a left-inner (or a right-inner) operator can be embedded into an inner one by the augmentation of the local isometry to unitary. More precisely, suppose $\begin{bmatrix} A & B \\ C & D \end{bmatrix}$ is an isometric realization of a left-inner operator; then there exist B_2 and D_2 such that $\begin{bmatrix} A & B & B_2 \\ C & D & D_2 \end{bmatrix}$ is unitary, thereby creating a second transfer operator T_2 with realization $\begin{bmatrix} A & B_2 \\ C & D_2 \end{bmatrix}$ such that $T_t := \begin{bmatrix} T & T_2 \end{bmatrix}$ is inner. Notice that the state dimensions, that is, the sequence of $\{\eta_k\}$'s of the total system, equal the state dimensions of the original. These are often called the *degree* of the system.

7.2 External Factorizations with Inner Denominators

Inner matrices allow us to derive new forms for the representation of a given causal operator T, which turn out to exhibit the natural state-space structure in a numerically appealing way and will often appear to be numerically useful in applications. They effectively lift the QR or LQ forms to the level of system operators, where the role of the orthogonal matrix Q is played by an inner operator. As a first observation, we see that the inverse of an inner operator – call it U – is anti-causal, since $U^{-1} = U'$, and the adjoint is automatically anti-causal: $U' = D' + B'Z'(I - A'Z')^{-1}C'$.

So a game we can play is as follows: given T, we can ask which inner U is such that its inverse/adjoint pushes T into anti-causality by multiplication to the right, to be denoted concisely by $TU' \in a$?; and a similar question when the inner factor is applied to the left: V inner such that $V'T \in a$. Moreover, we could ask whether there is a smallest possible U or V whose transpose pushes T to anti-causality from the right or the left.

The answer turns out to be pretty direct. Starting with the right-side case, let us choose for T a realization in an input normal form, hence one for which all the reachability matrices $\mathbf{R}_k$ are co-isometric ($\mathbf{R}_k\mathbf{R}_k' = I$). In that case, the resulting output normal form realization will have $\begin{bmatrix} A_k & B_k \end{bmatrix}$ co-isometric as well: $A_kA_k' + B_kB_k' = I$ for all k. Using the properties of Section 7.1, we may now construct an inner U by completion to a unitary realization, defining C_U and D_U such that

$$U \sim_c \begin{bmatrix} A & B \\ C_U & D_U \end{bmatrix} \tag{7.9}$$

is unitary. The property sought is then

Proposition 7.3 *TU' is anti-causal.*

Proof

By direct computation, we find

$$\begin{aligned} TU' &= (D + C(I - ZA)^{-1}ZB)(D_U' + B'Z'(I - A'Z')^{-1}C_U') \\ &= DD_U' + C(I - ZA)^{-1}ZBD_U' + DB'Z'(I - A'Z')^{-1}C_U' \\ &+ \ C(I - ZA)^{-1}ZBB'Z'(I - A'Z')^{-1}C_U'. \end{aligned} \tag{7.10}$$

The last quadratic term splits, thanks to partial fraction decomposition and $AA' + BB' = I$, as follows:

$$\begin{aligned} &C(I - ZA)^{-1}ZBB'Z'(I - A'Z')^{-1}C_U' \\ &\quad = C[(I - ZA)^{-1}ZA + I + A'Z'(I - A'Z')^{-1}]C_U'. \end{aligned} \tag{7.11}$$

Now using the unitarity of the realization for U, we see that the causal term in $C(I - ZA)^{-1}$ cancels, leaving

$$TU' = (CC_U' + DD_U') + (CA' + DB')Z'(I - A'Z')^{-1}C_U' \tag{7.12}$$

anti-causal. □

Hence, $\Delta_L' := TU'$ is anti-causal, and

$$\Delta_L \sim_c \begin{bmatrix} A & AC' + BD' \\ C_U & C_UC' + D_UD' \end{bmatrix} = \begin{bmatrix} A & B \\ C_U & D_U \end{bmatrix} \begin{bmatrix} I & C' \\ 0 & D' \end{bmatrix}, \tag{7.13}$$

and we have obtained a *right external factorization* for T:

$$T = \Delta_L' U = \Delta_L'(U')^{-1}. \tag{7.14}$$

Hence, we obtain a representation for the causal T as the ratio of two anti-causal operators, of which the right one is unitary. This is what we call a *right external factorization*, often called *right coprime factorization* in the literature, although the latter term implies coprimeness between the factors, which is not necessary for such a factorization to exist but is often assumed. U is an inner factor whose inverse pushes T to anti-causality from the right, and it is, as we have constructed it, “minimal” in doing so, meaning in this context that it has the smallest possible state dimension at each index k to do so. Proof for this requires some more work, which for completeness we shall do later in this chapter, in an advanced section on Bezout equations.

Similarly, and starting from an output normal form, we could obtain a *left external factorization*. This produces

$$T = V\Delta_R' \tag{7.15}$$

with V causal unitary and Δ_R' anti-causal (Δ_R causal). One starts out, dually, with a realization in an output normal form and $\begin{bmatrix} A \\ C \end{bmatrix}$ isometric. (Warning: A and C are different than in the previous, input normal form!) Again, one may complete $\begin{bmatrix} A \\ C \end{bmatrix}$ to unitary, thereby defining a unitary $V = D_V + C(I - ZA)^{-1}ZB_V$. We find the realization for Δ_R by working out the product

$$\begin{aligned} V'T &= \left(D_V' + B_V'Z'(I - A'Z')^{-1}C'\right)\left(D + C(I - ZA)^{-1}ZB\right) \\ &= B_V'(I - ZA')^{-1}Z'(A'B + C'D) + (D_V'D + B_V'B) \\ &\quad + (D_V'C + B_V'A)(I - ZA)^{-1}ZB \\ &= B_V'Z'(I - A'Z')^{-1}(A'B + C'D) + (D_V'D + B_V'B), \end{aligned} \tag{7.16}$$

where the causal term disappears again because of the unitarity of V and its realization. This gives as realization for Δ_R (assuming $A'A + C'C = I$!),

$$\Delta_R \sim \begin{bmatrix} A & B_V \\ B'A + D'C & B'B_V + D'D_V \end{bmatrix} = \begin{bmatrix} I & \\ B' & D' \end{bmatrix} \begin{bmatrix} A & B_V \\ C & D_V \end{bmatrix}. \tag{7.17}$$

System-Theoretic Significance

Given a causal and quasi-separable LTV system T (that is: having a time-variant realization with finite-dimensional state vectors), we have already figured out that we could determine realizations by choosing matching bases for reachability and the observability spaces. For example, we could choose an orthonormal basis for each reachability space (i.e., the range of H_k' at each index k) and determine the corresponding observability space and a realization according to the precepts of Chapter 4. This produces A, B, C and D in the input normal form, that is, such that $AA' + BB' = I$. We can show, as a central property,

Proposition 7.4 *Given a minimal realization for T in the input normal form, orthonormal bases for the reachability spaces are given by the columns of $\mathbf{R}' := B'Z'(I - A'Z')^{-1}$, and the causal part of $T\mathbf{R}'$ is, with Π_f projection on the present and future, given by*

$$\Pi_f\left[\left(D + C(I - ZA)^{-1}ZB\right)B'Z'(I - A'Z')^{-1}\right] = C(I - ZA)^{-1}, \tag{7.18}$$

that is, the response one gets with zero as (present and) future inputs, with the latter producing the matching observability bases $\mathbf{O}$ ($\mathbf{O}_k$ for each k).

Proof

The reachability spaces are by definition the ranges of the adjoint of the Hankel H_k' at each k. We have, by the construction of the realization,

$$H_k = \begin{bmatrix} C_k \\ C_{k+1}A_k \\ C_{k+2}A_{k+1}A_k \\ \vdots \end{bmatrix} \begin{bmatrix} \cdots & A_{k-1}A_{k-2}B_{k-3} & A_{k-1}B_{k-2} & B_{k-1} \end{bmatrix} = \mathbf{O}_k\mathbf{R}_k, \tag{7.19}$$

with this factorization being minimal and $\mathbf{R}_k'$ chosen to be isometric. The basis for H_k' is therefore just $\mathbf{R}_k'$, which is the kth block column of $B'Z'(I - A'Z')^{-1}$. The corresponding observability basis is then found by "decomposition in parts" of the quadratic term in (7.18):

$$(I - ZA)^{-1}ZBB'Z'(I - A'Z')^{-1} = (I - ZA)^{-1} + A'Z'(I - A'Z')^{-1}, \tag{7.20}$$

followed by the application of Π_f to yield the result. □

7.3 Coprimeness and the Bezout Equations*

To find the two canonical external forms one puts the realization in the input and output normal forms, respectively, which in each case requires the solution of a Lyapunov–Stein equation, a forward and a backward recursion, respectively, when just a general minimal realization is given to start with. Now, we want to show that it is advantageous to combine the two realizations, which will also allow us to assess whether the factorization is indeed coprime. Let us therefore derive a realization for what we called the right canonical external form $T = \Delta_L' U$, with U inner and Δ_L causal, using a minimal representation in the output normal form $\{A, B, C, D\}$.

Minimality of the system realization implies the nonsingularity of the reachability Gramian, let us call it for brevity $G = RR'$, in which R is diagonal with invertible square blocks. G satisfies the Lyapunov–Stein equation

$$G_{k+1} = B_k B_k' + A_k G_k A_k'. \tag{7.21}$$

Hence $[R_{k+1}^{-1} A_k R_k \;\; R_{k+1}^{-1} B_k]$ is co-isometric, so that a local state transformation $R_{k+1}^{-1} \cdot R_k$ turns the original output normal form into an input normal form. If $\begin{bmatrix} A & B \\ C_U & D_U \end{bmatrix}$ is a (nonunitary) realization for U, then the corresponding unitary realization, written in local terms, is

$$U \sim \begin{bmatrix} R_{k+1}^{-1} A_k R_k & R_{k+1}^{-1} B_k \\ C_{U,k} R_k & D_{U,k} \end{bmatrix}. \tag{7.22}$$

Similarly (and dual) as before, we find, for B_b and D_b to be specified further,

$$\Delta_L \sim \begin{bmatrix} A & B_b \\ C_U & D_b \end{bmatrix} \sim \begin{bmatrix} R_{k+1}^{-1} A_k R_k & R_{k+1}^{-1} B_{b,k} \\ C_{U,k} R_k & D_{b,k} \end{bmatrix}, \tag{7.23}$$

in which

$$\begin{bmatrix} B_b \\ D_b \end{bmatrix} = \begin{bmatrix} AG & B \\ C_U G & D_U \end{bmatrix} \begin{bmatrix} C' \\ D' \end{bmatrix}, \tag{7.24}$$

and the global recursion for G is

$$G^{\langle -1 \rangle} = BB' + AGA', \tag{7.25}$$

which also expresses orthogonality of the right factor U in $T = \Delta_L' U$. (The correctness of the given realizations can be checked directly but follows also from a procedure dual to the procedure just derived.)

To show coprimeness, we need to do more work on the realizations derived in Section 7.2; this will lead to the famous Bezout equations. The trick is to consider at first the realizations for the combinations $\begin{bmatrix} V \\ \Delta_R \end{bmatrix}$ and $\begin{bmatrix} -\Delta_L & U \end{bmatrix}$, which all share the same state transition matrix A (motivation for the minus sign is tradition and will soon appear):

$$\begin{bmatrix} V \\ \Delta_R \end{bmatrix} \sim_c \left[\begin{array}{cc} A & B_V \\ \hline C & D_V \\ C_a & D_a \end{array} \right], \quad \begin{bmatrix} -\Delta_L & U \end{bmatrix} \sim_c \left[\begin{array}{c|cc} A & -B_b & B \\ C_U & -D_b & D_U \end{array} \right]. \tag{7.26}$$

First, we show that $\begin{bmatrix} V \\ \Delta_R \end{bmatrix}$ has a causal left inverse (is left outer) and that (dually and for similar reasons) $\begin{bmatrix} -\Delta_L & U \end{bmatrix}$ has as causal right inverse (is right outer).

Proposition 7.5 $\begin{bmatrix} V \\ \Delta_R \end{bmatrix}$ *is left outer, and* $\begin{bmatrix} -\Delta_L & U \end{bmatrix}$ *is right outer.*

Proof

(This proof is only valid for finitely indexed systems. However, the property is more generally valid also for infinitely indexed systems. The simple finite proofs given here are interesting in their own right.) A finitely indexed system is left-outer if and only if its instantaneous operator D has a left-inverse (theorem 6.3). In the present case, the condition reduces to

$$\begin{bmatrix} D_V \\ D_a \end{bmatrix} = \begin{bmatrix} 0 & I \\ B' & D' \end{bmatrix} \begin{bmatrix} B_V \\ D_V \end{bmatrix}, \tag{7.27}$$

which is left invertible, or equivalently, its columns form a basis, meaning $\begin{bmatrix} D_V \\ D_a \end{bmatrix} X = 0 \Rightarrow X = 0$, with X being any conformal block diagonal matrix. (One may choose the components of X to be just vectors: $X = \text{diag}[X_k]$.) So, assume $\begin{bmatrix} D_V \\ D_a \end{bmatrix} X = 0$; then $D_V X = 0$ and $B' B_V X = 0$. Now consider any index k and

$$\begin{aligned} G_{k+1} B_{V,k} X_k &= B_k B_k' B_{V,k} X_k + A_k G_k A_k' B_{V,k} X_k \\ &= 0 - A_k G_k C_k' D_{V,k} X_k = 0 \end{aligned} \tag{7.28}$$

because, by orthogonality, $A_k' B_{V,k} + C_k' D_{V,k} = 0$. Since the realization is assumed reachable, G_{k+1} is nonsingular; hence, $B_{V,k} X_k = 0$ in addition to $D_{V,k} X_k = 0$ and also $X_k = 0$ since $\begin{bmatrix} B_{V,k} \\ D_{V,k} \end{bmatrix}$ forms a basis. The dual proof holds for $\begin{bmatrix} -\Delta_L & U \end{bmatrix}$. □

Hence, $\begin{bmatrix} V \\ \Delta_R \end{bmatrix}$ has a left inverse $\begin{bmatrix} M & N \end{bmatrix}$ with M and N causal operators, by definition of left-outer:

$$MV + N\Delta_R = I. \tag{7.29}$$

Dually and using a similar proof, there will also be causal $-X$ and Y such that

$$\Delta_L X + UY = I. \tag{7.30}$$

Putting these equations together and observing that dimensions fit, we may write

$$\begin{bmatrix} M & N \\ -\Delta_L & U \end{bmatrix} \begin{bmatrix} V & -X \\ \Delta_R & Y \end{bmatrix} = \begin{bmatrix} I & ? \\ 0 & I \end{bmatrix}. \tag{7.31}$$

Would it be possible to (1) give minimal realizations for these four new systems M, N, X and Y and (2) assure that the entry marked as "?" is zero? We have, of course, $-\Delta_L V + U\Delta_R = 0$ since $T = V\Delta_R' = \Delta_L' U$, so we show next that we can choose X and Y minimal such that $-MX + NY = 0$. This is done in Proposition 7.7. But we need the following lemma first.

Lemma 7.6

$$\begin{bmatrix} -D_b & D_U \end{bmatrix} \begin{bmatrix} D_V \\ D_a \end{bmatrix} = 0. \tag{7.32}$$

Proof

This is a direct consequence of the causality of $\begin{bmatrix} -\Delta_L & U \end{bmatrix}$ and $\begin{bmatrix} V \\ \Delta_R \end{bmatrix}$, and the fact that their product is zero. (It can also be verified directly on the realizations: good exercise!) □

Proposition 7.7 $\begin{bmatrix} V \\ \Delta_R \end{bmatrix}$ *has a left causal pseudo-inverse with realization (using* $n_a := D_V D_V' + D_a' D_a$*)*

$$\begin{bmatrix} M & N \end{bmatrix} \sim \left[\begin{array}{c|cc} A - B_V n_a^{-1}(D_V' C + D_a' C_a) & B_V n_a^{-1} D_V' & B_V n_a^{-1} D_a' \\ \hline -n_a^{-1}(D_V' C + D_a' C_a) & n_a^{-1} D_V' & n_a^{-1} D_a' \end{array} \right]. \tag{7.33}$$

Dually,

$$\begin{bmatrix} -\Delta_L & U \end{bmatrix} \sim \left[\begin{array}{c|cc} A & -B_b & B \\ C_U & -D_b & D_U \end{array} \right] \tag{7.34}$$

has a right causal pseudo-inverse (using $n_b := D_b D_b' + D_U D_U'$*) given by*

$$\begin{bmatrix} -R \\ S \end{bmatrix} \sim \left[\begin{array}{c|c} A - (B_b D_b' + B D_U') n_b^{-1} C_U & (B_b D_b' + B D_U') n_b^{-1} \\ \hline D_b' n_b^{-1} C_U & -D_b' n_b^{-1} \\ -D_U' n_b^{-1} C_U & D_U' n_b^{-1} \end{array} \right]. \tag{7.35}$$

Proof

Proposition 7.5 already established that $\begin{bmatrix} V \\ \Delta_R \end{bmatrix}$ is left-outer, which guarantees the existence of causal pseudo-inverses (see Chapter 6). One obvious choice for the left pseudo-inverse of the instantaneous term $\begin{bmatrix} D_V \\ D_a \end{bmatrix}$ is $n_a^{-1} \begin{bmatrix} D_V' & D_a' \end{bmatrix}$. Filling in the expression for the pseudo-inverse produces the formula for $\begin{bmatrix} M & N \end{bmatrix}$. A dual reasoning holds for $\begin{bmatrix} -\Delta_L \\ U \end{bmatrix}$. □

As a result, we obtain the Bezout relation.

Proposition 7.8

$$\begin{bmatrix} M & N \\ -\Delta_L & U \end{bmatrix} \begin{bmatrix} V & -R \\ \Delta_R & S \end{bmatrix} = I, \tag{7.36}$$

in which all M, N, R and S are causal.

Proof

We already established much of the claimed relation, except for $-MR + NS = 0$. This now follows from the explicit construction in Proposition 7.7. Indeed, we find

1. $-D_M D_R + D_N D_S = n_a^{-1}(-D'_V D'_b + D'_a D'_U) n_b^{-1} = 0$ by Lemma 7.6, and
2. let $X := D_X + C_X(I - ZA_X)^{-1} Z B_X$ and likewise Y be two causal transfer functions; then a sufficient condition for $XY = 0$ is that besides $D_X D_Y = 0$, also either $\begin{bmatrix} B_X \\ D_X \end{bmatrix} C_Y = 0$ or $B_X \begin{bmatrix} D_Y & C_Y \end{bmatrix} = 0$. In the present case, both are satisfied, for example,

$$\begin{bmatrix} B_X \\ D_X \end{bmatrix} C_Y \sim B_V n_a^{-1} \begin{bmatrix} D'_V & D'_a \end{bmatrix} \begin{bmatrix} D'_b \\ -D'_U \end{bmatrix} n_b^{-1} C_U = 0 \tag{7.37}$$

by Lemma 7.6. □

A consequence of the Bezout equations is that, for example, V and Δ_R are *right coprime*, in the sense that there cannot be a nondiagonal inner factor v common to both, that is, such that $V = \hat{V}v$ and $\Delta_R = \hat{\Delta}_r v$, with both $\hat{V}$ and $\hat{\Delta}_R$ causal. Indeed, suppose there were such a v; then we would have

$$M\hat{V} + N\hat{\Delta}_R = v', \tag{7.38}$$

showing that v' also is causal. But only diagonals are causal and anti-causal at the same time, making v necessarily diagonal unitary. Similarly, U and Δ_L are left-inner coprime.

7.4 The Global Geometric Picture*

Let us start out with a left external coprime factorization $T = V\Delta'_R$, with the dimensions of V, $\mathbf{n} \times \mathbf{r}$, $\mathbf{n} := (n_k)_{k=:}$ and $\mathbf{r} := (r_k)_{k=:}$, and make the following observation:

Proposition 7.9 *The orthogonal complement of the range of the Hankel H_k is $V[\ell_2^{\mathbf{r}}]_{k:\infty}$.*

Proof

We have, since V is unitary,

$$\ell_2^{\mathbf{n}} = V\ell_2^{\mathbf{r}} = V[\ell_2^{\mathbf{r}}]_{-\infty:k-1} \oplus V[\ell_2^{\mathbf{r}}]_{k:\infty} = \text{range} H_k \oplus V[\ell_2^{\mathbf{r}}]_{k:\infty}. \tag{7.39}$$

□

So we see that $V[\ell_2^{\mathbf{r}}]_{k:\infty} = \mathcal{M}_k$, which we call *the output null space of* T. This subspace has interesting shift-invariance properties, since

$$V[\ell_2^{\mathbf{r}}]_{k:\infty} \supset V[\ell_2^{\mathbf{r}}]_{k+1:\infty} \supset V[\ell_2^{\mathbf{r}}]_{k+2:\infty} \supset \cdots . \tag{7.40}$$

In order to properly explore shift invariance, we need some extra machinery that we introduce now.

Working with Global Subspaces

In our treatments until now, we looked at the behavior of the Hankel map corresponding to a causal transfer operator T at a selected but arbitrary index point k, taken as the actual present. To identify a time-variant system globally, one has to consider what happens at every index point k, and, as we saw in Chapter 4, compare the evolution of spaces from each point k with the next point $k+1$ in order to derive a state realization. At time index k, there is a Hankel map H_k whose range is the observability space $\mathcal{O}_k$, while the range of the adjoint H_k' is the reachability space $\mathcal{R}_k'$. Each minimal realization for T provides a basis for $\mathcal{O}_k$ given by $\mathbf{O}_k := \text{col}[C_k, C_{k+1}A_k, C_{k+2}A_{k+1}A_k, \ldots]$, and a basis for the reachability space at k given by $\mathbf{R}_k' := [\cdots\ B_{k-2}A_{k-1}\ B_{k-1}]'$, while $H_k = \mathbf{O}_k\mathbf{R}_k$. What is needed is a way to deal with these spaces and their relations in a global way.

Intermezzo: Frobenius Spaces

Let us consider just an $n \times m$ matrix T and let us study its effect on a collection of vectors of dimension m, say $u := [u^1, u^2, \ldots, u^k]$. Applying T on this concatenated (row) stack of vectors, we find

$$Tu = \begin{bmatrix} Tu^1 & Tu^2 & \cdots & Tu^k \end{bmatrix}. \tag{7.41}$$

In other words, the results just stack up independently as well. If we want to identify some properties of the operator T by experimenting on n vectors independently, we would apply T on each and see what it does on each individually. When u was supposed to be just a single vector, we defined the operator norm of T as

$$\|T\| = \sup_{u \neq 0} \frac{\|Tu\|_2}{\|u\|_2} \tag{7.42}$$

derived from the Euclidean (inner-product) norm $\|\cdot\|_2$ for vectors of $\mathbf{R}^m$. Viewing now T as acting on a stack u instead of a single vector, what should be the global definition of the norm on the stack in order to recover the correct $\|T\|$? It turns out that the correct inner-product norm for the stack u that will achieve this feat is just the stack of norm squares: $\|u\|_F^2 = \sum_{i=1:k} \|u^i\|_2^2$. The norm $\|u\|_F$ is called the *Frobenius norm* on the matrix u. It is easy to see (good exercise!) that this definition of norm on the stack u reproduces the operator norm on T, and formula (7.42) remains valid, now using the Frobenius norm

$$||T|| = \sup_{u \neq 0} \frac{||Tu||_F}{||u||_F}. \tag{7.43}$$

Remark that the Frobenius norm is just a special type of the Euclidean norm, namely the Euclidean norm of the overall vector $\text{col}[u^1, u^2, \ldots, u^k]$, as if the column vectors had been stacked column-wise and living in the space $\mathbf{R}^{mk}$. u lives in what we call a *Frobenius space*, which is:

1. a true Euclidean space,
2. but with some special properties. In particular, $u^1 \perp u^2 \perp \cdots \perp u^k$ when properly embedded in the Frobenius space, reflecting the independence of these vectors.

Notice also that the Frobenius norm of u is nothing but the square root of the sum total of all entries norm squared, sometimes called the *total energy* in u of the computational schema. □
End of the Intermezzo

At each time index point k, we now want to consider inputs and outputs for which k is the present. Stacking the inputs (starting from any far away point in the past until any far away point in the future), we get stacks of the type $u = [\cdots\ u^{-1}\ u^0\ u^1\ \cdots]$ on which we can apply the operator T, and we can observe, likewise, the corresponding stacked outputs $y = [\cdots\ y^{-1}\ y^0\ y^1\ \cdots]$, with $y = Tu$. On these spaces, we may use the Frobenius norm, as discussed in the intermezzo. We call the space of the stacked inputs $\mathcal{X}_{F,\mathbf{m}}$, that is, stacks of input sequences with the dimension sequence $\mathbf{m}$ and likewise $\mathcal{X}_{F,\mathbf{n}}$ for the output sequences. (We may drop the dimensions if they are clear from the context.) The division in strict past and future in these global spaces is reflected in the structure of u. Consider now the space of all u's in the stack that belong to the strict past with respect to their respective index point k – in other words, all strictly upper input stacks. Similarly, all input signals living in the present plus future with respect to their index form together a lower block triangular Frobenius space. A similar subdivision can be made for the output spaces as well. We call the Frobenius space of the strict past inputs $\mathcal{U}^-_{F,\mathbf{m}}$, $\mathcal{U}_{F,\mathbf{m}}$ assembles all the uppers, and $\mathcal{D}_{F,\mathbf{m}}$ assembles all the diagonals with individual dimensions $m_k \times 1$), while present plus future inputs are denoted $\mathcal{L}_{F,\mathbf{m}}$. Because of the inner-product norm, we have the orthogonal decomposition

$$\mathcal{X}_{F,\mathbf{m}} = \mathcal{U}^-_{F,\mathbf{m}} \oplus \mathcal{L}_{F,\mathbf{m}}. \tag{7.44}$$

Likewise for the output space. Here is how a (global) future input looks:

$$u = \begin{bmatrix} \ddots & & & & \\ \ddots & u_{-1}^{-1} & & & \\ \ddots & u_0^{-1} & \boxed{u_0^0} & & \\ \ddots & u_1^{-1} & u_1^0 & u_1^1 & \\ \cdot^{\cdot^{\cdot}} & \ddots & \ddots & \ddots & \ddots \end{bmatrix}. \tag{7.45}$$

Some remarks:

- All column dimensions in this stack are 1; row dimensions are $\mathbf{m}$.
- There is no harm in letting the schema extend to infinity both sides. Strictly speaking, that ends up producing a Hilbert space instead of a (finite-dimensional) Euclidean space, even though $\mathbf{m}$ may have finite support, but in the sequel, we shall only handle finitely indexed objects in it.
- The action of a causal T on $\mathcal{L}_F$ will be of the type: multiplication of a lower triangular operator with a lower triangular matrix of stacked inputs.
- Shifts may be applied both sides of a stack. When a shift Z is applied to the left-hand side, it produces a causal shift downward, that is, the normal causal shift on any signal sequence (also shifting the index sequence $\mathbf{m}$ to $Z\mathbf{m}$). However, the interesting new shift possibility is when Z is applied to the right-hand side: it then shifts an input sequence *to the left*, that is, partially into the strict future connected to the previous index point. So we see that the range of T applied to $\mathcal{L}_{F,\mathbf{m}}$ (which can handily be denoted as $\bigvee T\mathcal{L}_{F,\mathbf{m}}$) contains $\bigvee T\mathcal{L}_{F,\mathbf{m}}Z$, and

$$\bigvee T\mathcal{L}_{F,\mathbf{m}} \ominus \bigvee T\mathcal{L}_{F,\mathbf{m}}Z \subset \mathcal{D}_{F,\mathbf{n}}. \tag{7.46}$$

In case $T = Z$ (applied to the left-hand side!), then this subspace is just the block diagonals of dimension $\mathbf{n}$, $\mathcal{D}_{F,\mathbf{n}}$.

The next important result that puts these global spaces immediately in the limelight is given in Proposition 7.10. In connection with T acting globally on $\mathcal{X}_{F,\mathbf{m}}$, there is the global Hankel operator H, which acts on the global $u \in \mathcal{U}_{\mathbf{m}}^-$, and hence separately as H_k on each strict past u_{sp}^k. H has its range in $\mathcal{L}_{F,\mathbf{n}}$. Given a minimal realization $\{A, B, C, D\}$, this range has a remarkable form, assuming the sequence of dimensions of the state to be η.

Proposition 7.10 *The range of the global Hankel operator is $C(I - ZA)^{-1}\mathcal{D}_{F,\eta} \subset \mathcal{L}_{F,\mathbf{n}}$. The columns of $\mathbf{O} := C(I - ZA)^{-1}$ form a basis for this range ($T\mathcal{U}_{F,\mathbf{m}}^-$), and they form an orthonormal basis if and only if the realization is in the output normal form (i.e., $A'A + C'C = I$).*

Proof

Let us, without loss of generality as far as the proof goes, assume that the support for T runs from $k = 0$ to $k = \ell$, and let us use the abbreviated notation for the continuous product $A^{\ell \geq k} = A_\ell \cdots A_k$); then we can just check

$$C(I - ZA)^{-1} = \begin{bmatrix} \boxed{|} & & & & & \\ | & C_1 & & & & \\ | & C_2 A_1 & C_2 & & & \\ \vdots & \vdots & \vdots & \ddots & & \\ | & C_{\ell-1} A^{\ell-2 \geq 1} & C_{\ell-1} A^{\ell-2 \geq 2} & \cdots & C_{\ell-1} & \\ | & C_\ell A^{\ell-1 \geq 1} & C_\ell A^{\ell-1 \geq 2} & \cdots & C_\ell A_{\ell-1} & C_\ell \end{bmatrix} = \mathbf{O}, \tag{7.47}$$

and assuming that the basis chosen for the observability space at each k is orthonormal, that is, $A_k'A_k + C_k'C_k = I$, it follows, by backward recursion, $\Pi_0(\mathbf{O}'\mathbf{O}) = I$. To check that $\mathcal{D}_{F,\eta}$ generates the range of H when $\mathbf{O}$ is applied to it, one has to show that each entry $x_k^k \in [\mathcal{D}_{F,\eta}]_k^k$ generates the range of each individual H_k. But this range is indeed spanned by $\mathbf{O}^k = \text{col}[C_k \; C_{k+1}A_k \; \cdots]$, that is, the kth column of $\mathbf{O}$. □

The space $\mathbf{O}\mathcal{D}_{F,\eta}$ is the *space of natural responses of the system*, that is, the responses from states only, putting future inputs equal to zero – assuming the linearity of course (otherwise the notion has no meaning.) It has a remarkable shift-invariance property, as can immediately be seen from Eq. (7.47): if one shifts the matrix one index notch to the right and cuts out the noncausal part from the shifted version (called a "restricted shift"), one obtains a subspace of the original (using the diagonal downward shift on A):

$$\Pi_{\mathrm{f}}\mathbf{O}^{\rightarrow} = \mathbf{O}A^{\langle 1 \rangle}. \tag{7.48}$$

(This property should not be surprising, as it is used to define A in realization theory!)

Now, consider the external coprime factorization $T = V\Delta_R'$ and assume the realization $\{A, B, C, D\}$ to be in the output normal form. V is obtained by the orthogonal completion of $\begin{bmatrix} A \\ C \end{bmatrix}$. V is a unitary operator with realization $\begin{bmatrix} A & B_V \\ C & D_V \end{bmatrix}$. V consists of orthonormal columns and is, in this case,

$$V = \begin{bmatrix} \boxed{D_{V,0}} & & & & & \\ C_1 B_{V,0} & D_{V,1} & & & & \\ C_2 A_1 B_{V,0} & C_2 B_{V,1} & D_{V,2} & & & \\ \vdots & \vdots & \vdots & \ddots & & \\ C_{\ell-1} A^{\ell-2 \geq 1} B_{V,0} & C_{\ell-1} A^{\ell-2 \geq 2} B_{V,1} & C_{\ell-1} A^{\ell-2 \geq 3} B_{V,2} & \cdots & D_{V,\ell-1} & \\ C_\ell A^{\ell-1 \geq 1} B_{V,0} & C_\ell A^{\ell-1 \geq 2} B_{V,1} & C_\ell A^{\ell-1 \geq 3} B_{v,2} & \cdots & C_\ell B_{V,\ell-1} & D_{V,\ell} \end{bmatrix}. \tag{7.49}$$

Remarks:

1. $\begin{bmatrix} B_{V,k} \\ D_{V,k} \end{bmatrix}$ are found by the unitary completion of $\begin{bmatrix} A_k \\ C_k \end{bmatrix}$; in particular, one can choose $\begin{bmatrix} B_{V,0} \\ D_{V,0} \end{bmatrix} = \begin{bmatrix} 0 & I \\ I & 0 \end{bmatrix}$ since $\begin{bmatrix} A_0 \\ C_0 \end{bmatrix} = \begin{bmatrix} | \\ | \end{bmatrix}$. This fixes the dimensions of V.
2. One must keep the difference between the regular Euclidean inner product on columns of a matrix, seen as vectors in a vector space, and the Frobenius inner product defined on the stack of columns in mind. Columns of $\mathbf{O}$ belonging to different column blocks are not orthogonal to each other for the Euclidean inner product, but they belong to different, orthogonal subspaces in the Frobenius setting. (However, they are orthogonal in the normal sense in the case of the input normal form.) On the other hand, all columns of V over all k are orthonormal for the regular Euclidean norm.
3. As operators on $\mathcal{L}_{F,\mathbf{m}}$, T and V generate subspaces of $\mathcal{L}_{F,\mathbf{n}}$ that consist of what we can call "sliced spaces," that is, they consist of a stack of individual subspaces

and for each index k, the result of the application of T or V on the corresponding present and future $[\ell_2^{\mathbf{m}}]_{k:\infty}$. Let us work this out further.

Let $\ell_2^{\mathbf{n}}$ be the global output space of T and V; then the columns of T and V located in lower (block) triangular matrices generate subspaces that have a characteristic *invariance* property. More precisely, consider the sequence of column bases subsequently generated by the columns of V (or for any other causal operator T) as they progress into the future:

$$\bigvee V^{0:\ell} \supset \bigvee V^{1:\ell} \supset \cdots \supset \bigvee V^{\ell}. \tag{7.50}$$

They form what has been called a *nest*, that is, a sequence of subspaces for which the next subspace is contained in the previous. If one defines, for each k, a subspace $\mathcal{W}_k := \bigvee V^{k:\ell} \ominus \bigvee V^{k+1,\ell}$, then the range of V, which is obviously also $\bigvee V^{0:\ell}$, is the direct sum of these subsequent spaces. $\mathcal{W}_k$ is called a *wandering subspace*, and we have

$$\text{range } V = \bigvee V^{0:\ell} = \bigoplus_{k=0:\ell} \mathcal{W}_k. \tag{7.51}$$

Since V is an orthogonal or unitary operator, the block column k forms an orthonormal basis $\mathcal{W}_k$. The converse works as well: we may choose an individual orthonormal basis for each $\mathcal{W}_k$, and, putting these together in sequence in a new matrix $\widehat{V}$, we have $\widehat{V} = VD$, in which D is a unitary block diagonal matrix. One may say that $\widehat{V}$ and V are *essentially equal*, or, to put it differently, it will make no difference if either of these is used in the external factorization.

The interesting point is that the nest is the characteristic object, rather than V itself, because any orthonormal basis for the nest that respects the nest order can serve as V, and the set of block columns of each such V are nothing more than a basis for the nest. This is the gist of what is known as the *generalized Beurling theorem for nest algebras* – here in its finitely indexed version and summarized in Proposition 7.12 (matrix Beurling) that considers subspaces $\mathcal{M}$ that are generated by bases at each individual index point.

Definition 7.11 A sliced space in $\mathcal{L}_{F,\mathbf{n}}$ is a space that is the direct sum of slices $\mathcal{W}_i$, in which each slice $\mathcal{W}_i$ is a subspace of $\ell_2^{\mathbf{k}}$, with $\mathbf{k}$ being a sequence of indices.

A sliced space is automatically invariant for multiplication with a compatible diagonal operator D on the right, meaning that $\mathcal{M}D \subset \mathcal{M}$. In fact, right D-invariance is the same as "sliced." Some sliced spaces (like the one generated by V before) have an additional property, namely they are also right-invariant for the causal shift: $\mathcal{M}Z \subset \mathcal{M}$. These are the subject of the celebrated Beurling theorem, now re-formulated for the matrix setting:

Proposition 7.12 *Any (right) shift-invariant sliced subspace $\mathcal{M} \subset \mathcal{L}_{F,\mathbf{n}}$ has the form $U\mathcal{L}_{F,\mathbf{k}}$, wherein U is causal isometric and $\mathbf{k}$ is some index sequence. Conversely, any causal and isometric U mapping $\mathcal{L}_{F,\mathbf{k}} \to \mathcal{L}_{F,\mathbf{n}}$ defines an index sequence and a right shift-invariant subspace $\mathcal{M} = U\mathcal{L}_{F,\mathbf{k}}$.*

Proof

Define the wandering subspace $\mathcal{W}_k := \mathcal{M} \ominus \mathcal{M}Z$ (defined as in the discussion before) and construct an orthonormal basis U^k for each slice $\mathcal{W}_k$ (also as is the case for V^k above). By construction, $U'U = I$. The converse is even more obvious, given the definition and the construction detailed in the beginning of this section. □

The V we constructed earlier starting from an isometric $\begin{bmatrix} A \\ C \end{bmatrix}$ was more than isometric: it was actually unitary. The isometric situation of the matrix Beurling proposition is indeed more general. However, an isometric U can be completed to a unitary one by the completion of the basis $V = \begin{bmatrix} W & U \end{bmatrix}$. This has to be done with some care: a unitary V has a unitary realization, and there is a relationship between the state and the respective input and output dimensions: $\eta_i + k_i = \eta_{i+1} + n_i$ at index i. In the present case, n_i is fixed, and, if a minimal unitary extension is desired, also η_i and η_{i+1} are fixed, fixing $\mathbf{k}$ recursively; since $\sum \eta_i = \sum \eta_{i+1}$ over all relevant i, hence $\sum \mathbf{k}_i = \sum n_i$ as it should be for a unitary operator. (This reasoning breaks down when any of the sums is infinite, giving rise to more possibilities!) An isometric U will have its $\mathbf{k}_i$ smaller than or equal to $\eta_{i+1} - \eta_i + n_i$, with W compensating for the remaining dimensions. Notice that in the isometric case and with W complementation, we have $W'U = 0$; hence, W generates the co-kernel of U in $\mathcal{L}_{F,\mathbf{n}}$. In the present theory, this co-kernel does not play a role directly, but it will do so in the inner–outer theory to be covered in later chapters. Important is to note that in the case of $V : \mathcal{L}_{F,\mathbf{k}} \to \mathcal{L}_{F,\mathbf{n}}$, it is *not true* that $\mathbf{k} = \mathbf{n}$ (and certainly not $\mathbf{k} = \mathbf{m}$), because the state dimensions play a role in the definition of $\mathbf{k}$. For example, if $V = Z$, then $\mathbf{k} = Z^{-1}\mathbf{n}$, but that is definitely not true in general. This situation is essentially different from what happens in the LTI case.

Definition 7.13 A right DZ-invariant subspace $\mathcal{M} = U\mathcal{L}_{F,\mathbf{k}}$ is called *full range* if U can be chosen inner (i.e., causal and unitary).

7.5 Discussion Items

- Z is an inner factor, maybe the most trivial one except for unitary diagonal matrices. What are realizations for Z^k?
- Finite-dimensional inner factors: a lower triangular finitely indexed matrix with scalar non-empty entries is necessarily diagonal (do you see that?). What are the consequences for inner factors? In particular, can a full unitary matrix be causal?
- Elementary examples: external factorizations are somewhat peculiar in our setup, in particular with semi-infinite or infinite indices, which are worth considering in an exploratory way. Look at related examples to see the peculiarities. For example, try the left and right external factorizations of the following causal matrices:

$$\begin{bmatrix} \boxed{1} & & & \\ 1/2 & 1 & & \\ & 1/2 & 1 & \\ & & 1/2 & 1 \end{bmatrix}, \quad \begin{bmatrix} \boxed{1} & & & \\ 1/2 & 1 & & \\ & 1/2 & 1 & \\ & & \ddots & \ddots \end{bmatrix},$$

$$\begin{bmatrix} \boxed{1} & & & \\ 2 & 1 & & \\ & 2 & 1 & \\ & & 2 & 1 \end{bmatrix}, \quad \begin{bmatrix} \boxed{1} & & & \\ 2 & 1 & & \\ & 2 & 1 & \\ & & \ddots & \ddots \end{bmatrix}, \tag{7.52}$$

where the second and last matrices are half infinite.

7.6 Notes

1. The external factorization treated in this chapter has the peculiarity that it uses inner functions as one of the factors, namely, the factor that in the LTI theory characterizes the poles of the system. That choice restricts the usefulness of the theory for infinitely indexed and LTI systems to systems that are u.e.s. An alternative is available and detailed in Chapter 10. In Chapter 10, we show how external factorizations consisting of polynomials in the shift – both in the numerator and in the denominator – can be obtained. This theory, which turns out to be simpler than the theory of the present chapter, uses *unilateral series calculus*, in which series expansions are allowed in the causal shift Z plus at most a finite number of terms in the anti-causal shift Z'. Such series (of course, including polynomials in the shifts) can be finitely multiplied with each other, so that arbitrary causal but unstable transfer functions can be dealt with. More details are to be found in Chapter 10.
2. In the case of scalar LTI transfer functions, the external factorization looks as follows (a typical example of degree 2 with $|a|, |b| < 1$ and the $\alpha_{2:0}$ arbitrary):

$$\frac{\alpha_2 z^2 + \alpha_1 z + \alpha_0}{(1 - az)(1 - bz)} = \frac{\alpha_2 z^2 + \alpha_1 z + \alpha_0}{(z - \overline{a})(z - \overline{b})} \cdot \frac{(z - \overline{a})(z - \overline{b})}{(1 - az)(1 - bz)}, \tag{7.53}$$

   with the second factor being the inner factor. The interest in using the inner type of factorization instead of the polynomial is that the former gives orthonormal bases for observability and reachability spaces. Otherwise, and, for example, for control purposes, the polynomial form may be preferred because of its simplicity.
3. Good examples of the practical use of external factorizations are in algorithms for efficiently solving systems of equations. Fully worked-out cases are given in Chapter 11 and 12. For another application of external factorizations for time-variant systems, now in the control area, see the work of W. N. Dale and M. C. Smith and colleagues in [19].
4. What happens with the left and right coprime factorizations of a causal operator T when the indices run from $-\infty$ to $+\infty$? First of all, as the operators are now all infinite dimensional, one has to put some boundedness assumptions on T and the diagonal operators A, B, C and D. One common strategy is to just assume that T is bounded as a map from an input $\ell_2^{\mathbf{m}}$ space to an output $\ell_2^{\mathbf{n}}$ space. Next, one can, for example, and just as before, choose a realization for T in an input normal form, by choosing an orthonormal basis for the co-range of each. In order to obtain realizations with finite matrices, one assumes that such a basis is finite dimensional

for all H_k, which are all bounded operators since they are all suboperators of T. This leads, as in the finitely indexed case, to an A and a B that are contractive (and hence bounded), but it may very well happen that A so obtained is not u.e.s. One therefore often puts the *extra requirement* that the resulting A is indeed u.e.s. so that the inverse $(I - ZA)^{-1}$ exists as a bounded operator. One can then show that the resulting reachability operators $\mathbf{R}_k = (I - ZA)^{-1}ZB$ exist and are uniformly bounded. To complete the realization in an input normal form, a further assumption is that the complementary C operator is also uniformly bounded, and one may, of course, trust that also D is bounded, because this turns out to follow from the boundedness of T. In this way, one obtains a realization in an input normal form in which all four realization operators are bounded, and A is, moreover, u.e.s.

Conversely, any realization with bounded A, B, C, D and A u.e.s. produces a bounded, causal transfer function $T = D + C(I - ZA)^{-1}ZB$ just as in the finitely indexed case, and the external factorizations as detailed in Section 7.2 of this chapter through exactly as described, with the only difference being an extra set of conditions to make things work for the infinite-dimensional matrices and operators involved.

5. What we call "external factorization" is usually called "coprime factorization" in the literature. The reason to introduce a new term is that an external factorization as we conceive it does not have to be coprime in the usual algebraic sense – "coprime" means the factors have no common nontrivial divisor within their class. (In the matrix case, one must distinguish between the right and left divisors.) Coprime factorizations – and by extension, external factorizations – have played an important role in the development of dynamical system theory in the wake of the seminal book of Kalman, Falb and Arbib [49]. It was recognized early on that in the LTI case, rational matrix functions and, in particular, polynomial matrices in a single variable z provide the algebraic framework needed to characterize important objects related to system theory such as kernels, state characterization, state equivalence and canonical forms for the state transition operator. This realization provided a valuable link with the pre–state-space approach, which was entirely based on rational matrix functions, and in particular the characterization of their dynamic properties through Smith–Macmillan forms and other algebraic devices based on algebraic structures called "rings," "principal ideal domains" and "modules" (as a multiport generalization of polynomial or rational algebras). All these efforts have led to many results in electrical engineering, in particular in control theory and network theory, many of which can be reinterpreted in the context of LTV systems.

   Unfortunately, even though module theory has produced many nice and important results, it does not provide the correct framework for time-variant or nonlinear systems. This is mainly due to the fact that the general shift operator Z does not commute with "instantaneous" operators, which in our case are the diagonal operators A, B, C and D. Hence, module theory is not applicable in the time-variant or quasi-separable context, but, as already mentioned, the relevant structure is a "nest algebra" originally proposed by Ringrose in [58] and further developed by Arveson [7]. Remarkable now is that most algebraic properties needed to solve the various system-related problems still work in that setting, at the "cost" of using methods that are more elementary than those used in ring and module theory. The biggest casualty of the reduced algebraic structure is the lack of an eigenvalue theory, but most properties and techniques needed for system analysis remain: coprime factorization, inner–outer factorization (the topic of Chapter 8), reachability, observability and then more elaborate topics, such as Bezout equations, embedding, interpolation and model reduction. The reason why most properties needed still work is to be found in their "geometric foundation," which can be properly explored by simple matrix calculus. This is the approach we have been taking and, we hope, shall become pretty convincing in the remainder of the book.

8 Inner–Outer Factorization

Inner–outer factorization is likely the most important operation in system theory. It certainly solves major problems in almost all areas of interest: system inversion, estimation theory and control. On the face of it, it seems not much more than the recursive application of a QR-type factorization or SVD on a system description, aiming at determining important system subspaces, which then play a central role in the solution of various problems. Its power is due to the fact that it leads to a recursion of low numerical complexity, combined with stable numerical operations. In this chapter, we concentrate on the algorithm itself, leaving its use in major problems and applications to further chapters. To introduce the theory and illustrate the procedure, we start with working out a standard 4×4 sample case of numerical inversion in detail. We then move on to the underlying theory and end up with "geometric" considerations that will improve our general insights into system properties.

Menu

Hors d'oeuvre
QR Factorization Revisited

First Course
A Prototype Outer–Inner Factorization

Second Course
Quasi-separable Outer–Inner Factorization

Dessert
Infinitely Indexed Systems and the Extension to Nonlinear Systems

8.1 Preliminary: Echelon Forms, QR and LQ Factorizations

Let us start out with a simple algebraic exercise: rotating a unitary (column) vector $u = \begin{bmatrix} u_1 \\ u_2 \end{bmatrix}$ of dimension $1 + n$, with $\|u\|^2 = |u_1|^2 + \sum_{i=1}^{n} |(u_2)_i|^2 = 1$ and with a nonnegative real first element u_1 to the positive first axis, which we call e_1 generically (ignoring its dimension); e_1 is then here a vector of dimension $1 + n$. The following orthogonal (or unitary in the complex case) matrix does the trick:

$$Q_u := \begin{bmatrix} u_1 & -u_2' \\ u_2 & I - u_2 \frac{1}{1+u_1} u_2' \end{bmatrix}, \tag{8.1}$$

and we shall have $u = Q_u e_1$ and $Q_u' u = e_1$, which is easily verified directly. The matrix Q_u has $\det Q_u = 1$; it is a generalized rotation matrix. One can actually show that it can be produced by a sequence of elementary rotations (often called "Givens rotations"), but a direct application of such a matrix to an arbitrary vector – say $x = \begin{bmatrix} x_1 \\ x_2 \end{bmatrix}$ with x_1 scalar – produces the following "efficient" computation:

$$Q_u x = \begin{bmatrix} u_1 x_1 - u_2' x_2 \\ x_2 + u_2(x_1 - \frac{u_2' x_2}{1+u_1}) \end{bmatrix}, \tag{8.2}$$

in which the inner product $u_2' x_2$ should be executed only once.

A more general nonzero vector $a = \begin{bmatrix} a_1 \\ a_2 \end{bmatrix}$ of dimension $1 + n$ can, of course, also be rotated to the direction of the first unit vector e_1. It pays to do that carefully. In the general complex case, let $\|a\|$ be the Euclidean norm of the vector and $a_1 = |a_1| e^{j\phi}$ (with $j = \sqrt{-1}$); then $u = a \frac{e^{-j\phi}}{\|a\|}$ will be like u before, and the effect of Q_u on a will be $Q_u a = e_1 \|a\| e^{j\phi}$. (It is useful to retain the norm of a and the phase of the first element for further use.) Let, for a general nonzero vector a, Q_a be defined consistently as $Q_a := Q_{a \frac{e^{-j\phi}}{\|a\|}}$.

Suppose next that one disposes of a collection $A = [a_1 \ \cdots \ a_m]$ of vectors of dimension $1 + n$ (typically columns of a matrix), and suppose that we are entitled to apply rotations to them (i.e., to the left). Suppose a_k is the nonzero vector with the smallest k and that its first element is $a_{k,1} = |a_{k,1}| e^{j\phi_1}$ (in case $a_{k,1} = 0$ one puts $\phi_1 = 0$). Applying Q_{a_k} to the stack now produces the following typical form:

$$\begin{aligned} & Q_{a_k}' \begin{bmatrix} a_1 & \cdots & a_{k-1} & a_k & a_{k+1} & \cdots & a_m \end{bmatrix} \\ &= \begin{bmatrix} 0 & \cdots & 0 & e_1 \|a_k\| e^{j\phi_1} & Q_{a_k}' a_{k+1} & \cdots & Q_{a_k}' a_m \end{bmatrix} \\ &= \begin{bmatrix} 0 & \cdots & 0 & \|a_k\| e^{j\phi_1} & * & \cdots & * \\ 0 & \cdots & 0 & 0 & b_{k+1} & \cdots & b_m \end{bmatrix}, \end{aligned} \tag{8.3}$$

where the "$*$" indicate entries that have been modified (and will remain unchanged further on), and the $\begin{bmatrix} b_{k+1} & \cdots & b_m \end{bmatrix}$ is a new collection of vectors, now of dimension n, and on which the procedure can be repeated without producing new fill-ins in the zero elements obtained so far, now with one dimension less. (Some of the zeros shown above may disappear, e.g., when a_1 is nonzero.) Continuing this way, now on the b's and realizing that products of rotation matrices remain orthogonal or unitary, after a number of steps, one obtains the so-called *echelon* form

$$A = \begin{bmatrix} Q_1 & Q_2 \end{bmatrix} \begin{bmatrix} R_1 \\ 0 \end{bmatrix}, \tag{8.4}$$

in which

$$R_1 = \begin{bmatrix} 0 & \cdots & 0 & R_{1,k_1} & \cdots & * & * & \cdots & * & * & \cdots \\ 0 & \cdots & 0 & 0 & \cdots & 0 & R_{2,k_2} & \cdots & * & * & \cdots \\ & & & & \vdots & & & & & & \\ 0 & \cdots & 0 & 0 & \cdots & 0 & 0 & \cdots & 0 & R_{\alpha,k_\alpha} & \cdots \end{bmatrix}, \tag{8.5}$$

where α is the rank of A and Q is orthogonal or unitary. The columns of Q_1 form a basis for the range of A, while the columns of R_1' form a basis for the co-range (i.e., the range of A'), and the columns of Q_2 for the co-kernel.

A similar, even more powerful, result could have been obtained by SVD of A:

$$A = \begin{bmatrix} U_1 & U_2 \end{bmatrix} \begin{bmatrix} \Sigma & 0 \\ 0 & 0 \end{bmatrix} \begin{bmatrix} V_1' \\ V_2' \end{bmatrix} = \begin{bmatrix} U_1 & U_2 \end{bmatrix} \begin{bmatrix} \Sigma V_1' \\ 0 \end{bmatrix} \tag{8.6}$$

at the cost of more computations. Here $\Sigma = \begin{bmatrix} \sigma_1 & & \\ & \ddots & \\ & & \sigma_\alpha \end{bmatrix}$ are the singular values of A in order: $\sigma_1 \geq \sigma_2 \geq \cdots \geq \sigma_\alpha > 0$; the columns of U_1 form a basis for the range of A, while the columns of V_1 and $V_1\Sigma$ form a basis for the co-range.

The example in Section 8.2 (and many in subsequent chapters) uses a variant of the QR algorithm just presented, namely an algorithm that starts at the bottom-right corner and produces a factorization in reverse order, which we will call generically of "LQ-type" with L again an echelon matrix in the column form and Q an orthogonal or unitary matrix. The procedure now starts out with a collection of rows rather than columns; it is dual to the preceding:

$$A = \begin{bmatrix} 0 & L_2 \end{bmatrix} \begin{bmatrix} Q_1 \\ Q_2 \end{bmatrix}. \tag{8.7}$$

Here L_2 is obtained by compressing toward the last column starting with the bottom row (skipping it when zero); it will look like

$$\begin{bmatrix} \vdots & & \vdots & \vdots \\ L_{\alpha,k_\alpha} & & * & * \\ 0 & & * & * \\ \vdots & & \vdots & \vdots \\ 0 & & L_{2,k_2} & * \\ 0 & & 0 & * \\ \vdots & \cdots & \vdots & \vdots \\ 0 & & 0 & L_{1,k_1} \\ 0 & & 0 & 0 \\ \vdots & & \vdots & \vdots \\ 0 & & 0 & 0 \end{bmatrix} \tag{8.8}$$

with these columns now forming a (nonorthogonal) basis for the range of A. Also in this case, a more accurate result can be obtained through SVD, when needed.

This LQ form (starting from the bottom right and moving to the left and upward) may appear awkward (why not start from the top left and move down?), but it is motivated by history, namely the need to compromise between various representation types like $\begin{bmatrix} A & B \\ C & D \end{bmatrix}$ realizations and various habits like causal operators being of "lower" type, and the habit of numerical analysts to work on the columns rather than the rows of a matrix. All this is not essential; what is important is the efficient computation of bases for various ranges and kernels, and it often does not matter what form these bases have (echelon or other).

Terminology. We use the term "QR factorization" for a form where Q is a unitary left factor and R is a right factor with independent nonzero rows, while an "LQ factorization" is a factorization where the unitary Q is a right factor and the nonzero columns of L are independent. (This terminology is however nonstandard in the literature.)

There is more to say about these various forms of orthogonalization. In $A = L_2Q_2$, with L_2 in the column echelon form, the rows of Q_2 form a minimal orthonormal basis for the rows of A, or, equivalently, the columns of Q_2' form an orthonormal basis for the co-range of A, while the columns of L_2 form a basis for the range of A. Because of the echelon form, this latter basis is ordered and is obtained in a bottom-up fashion, since L_2 is in the upper form. If, on the contrary, one computes the basis in top-down order, then an equally valid LQ factorization is obtained, with L in the lower echelon form. One may, of course, obtain such orthogonalizations directly (this is known as a *Gram–Schmidt orthogonalization*), which amounts to recursively computing AQ' starting either from the bottom or from the top. A full LQ factorization produces a basis for the kernel of A as well. In the case of $A = \begin{bmatrix} 0 & L_2 \end{bmatrix} \begin{bmatrix} Q_1 \\ Q_2 \end{bmatrix}$, with Q unitary, the columns of Q_1' provide for an orthonormal basis of the kernel of A. One may say that the LQ factorization is the most elementary example of an outer–inner factorization, to be discussed in Section 8.2.

8.2 A Prototype Example of Outer–Inner Factorization

Let us consider the (lower) LQ factorization of a 4×4 lower block triangular (i.e., causal) matrix *given by means of its realization* $\{A,B,C,D\}$. Writing out the realization entry-wise, the matrix is defined *implicitly* as

$$T = \begin{bmatrix} \boxed{D_0} & & & \\ C_1B_0 & D_1 & & \\ C_2A_1B_0 & C_2B_1 & D_2 & \\ C_3A_2A_1B_0 & C_3A_2B_1 & C_3B_2 & D_3 \end{bmatrix} \tag{8.9}$$

(see Fig. 8.1). Our goal will be to find a factorization of the type $T = \begin{bmatrix} 0 & T_o \end{bmatrix} \begin{bmatrix} V_1 \\ V_2 \end{bmatrix} = T_oV_2$ such that V is causal unitary, T_o is in lower column echelon form and left invertible, V_2 is lower co-isometric, with the right-upper inverse V_2', *and whereby all*

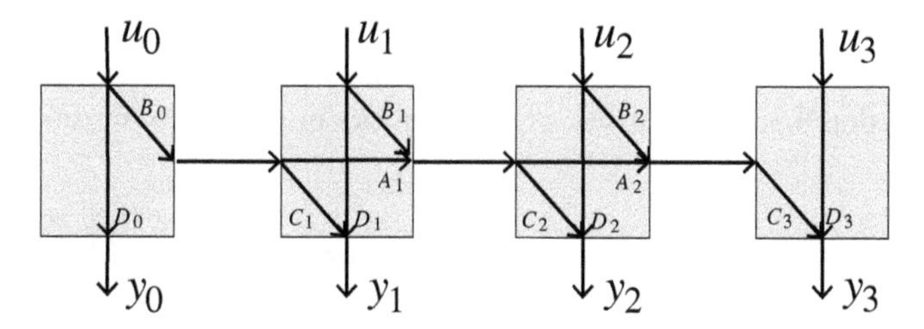

Figure 8.1 The given realization.

these quantities are given by realizations. We shall indeed discover that realizations for T_o and V are just as simple as the realization for T, and they can easily be derived from it via a linear forward recursion. A first observation is that in an LQ factorization in the echelon form of T, the R factor turns out to be lower-block triangular, because the left-to-right recursive operation respects the lower triangular form. But much more is the case.

The LQ factorization procedure works column by column starting with the first. (Actually, it works on the main block diagonals downward, but that will soon be apparent.) All blocks of the first column, except for the first, share B_0:

$$T^0 = \begin{bmatrix} D_0 \\ C_1 B_0 \\ C_2 A_1 B_0 \\ C_3 A_2 A_1 B_0 \end{bmatrix} = \begin{bmatrix} I & 0 \\ 0 & C_1 \\ 0 & C_2 A_1 \\ 0 & C_3 A_2 A_1 \end{bmatrix} \begin{bmatrix} D_0 \\ B_0 \end{bmatrix}. \tag{8.10}$$

How to proceed to get the first block column of T_o? Remark that getting T_o left outer (i.e., with a left pseudo-inverse, or, more precisely, in the causal column echelon form), it is necessary and sufficient that D_{o0} is made left invertible (Theorem 6.3). D_0 is a single diagonal entry on block row 0, and its reduction from the right-hand side will remain unaffected by later steps undertaken on subsequent block columns, so D_{o0} is solely dependent on D_0 and nothing else; this observation *determines* its column dimension. An LQ factorization of $\begin{bmatrix} B_0 \\ D_0 \end{bmatrix}$ achieves this and is done traditionally starting from the bottom, compressing to the right. (See also the discussion at the end of Section 8.1.) Therefore, we reverse their order and use new names for the blocks of the right-hand side,[1]

$$\begin{bmatrix} B_0 \\ D_0 \end{bmatrix} := \begin{bmatrix} 0 & Y_1 & B_{o0} \\ 0 & 0 & D_{o0} \end{bmatrix} \begin{bmatrix} Q_1^{(0)} \\ Q_2^{(0)} \\ Q_3^{(0)} \end{bmatrix} \text{ (= upper LQ factorization),} \tag{8.11}$$

[1] The symbols $\{Y_1, B_{o0}, \ldots\}$ may appear arbitrary at this point but are duly motivated by the result, which recursively appears.

in which Y_1 and D_{o0} are left invertible, $Q^{(0)}$ is unitary and all quantities of the right-hand side are computed from the given left-hand side quantities by the LQ algorithm. (In Section 8.3, we shall see what the blocks in Q actually mean in terms of realizations.)

Remark: An alternative would be to use an LQ factorization on $\begin{bmatrix} D_0 \\ B_0 \end{bmatrix}$, which would lead to an L factor in the column echelon form and a formalism that is identical to what follows. It does not really matter what form the L factor has, as long as its nonzero component is left invertible (i.e., consists of independent columns). Also, an SVD evidently does the trick, with more numerical accuracy and additional computational cost.

Applying $Q^{(0)\prime}$ to the first block column T^0 (i.e., to the right) of T now produces (after exchanging the order of B_0 and D_0)

$$T^0 Q^{(0)\prime} = \begin{bmatrix} 0 & 0 & D_{o0} \\ 0 & C_1Y_1 & C_1B_{o0} \\ 0 & C_2A_1Y_1 & C_2A_1B_{o0} \\ 0 & C_3A_2A_1Y_1 & C_3A_2A_1B_{o0} \end{bmatrix}. \tag{8.12}$$

The columns of the last (third) subcolumn form a basis, because D_{o0} is left invertible. Let us permute the second and third columns with a permutation matrix P_0 (the latter will become part of the final left factor) leaving the zero columns to the left. Applying $\begin{bmatrix} Q^{(0)\prime}P_0 & & & \\ & I & & \\ & & I & \\ & & & I \end{bmatrix}$ to the right of the whole matrix T does not change the remaining columns (no fill ins!), and the process can be repeated on the block rows starting at row 2, leaving the first two columns unchanged *and taking into account the remainder of the operation on the first block column represented by the third column starting with* C_1Y_1 *on the second block row.*

The remainder, combined with the rest of the matrix that has been left untouched is now

$$\left[\begin{array}{cc|cc} C_1Y_1 & D_1 & 0 & 0 \\ \hline C_2A_1Y_1 & C_2B_1 & D_2 & \\ C_3A_2A_1Y_1 & C_3A_2B_1 & C_3B_2 & D_3 \end{array}\right], \tag{8.13}$$

showing that the column starting with C_1Y_1 *had to be moved to the next stage*. One may now leave the first block row intact and move to the second block row and column, noticing that the dimensions of the diagonal element $[C_1Y_1 \;\; D_1]$ (and hence of the second block column) have been forced to change.

The next step is now to reduce the new first diagonal element $\begin{bmatrix} C_1Y_1 & D_1 \end{bmatrix}$, taking into account that the effect has to be propagated down the combined column and that this combined column has a common right factor $\begin{bmatrix} A_1Y_1 & B_1 \end{bmatrix}$ (as before, but this is the "generic" step!):

$$\begin{bmatrix} C_1Y_1 & D_1 \\ \hline C_2A_1Y_1 & C_2B_1 \\ C_3A_2A_1Y_1 & C_3A_2B_1 \end{bmatrix} = \begin{bmatrix} 0 & I \\ \hline C_2 & 0 \\ C_3 & 0 \end{bmatrix} \begin{bmatrix} A_1Y_1 & B_1 \\ C_1Y_1 & D_1 \end{bmatrix}. \tag{8.14}$$

Compute therefore a new LQ factorization, with a new $Q^{(1)}$ structured as before

$$\begin{bmatrix} A_1Y_1 & B_1 \\ C_1Y_1 & D_1 \end{bmatrix} = \begin{bmatrix} 0 & Y_2 & B_{o1} \\ 0 & 0 & D_{o1} \end{bmatrix} Q^{(1)}. \tag{8.15}$$

Application of this $Q^{(1)\prime}$ to the right of the new first column (it has to be verified that this will not change any of the other elements that have already been set aside: there are no "fill ins") produces the new second subcolumn

$$\begin{bmatrix} 0 & 0 & D_{o1} \\ 0 & C_2Y_2 & C_2B_{o1} \\ 0 & C_3A_2Y_1 & C_3A_2B_{o1} \end{bmatrix} \tag{8.16}$$

and, again, the columns of the third subcolumn form a basis. Again, we permute relevant columns and move the zero column to the far left. Let us check the overall result, after the application of the two subsequent Q's:

$$\left[\begin{array}{cc|cc|ccc} 0 & 0 & D_{o0} & 0 & 0 & 0 & 0 \\ 0 & 0 & C_1B_{o0} & D_{o1} & 0 & 0 & 0 \\ 0 & 0 & C_2A_1B_{o0} & C_2B_{o1} & C_2Y_2 & D_2 & 0 \\ 0 & 0 & C_3A_2A_1B_{o0} & C_3A_2B_{o1} & C_3A_2Y_2 & C_3B_2 & D_3 \end{array}\right]. \tag{8.17}$$

The next operation takes place on block subcolumns 3 and 4, with again a new Q, similar to the previous

$$\begin{bmatrix} A_2Y_2 & B_2 \\ C_2Y_2 & D_2 \end{bmatrix} = \begin{bmatrix} 0 & Y_3 & B_{o2} \\ 0 & 0 & D_{o2} \end{bmatrix} Q^{(2)}, \tag{8.18}$$

and produces

$$\left[\begin{array}{ccc|ccc|cc} 0 & 0 & 0 & D_{o0} & 0 & 0 & 0 & 0 \\ 0 & 0 & 0 & C_1B_{o0} & D_{o1} & 0 & 0 & 0 \\ 0 & 0 & 0 & C_2A_1B_{o0} & C_2B_{o1} & D_{o2} & 0 & 0 \\ 0 & 0 & 0 & C_3A_2A_1B_{o0} & C_3A_2B_{o1} & C_3B_{o2} & C_3Y_2 & D_3 \end{array}\right]. \tag{8.19}$$

The final step is simpler, with a last Q for which $\begin{bmatrix} C_3Y_2 & D_3 \end{bmatrix} = \begin{bmatrix} 0 & D_{o3} \end{bmatrix}$ with D_{o3} left invertible, giving finally

$$\left[\begin{array}{cccc|cccc} 0 & 0 & 0 & 0 & D_{o0} & 0 & 0 & 0 \\ 0 & 0 & 0 & 0 & C_1B_{o0} & D_{o1} & 0 & 0 \\ 0 & 0 & 0 & 0 & C_2A_1B_{o0} & C_2B_{o1} & D_{o2} & 0 \\ 0 & 0 & 0 & 0 & C_3A_2A_1B_{o0} & C_3A_2B_{o1} & C_3B_{o2} & D_{o3} \end{array}\right]. \tag{8.20}$$

The operations so far have produced a global LQ factorization

$$T = \begin{bmatrix} 0 & T_o \end{bmatrix} Q, \tag{8.21}$$

in which T_o is left invertible and Q, the product of all the orthogonal transformations after embedding, is orthogonal or unitary. In addition, T_o has partly inherited the state-space structure of T. Hence, we have obtained $T_o = D_o + C(I - ZA)^{-1}ZB_o$. As already explained in Chapter 6, a left inverse for T_o will have the same state structure as T_o itself, and because all the D_{ok} are by themselves left invertible, a realization for a left inverse is simply $T_o^+ = D_o^+ - D_o^+ C(I - Z\Delta)^{-1} ZB_o D_o^+$, in which $\Delta = A - B_o D_o^+ C$. Notice that the realizations of T_o and T_o^+ are not necessarily minimal, it may even be that T_o is purely block diagonal.

What about the overall Q? Backtracking, we see that each Q just operates on the first block column of the subsequent matrices, corresponding to subsequent block columns of the original. However, part of the result is propagated further on, but in a limited way: just to the next block column. This is a strong indication that also the Q matrix has a limited state-space structure. It is very well possible to track the structure of Q down in detail from the previous, but there is an attractive shortcut, using the diagonal algebra defined in chapter 3 (section 3.8), presented in Section 8.3.

8.3 The General Matrix Case

The procedure we went through on a 4×4 block example generalizes to causal systems, using the general system realization formalism we developed so far. This produces the following basic theorem that we give in full detail for later reference.

Theorem 8.1A (Outer–Inner Factorization) *Let $T = D + C(I - ZA)^{-1}ZB$ be a causal, finitely indexed transfer function with minimal realization $\{A, B, C, D\}$; then T can be factored into $T = \begin{bmatrix} 0 & T_o \end{bmatrix} \begin{bmatrix} W \\ V \end{bmatrix}$, in which T_o is left-outer (i.e., is causal and has a causal left inverse); $\begin{bmatrix} W \\ V \end{bmatrix}$ is causal unitary and has a unitary realization $\begin{bmatrix} D_W \\ D_V \end{bmatrix} + \begin{bmatrix} C_W \\ C_V \end{bmatrix} (I - ZA_V)^{-1} ZB_V$; T_o has a (not necessarily minimal) realization $T_o = D_o + C(I - ZA)^{-1}ZB_o$; the relationship between the given realization of T and the realizations of T_o and $\begin{bmatrix} W \\ V \end{bmatrix}$ is given by the forward recursion*

$$\begin{bmatrix} AY & B \\ CY & D \end{bmatrix} \left[\begin{array}{c|c|c} C'_W & A'_V & C'_V \\ D'_W & B'_V & D'_V \end{array} \right] = \left[\begin{array}{c|c|c} 0 & Y^{\langle -1 \rangle} & B_o \\ 0 & 0 & D_o \end{array} \right], \tag{8.22}$$

in which both D_o and Y are left invertible block diagonal matrices. The initial value for the intermediate Y is "empty," and Y as well as all new A_V, B_V, C_V, D_V, C_W, D_W, D_o and B_o are then further determined by the forward recursion (8.22), which consists of a sequence of local block (upper) LQ factorizations.

Theorem 8.1B (Inner–Outer Factorization) *Let $T = D + C(I - ZA)^{-1}ZB$ be a causal, finitely indexed transfer function with minimal realization $\{A, B, C, D\}$; then T*

can be factored into $T = \begin{bmatrix} W & V \end{bmatrix} \begin{bmatrix} 0 \\ T_o \end{bmatrix}$, *in which* T_o *is right-outer (i.e., is causal and has a causal right inverse);* $\begin{bmatrix} W & V \end{bmatrix}$ *is causal unitary and has a unitary realization* $\begin{bmatrix} D_W & D_V \end{bmatrix} + C_V(I - ZA_V)^{-1}Z\begin{bmatrix} B_W & B_V \end{bmatrix}$*;* T_o *has a (not necessarily minimal) realization* $T_o = D_o + C_o(I - ZA)^{-1}ZB$*; the relationship between the given realization of* T *and the realizations of* T_o *and* $\begin{bmatrix} W & V \end{bmatrix}$ *is given by the backward recursion*

$$\begin{bmatrix} B_W' & D_W' \\ A_V' & C_V' \\ B_V' & D_V' \end{bmatrix} \begin{bmatrix} YA & YB \\ C & D \end{bmatrix} = \begin{bmatrix} 0 & 0 \\ Y^{\langle +1 \rangle} & 0 \\ C_o & D_o \end{bmatrix}, \tag{8.23}$$

in which both D_o *and* Y *are right invertible block diagonal matrices. The initial value for the intermediate* Y *is "empty," and* Y *as well as all new* A_V, B_V, C_V, D_V, B_W, D_W, D_o *and* C_o *are then further determined by the backward recursion (8.23), which consists of a sequence of local block QR factorizations.*

Warning: W, V and T_o in the two parts of the theorem are, of course, different. If needed, they can be distinguished by "left" and "right" as in $T = V_\ell T_{o,r} = T_{o,\ell} V_r$.

Proof

We produce the result constructively and give a complete proof for the outer–inner case; the inner–outer case is dual, but its salient features are discussed at the end of the proof. Let v be a causal unitary system with unitary realization $v \sim_c \begin{bmatrix} A_v & B_v \\ C_v & D_v \end{bmatrix}$ and consider Tv':

$$Tv' = [D + C(I - ZA)^{-1}ZB][D_v' + B_v'Z'(I - A_v'Z')^{-1}C_v']. \tag{8.24}$$

Let us determine the conditions on v under which Tv' is causal. For this, it is necessary to split the "quadratic term" $(I - ZA)^{-1}ZBB_v'Z'(I - A_v'Z')^{-1}$. It is easy to see (and we did this "decomposition in parts" already before) that it is equal to $(I - ZA)^{-1}ZAY + Y + YA_v'Z'(I - A_v'Z')^{-1}$ with

$$Y^{\langle -1 \rangle} = BB_v' + AYA_v'. \tag{8.25}$$

(Proof is derived by pre- and postmultiplication with $(I - ZA)$ and $(I - A_v'Z')$ respectively.) Hence,

$$\begin{aligned} Tv' = {} & (DD_v' + CYC_v') + C(I - ZA)^{-1}Z(BD_v' + AYC_v') \\ & + (DB_v' + CYA_v')Z'(I - A_v'Z')^{-1}C_v'. \end{aligned} \tag{8.26}$$

This product will be causal if $DB_v' + CYA_v' = 0$, a condition we must enforce. Assuming this to be the case, we define further new block diagonal matrices $d_o := DD_v' + CYC_v'$ and $b_o := BD_v' + AYC_v'$, making $Tv' = d_o + C(I - ZA)^{-1}b_o$. Bringing these relations in matrix form gives

$$\begin{bmatrix} AY & B \\ CY & D \end{bmatrix} \left[\begin{array}{c|c} A_v' & C_v' \\ B_v' & D_v' \end{array}\right] = \left[\begin{array}{c|c} Y^{\langle -1 \rangle} & b_o \\ 0 & d_o \end{array}\right]. \tag{8.27}$$

This equation consists of a sequence of incomplete (or partial) block LQ factorizations of $\begin{bmatrix} A_k Y_k & B_k \\ C_k Y_k & D_k \end{bmatrix}$ with "Q_k" $= \begin{bmatrix} A_{vk} & B_{vk} \\ C_{vk} & D_{vk} \end{bmatrix}$ and "R_k" $= \begin{bmatrix} Y_{k+1} & b_{ok} \\ 0 & d_{ok} \end{bmatrix}$. Let us assume that the system T is empty for indices $k \le k_\ell$, and putting Y_k = empty for $k \le k_\ell$, the recursion (8.27) yields causal systems v and t_o such that $T = t_o v$, with v unitary. However, at this point, neither t_o nor Y are necessarily left invertible (i.e., t_o left-outer), and hence further refinement is necessary, in particular to obtain both a Y and a $d_o \leftarrow D_O$ that are left invertible. This is done locally, for each k, in two steps:

1. Use a unitary transformation $[Q_k]_1$ to compress the columns of $[C_k Y_k \; D_k]$ into a left-invertible D_{ok} (i.e., with independent columns) and apply $[Q_k]_1'$ to the whole matrix. This defines the right-most columns of the result and an intermediate matrix denoted as a fill-in "$*$":

$$\begin{bmatrix} A_k Y_k & B_k \\ C_k Y_k & D_k \end{bmatrix} [Q_k]_1' = \begin{bmatrix} * & B_{ok} \\ 0 & D_{ok} \end{bmatrix}. \tag{8.28}$$

   This step also determines the dimensions of D_{ok} and a new matrix that we denote as B_{ok}.

2. Next compress the columns of "$*$" with a unitary matrix $\begin{bmatrix} [Q_k]_2 & \\ & I \end{bmatrix}$, leaving the last block rows undisturbed. This step now defines a new left-invertible matrix Y_{k+1} that satisfies the recursion

$$\begin{bmatrix} A_k Y_k & B_k \\ C_k Y_k & D_k \end{bmatrix} [Q_k]_1' \begin{bmatrix} [Q_k]_2 & \\ & I \end{bmatrix}' = \begin{bmatrix} 0 & Y_{k+1} & B_{ok} \\ 0 & 0 & D_{ok} \end{bmatrix}. \tag{8.29}$$

Cleaning up, we see that the overall Q_k decomposes in six blocks, which we can adorn with new names:

$$Q_k \leftarrow \begin{bmatrix} C_W & D_W \\ A_V & B_V \\ C_V & D_V \end{bmatrix}, \tag{8.30}$$

satisfying (dropping the index k)

$$\begin{bmatrix} AY & B \\ CY & D \end{bmatrix} \left[\begin{array}{c|cc} C_W' & A_V' & C_V' \\ D_W' & B_V' & D_V' \end{array} \right] = \left[\begin{array}{c|cc} 0 & Y^{\langle -1 \rangle} & B_o \\ 0 & 0 & D_o \end{array} \right]. \tag{8.31}$$

The more refined LQ factorization hence produces a more specialized $v = \begin{bmatrix} D_W \\ D_V \end{bmatrix} + \begin{bmatrix} C_W \\ C_V \end{bmatrix} (I - ZA)^{-1} ZB$ that satisfies the factorization equation $T = \begin{bmatrix} 0 & T_o \end{bmatrix} \begin{bmatrix} W \\ V \end{bmatrix}$, in which T_o is causally left invertible, because each D_{ok} is left invertible, and Y_k, now satisfying

$$Y_{k+1} = B_k B_{Vk}' + A_k Y_k A_{Vk}', \tag{8.32}$$

is also left invertible by construction.

The inner–outer case is obviously dual to the previous, using conjugates and time reversal. However, it is more practical to work directly on the original matrices, and this translates to what is given in the statement of the theorem. The construction for this case can be given along the lines of the inner–outer construction, now working on rows rather than on columns. □

Remarks

1. In the outer–inner case, we obtain $T = T_o V$, in which V is merely co-isometric ($VV' = I$) with, in addition, $VW' = 0$. It follows that also $TW' = 0$, and W' produces a basis for the kernel of T. In many applications, the kernel is sought of a causal system, and outer–inner factorization produces an efficient solution to this problem, provided the system has a low-complexity state description. Conversely, in the inner–outer case, W produces a system realization for the co-kernel.
2. There is a global expression for Y in both cases. In the outer–inner case, we may define the reachability operators of both T and V: $\mathcal{R}_T$ and $\mathcal{R}_V$, respectively. Solving the recursion from past to future, we find

$$Y = \mathcal{R}_T \mathcal{R}_V'. \tag{8.33}$$

   Since we sharpened the recursion to having Y left invertible, and noticing that $\mathcal{R}_T$ is assumed nonsingular by the minimality of the realization of T, we find that $\mathcal{R}_V$ is nonsingular as well, and V is, besides co-isometric, also reachable. This actually makes the co-isometric realization for V minimal. (This requires some proof but is consistent with the fact that a minimal co-isometric realization has a co-isometric reachability matrix, as is shown in Chapter 6.)
3. The outer–inner factorization is nothing but the forward recursive version of an LQ factorization, where the original transfer matrix T is lower block triangular or causal. It turns out that not only the L factor is block lower, also the Q factor is block lower, but the dimensions of both are adapted to fit the dimensions of respective ranges: the range of T for the L factor and the co-range of T (and kernel) for the Q factor produce an orthonormal representation for the co-range of T (remember: the co-range of T is the adjoint of the range of T', which is anti-causal when T is causal).

8.4 Geometric Interpretation

Let us consider the inner–outer case $T = VT_o$ and specialize to just one stage of the recursive computation – this is the case that occurs in quadratic optimization, as already discovered in Chapter 1. A similar treatment for outer–inner is dual and has to be performed on the dual system. Let us, for ease of notation, also drop the index (typically "k"). We look at what happens geometrically between the stage k and the stage $k+1$. We consider the linear case first and then show how the treatment can be generalized to nonlinear systems.

Let us abbreviate the state at k by x and at $k+1$ by x_+. The state x and input u get mapped to the next state x_+ and output y by the transition operator:

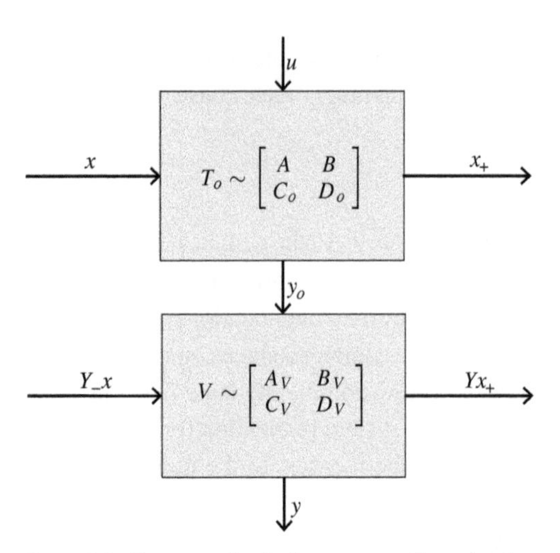

Figure 8.2 One stage in the inner–outer factorization.

$$\begin{bmatrix} x_+ \\ y \end{bmatrix} = \begin{bmatrix} A & B \\ C & D \end{bmatrix} \begin{bmatrix} x \\ u \end{bmatrix}. \tag{8.34}$$

The inner–outer factorization works backward. It starts with an intermediate quantity Y located at $k+1$ and produces three matrices Y_-, C_o and D_o via the block-QR decomposition

$$\begin{bmatrix} YA & YB \\ C & D \end{bmatrix} = \begin{bmatrix} A_V & B_V \\ C_V & D_V \end{bmatrix} \begin{bmatrix} Y_- & 0 \\ C_o & D_o \end{bmatrix}, \tag{8.35}$$

in which $\begin{bmatrix} A_V & B_V \\ C_V & D_V \end{bmatrix}$ is isometric and D_o as well as Y_- are right invertible ("flat" matrices). The corresponding signal-processing diagram is shown in Fig. 8.2. The outer part produces the next state and a "compressed" output y_o, while the output $\begin{bmatrix} Yx_+ \\ y \end{bmatrix}$ is a rotation of $\begin{bmatrix} Y_-x \\ y_o \end{bmatrix}$.

To interpret this further, let us apply the input $\begin{bmatrix} x \\ u \end{bmatrix}$ to Eq. (8.35). This gives

$$\begin{bmatrix} Yx_+ \\ y \end{bmatrix} = V \begin{bmatrix} Y_-x \\ C_o x + D_o u \end{bmatrix}. \tag{8.36}$$

In other words (and using Euclidean norms),

$$\|Yx_+\|^2 + \|y\|^2 = \|Y_-x\|^2 + \|C_o x + D_o u\|^2. \tag{8.37}$$

This expression exhibits a minimization challenge. Once the state x has been reached (at the beginning of stage k), the quantity $Y(x) := \|Y_-x\|^2$ is determined as well, and the output quantity $\|Yx_+\|^2 + \|y\|^2$ is minimized by choosing an input u such that $\|C_o x + D_o u\|^2$ is minimized, that is, by choosing

$$\widehat{u} := -D_o^{\dagger} C_o x. \tag{8.38}$$

In major applications (like optimal quadratic control or, dually, in Kalman filtering), D_o is not only right (or left) invertible but fully invertible (square and invertible), hence $\|Y_{-}x\|^2$ appears to reach a *global* "cost" minimum, starting from a given state x, because the same reasoning applies for x_+ as well, and so on down the chain that propagates the state, assuming, of course, that cost is expressed by quadratic norms as shown.

A further observation is that y_o measures in a very straight way the deviation from optimality. Indeed, we have immediately

$$y_o = D_o(u - \widehat{u}), \tag{8.39}$$

corresponding to the "excess cost" $\|y_o\|^2$ when a nonoptimal input u is used. The "outerness" of $\begin{bmatrix} A & B \\ C_o & D_o \end{bmatrix}$ reflects the strict convexity of the cost (8.37) as a function of u, given x (the realization does not have to be minimal though.)

What happens when D_o is not invertible? The minimization property remains valid, but now the optimizing input u is used to reduce a part of the state that is not propagated further and is hence unobservable. The model itself is then not minimal. (This issue often motivates the choice of minimal models in some major applications.)

Whatever the case, there is obviously a close connection between the inner–outer problem and quadratic optimization: one can be turned into the other, given a cost model using quadratic norms. The connection even extends to nonlinear models and is briefly discussed in Section 8.6.

Summarizing: In this section, we have established the fact that an arbitrary causal and regular operator T always admits a factorization $T = T_{ol}V_r$, in which T_{ol} is left-outer and V_r is right-inner. Such a factorization is called an outer–inner factorization. Dually, T admits an inner–outer factorization $T = V_\ell T_{or}$, with V_ℓ left-inner and T_{or} right-outer. When T is already left-outer, then the outer–inner factorization is trivial, with $V_r = I$.

8.5 Computing the LTI Case*

The LTI case is of importance for the LTV case, because the inner–outer factorization is recursive, and starting values for the recursion may have to be provided in important application cases. Such a starting value is often the solution of a fixed-point equation, as in the LTI case, where the outer–inner recursion becomes

$$\begin{bmatrix} AY & B \\ CY & D \end{bmatrix} = \begin{bmatrix} 0 & Y & B_o \\ 0 & 0 & D_o \end{bmatrix} \overline{V}, \tag{8.40}$$

in which we have put $\overline{V} := \begin{bmatrix} W \\ V \end{bmatrix}$, a unitary matrix. Equation (8.40) can now be interpreted as an ordinary matrix equation involving the LTI realization $\{A, B, C, D\}$. In this equation, all the matrices on the right-hand side are unknown and are derived from the data on the left-hand side, which, however, also contains an unknown matrix, namely Y

(Y is the "fixed point"!). In this section, we discuss a straight (noniterative) square-root solution based on a Schur eigenvalue decomposition of the inverse of the given state transition matrix. It turns out that a Schur eigenvalue decomposition and the solution of a single (to be constructed) Lyapunov–Stein equation solve the problem. Both parts of the algorithm have numerically stable and efficient solution methods. The treatment is a bit tricky and requires some knowledge of z-transform theory; see the remark at the end of the section.

The traditional way of solving the outer–inner equation is by canceling the unitary $\overline{V}$ out by multiplying the equation on the right-hand side with its conjugate, which leads to a quadratic equation for Y, called an algebraic *Riccati equation*. There is abundant literature on how to solve this equation, both in theory and numerically. However, and as we already know from the (linear) Lyapunov–Stein equation, it is not advisable to square an equation in order to solve it and certainly not when it is a matrix equation, and this for several reasons, among others, because many undesired solutions are introduced, and the numerical conditioning of the system is greatly compromised. The criticism is especially valid for this case, since the original equation can be solved directly efficiently and in a numerical stable way. Here, we treat only the square case with nonsingular D. The general case requires more casuistry and is fully treated in [30].

To solve the square case, assume the realization is minimal and D non-singular. Define $\Delta := A - BD^{-1}C$ and consider the rational transfer function $R(z) := D^{-1} - D^{-1}C(I - z\Delta)^{-1}zBD^{-1}$; then it is easy to see, by direct evaluation, that $R(z)T(z) = T(z)R(z) = I$; in other words, $R(z) = T(z)^{-1}$ as a rational function in z (and it possesses a unilateral Maclaurin series in z as well that converges in an open disc around $z = 0$). To find the property needed to solve the outer–inner equation, suppose that we have indeed succeeded in factoring $T(z) = T_o(z)V(z)$, with T_o outer and V inner; then the inversion gives $T(z)^{-1} = V(z)^{-1}T_o(z)^{-1}$, in which $V(z)^{-1}$ is analytic in the complement of the open unit disk, and $T_o(z)$ is analytic in the open unit disk. Or, to put it in more informal language, $V(z)^{-1}$ has poles in the open unit disk (and not on the unit circle, excluding $z = 0$ as well because of the nonsingularity of D), and, on the other hand, $T_o(z)^{-1}$ has all poles in the complement, that is, in $|z| \geq 1$. The inner–outer factorization hence separates poles of $T(z)^{-1}$ strictly inside from poles outside the unit disk and the spectrum of Δ splits accordingly. Moreover, it will appear during the construction that the nonsingularity of D has as a consequence that $V(z)^{-1}$ has a unilateral expansion (Maclaurin expansion) at $z = 0$, so that a unitary realization $\{A_V, B_V, C_V, D_V\}$ for $V(z)$ has both A_V and D_V nonsingular. In the derivation, we just admit this latter property as a fact and show that it leads to the solution.

Looking at the realization $\begin{bmatrix} \Delta & BD^{-1} \\ -D^{-1}C & D^{-1} \end{bmatrix}$ for the (unilateral) inverse of T, a factorization of the type $V^{-1}T_o^{-1}$ would follow from a block decomposition of the transition matrix Δ as

$$\Delta = Q' \begin{bmatrix} \alpha_{11} & \alpha_{21} \\ 0 & \alpha_{22} \end{bmatrix} Q, \tag{8.41}$$

in which α_{11} is a square matrix with all eigenvalues in $\{|z| > 1\}$ (one should realize that poles of the transfer function are inverses of the eigenvalues of the state transition matrix; they are "zeros" of $I - zA$!), α_{22} has all eigenvalues in $|z| \leq 1$, and $Q = \begin{bmatrix} Q_1 \\ Q_2 \end{bmatrix}$ is unitary, with the columns of Q_1' spanning the eigenspace corresponding to the eigenvalues of Δ in the open unit disk. Suppose, indeed, that $V^{-1}(z) \sim \begin{bmatrix} A_1 & B_1 \\ C_1 & D_1 \end{bmatrix}$ and $T_o(z)^{-1} \sim \begin{bmatrix} A_2 & B_2 \\ C_2 & D_2 \end{bmatrix}$; then (see Chapter 6)

$$V^{-1}T_o^{-1} \sim \left[\begin{array}{cc|c} A_1 & B_1C_2 & B_1D_2 \\ 0 & A_2 & B_2 \\ \hline C_1 & D_1C_2 & D_1D_2 \end{array}\right]. \tag{8.42}$$

The chain of consequences is then as follows:

1. State equivalence induced by Q' on the realization of T^{-1} gives

$$T^{-1} \sim \left[\begin{array}{cc|c} \alpha_{11} & \alpha_{12} & Q_1'BD^{-1} \\ 0 & \alpha_{22} & Q_2'BD^{-1} \\ \hline -D^{-1}CQ_1 & -D^{-1}CQ_2 & D^{-1} \end{array}\right], \tag{8.43}$$

where α_{11} has all eigenvalues with modulus greater than 1 and hence is invertible.

2. Hence, $\begin{bmatrix} A_1 \\ C_1 \end{bmatrix}$ could be chosen as $\begin{bmatrix} \alpha_{11} \\ -D^{-1}CQ_1 \end{bmatrix} := \begin{bmatrix} \alpha_{11} \\ \gamma \end{bmatrix}$.
3. However, this part of the realization of the candidate conjugate inner V^{-1} does not (yet) correspond to an isometric realization $\begin{bmatrix} A_V & B_V \\ C_V & D_V \end{bmatrix}$ for $V(z)$; the corresponding formal and unstable realization for $V^{-1}(z)$ is, assuming A_V invertible (see the remark at the end of this section),

$$V(z)^{-1} \sim \begin{bmatrix} A_V^{-\prime} & -A_V^{-\prime}C_V' \\ B_V'A_V^{-\prime} & D_V^{-1} \end{bmatrix}, \tag{8.44}$$

so in order to make the correspondence, one has to find a state transformation R so that

$$\begin{bmatrix} A_1 & B_1 \\ C_1 & D_1 \end{bmatrix} = \begin{bmatrix} R^{-1}A_V^{-\prime}R & -R^{-1}A_V^{-\prime}C_V' \\ B_V'A_V^{-\prime}R & D_V^{-1} \end{bmatrix} = \begin{bmatrix} \alpha_{11} & B_1 \\ \gamma & D_1 \end{bmatrix} \tag{8.45}$$

or $A_V = R^{-\prime}\alpha_{11}^{-\prime}R'$ and $B_V = R^{-\prime}\alpha_{11}^{-\prime}\gamma'$; hence, $M = R'R$ must solve the Lyapunov–Stein equation $A_VA_V' + B_VB_V' = I$:

$$M = \alpha_{11}^{-\prime}\gamma'\gamma\alpha_{11}^{-1} + \alpha_{11}^{-\prime}M\alpha_{11}^{-1}; \tag{8.46}$$

the realization for V^{-1} completes then by putting $D_V = (I + \gamma M^{-1}\gamma')^{-1/2}$ (one uses a Cholesky factorization!), and $C_V = -D_V\gamma R^{-1}$. Notice that α_{11} is nonsingular because all its eigenvalues are located in $\{|z| > 1\}$; hence,

$$\begin{bmatrix} A_1 & B_1 \\ C_1 & D_1 \end{bmatrix} = \begin{bmatrix} \alpha_{11} & \alpha_{11}M^{-1}\gamma'D_V' \\ \gamma & D_V^{-1} \end{bmatrix}. \tag{8.47}$$

4. The realization for T_o^{-1} then readily follows:

$$T_o^{-1} \sim \begin{bmatrix} \alpha_{22} & Q_2' B D^{-1} \\ -D_2 C Q_2 & D_2 \end{bmatrix}, \tag{8.48}$$

with $D_2 = D_1^{-1} D = D_V D$.

Translating to the original outer–inner square equation, all this amounts to

Proposition 8.2 $Y = Q \begin{bmatrix} \sigma \\ 0 \end{bmatrix}$*, with σ square nonsingular and Q unitary is the (essentially unique) solution of (8.40) where* $Q'\Delta Q = \begin{bmatrix} \alpha_{11} & \alpha_{12} \\ 0 & \alpha_{22} \end{bmatrix}$*, with α_{11} invertible and with eigenvalues strictly outside the unit disk, α_{11} of the same dimensions as σ, and $M := \sigma^{-\prime}\sigma^{-1}$ the unique (and nonsingular) solution of the Lyapunov–Stein fixed-point equation*

$$M = \gamma'\gamma + \alpha_{11}^{-\prime} M \alpha_{11}^{-1}, \tag{8.49}$$

in which $\gamma := D^{-1} C Q \begin{bmatrix} \alpha_{11}^{-1} \\ 0 \end{bmatrix}$.

("Essentially unique" means modulo a left unitary factor on σ.)

Remark: When $V = D_V + C_V(I - zA_V)^{-1} z B_V$ is inner with $\begin{bmatrix} A_V & B_V \\ C_V & D_V \end{bmatrix}$ unitary and D_V nonsingular, then two types of "inverse" of V may be considered. V has a unitary and anti-causal inverse given by its dual $V'(z) = D_V' + B_V' z^{-1}(I - A_V' z^{-1})^{-1} C_V'$, often called the "para-Hermitian conjugate" in the literature and denoted as $V_*(z)$. In this formalism, $z' = z^{-1}$; hence, when z is interpreted as a complex quantity, the expression $VV' = V'V = I$ is only numerically valid on the unit circle $|z|^2 = 1$ (or, alternatively, z^{-1} has to be interpreted in a "matrix algebra fashion" as Z'). However, it is true that $V(z)V'(z^{-1}) \equiv I$, when z is seen as a formal variable. Another important inverse of the causal $V(z)$ is a *causal* but unbounded transfer function $V(z)^{-1}$, which is formally (as a rational function of z) equal to $V'(z)$ but now produces the unilateral inverse $V(z)^{-1} = D_V^{-1} - D_V^{-1} C_V (I - z\Delta_V)^{-1} B_V D_V^{-1}$, where $\Delta_V = A_V - B_V D_V^{-1} C_V$ is unbounded. Indeed, from the unitarity of $\begin{bmatrix} A_V & B_V \\ C_V & D_V \end{bmatrix}$, one gets $\Delta_V = A_V^{-\prime}$ and

$$\begin{bmatrix} \Delta_V & B_V D_V^{-1} \\ -D_V^{-1} C_V & D_V^{-1} \end{bmatrix} = \begin{bmatrix} A_V^{-\prime} & -A_V^{-\prime} C_V' \\ B_V' A_V^{-\prime} & D_V^{-1} \end{bmatrix}, \tag{8.50}$$

showing the formal equivalence between the anti-causal dual and the unbounded causal inverse. From all this, the important role of the inner factor should be apparent: it characterizes the unstable part of the inverse of the system.[2]

[2] To add to the confusion, a variety of other conventions can be found in the literature. For example, "$V'(z)$" is sometimes used to indicate the (Hermitian) transpose of the coefficients in V but keeping the argument z unchanged. In that convention, one has, for the para-Hermitian conjugate, $V_*(z) = V'(1/z')$,

8.6 Nonlinear Generalization*

Inner–outer (and outer–inner) factorization extends to the nonlinear case, using a quadratic cost criterion as derived in Section 8.4 and some reachability assumptions on the propagation. We briefly sketch how the inner–outer theory proceeds for the nonlinear time-invariant case.

In the nonlinear case, we start with the transition operator (using the same repression of the index k)

$$\begin{bmatrix} x_+ \\ y \end{bmatrix} = \begin{bmatrix} f(x,u) \\ g(x,u) \end{bmatrix}, \tag{8.51}$$

where $x \in \mathbf{R}^{\eta}$, $x_+ \in \mathbf{R}^{\eta_+}$, $u \in \mathbf{R}^m$ and $y \in \mathbf{R}^n$, and let us suppose that there exist an optimal input $\widehat{u}(x)$ and an optimal cost function $Y_-(x)$, such that $\|Y_-(x)\|^2$ reaches a minimum, say in an open domain of $\mathbf{R}^{\eta}$ with, likewise, an existing optimal $Y(x_+)$ for the x_+ of interest. (In the reasoning, we fix the state x and restrict the potential inputs u to an open domain around the optimal $\widehat{u}$.)

We must require some additional specifications on $Y(x_+) = (Yf)(x,u))$ and $g(x,u)$ in order to avoid singularities. ("Yf" stands for the application of first f followed by Y.) We start out (as in the linear case) with a fixed x, and search for a $\widehat{u}(x)$ that minimizes the "cost" $\|g(x,u)\|^2 + \|Y(x_+)\|^2$, and we assume that this cost function has a single minimum in the domain of consideration, which is also a function of x, and is given by $\widehat{x}_+(x)$. This optimization will involve the differential equation that determines $\widehat{u}(x)$ as the solution of the equation (dropping a factor 2):

$$\begin{bmatrix} \partial_u((Yf)(x,\widehat{u})) & \partial_u(g(x,\widehat{u})) \end{bmatrix} \begin{bmatrix} Yf(x,\widehat{u}) \\ g(x,\widehat{u}) \end{bmatrix} = 0, \tag{8.52}$$

in which the differentials are evaluated at $(x,\widehat{u})$; hence, also at $\widehat{x}_+ = f(x,\widehat{u})$. (Given a matrix function $F(u) : \mathbf{R}^m \to \mathbf{R}^n$, the matrix $[\partial_u F]_{i,j} = \frac{\partial F_i}{\partial u_j}$ defines the approximate increment $\Delta F(u) \sim \partial_u(F)\Delta u$ induced by the vector $\Delta u_{1:m}$ of increments of u.) We require that Eq. (8.52) has full rank and leads to a unique solution. (The simplest and most common set of dimensional assumptions is $Y : \mathbf{R}^{\eta_+} \to \mathbf{R}^{\eta_+}$, and $\eta + m = \eta_+ + n$, consistent with full reachability.)

We now have, by definition, $\|Y_-(x)\|^2 = \|Y(\widehat{x}_+)\|^2 + \|g(x,\widehat{u})\|^2$ as the minimum cost for x, and hence there should exist an orthogonal matrix $V(x)$ such that

$$V(x)' \begin{bmatrix} Y(\widehat{x}_+) \\ g(x,\widehat{u}) \end{bmatrix} = \begin{bmatrix} Y_-(x) \\ 0 \end{bmatrix}. \tag{8.53}$$

Actually, we may *define* $V(x)$ to achieve this, and this can be done continuously in a neighborhood (abbreviated to NBH) of x (thanks to the original assumptions.) Having defined $V(x)$, $g_o(x,u)$ is defined next in the NBH of u by

$$g_o(x,u) = V_{1,2}(x)'Y_+(x) + V_{2,2}(x)'g(x,u). \tag{8.54}$$

and z is viewed as a complex variable. Note, however, that $V'(z)$ does not correspond to the conjugate transfer operator V' as in the convention we use here.

The unitary transformation $V(x)'$ on the image space $\mathbf{R}^{\eta_+} \oplus \mathbf{R}^n$ amounts to a change of coordinates, derived from the functions Y, f and g, and such that the range of Y_- is orthogonal to $\Delta u := u - \widehat{u}$ to the second order. This relation remains valid for all x in a sufficiently small NBH, provided differentiability is assured.

The nonlinear inner–outer factorization hence leads to the same diagram as shown in Fig. 8.2, with the respective transition matrices replaced by their nonlinear equivalents. Remains the issue of the definition of $Y(x+)$, which is assumed to exist at the beginning of the recursion. With the definition of the (additive) quadratic norm, the assumption posits that *starting from any admissible x, there always is a quadratically finite sequence of inputs that minimizes the total error*: it is a reachability assumption on the global properties of the system and the cost criterion. The property propagates backward. After all, the minimal cost amounts to a summation of all optimal individual costs starting from a given x to the final stage (or to $+\infty$ is there is no final stage):

$$\|Y_{k-1}(x_k)\|^2 = \sum_{i=k:\infty} \|g_i(\widehat{x}_i, \widehat{u}_i)\|^2, \tag{8.55}$$

in which the indices now stand for their stage in the recursion.

8.7 Items for Discussion

- As in Chapter 7 on external factorizations, it is interesting to work out some simple examples, and study inner–outer as well as outer–inner factorization of a few "simple" cases, in particular, the matrices

$$\begin{bmatrix} \boxed{1} & & & \\ 1/2 & 1 & & \\ & 1/2 & 1 & \\ & & 1/2 & 1 \end{bmatrix}, \begin{bmatrix} \boxed{1} & & & \\ 1/2 & 1 & & \\ & 1/2 & 1 & \\ & & \ddots & \ddots \end{bmatrix}, \begin{bmatrix} \boxed{1} & & & \\ 2 & 1 & & \\ & 2 & 1 & \\ & & \ddots & \ddots \end{bmatrix}. \tag{8.56}$$

  Several interesting phenomena appear that will motivate quite a few further developments! Compare also what happens in relation to the external factorizations discussed before.

- An often occurring transfer matrix has the form $T = \begin{bmatrix} T_{0,-1} & \boxed{T_{0,0}} \\ T_{1,-1} & T_{1,0} \end{bmatrix}$. What would its inner–outer and outer–inner factorization look like? (Hint: first try your hand on the simple example $T = \begin{bmatrix} 1 & \boxed{1} \\ 2 & 1 \end{bmatrix}$. Can T be left-outer? right-outer? outer? Find minimal realizations and perform the factorizations!) This example requires deft handling of empty entries and the existence of their right and left inverses!

- With respect to the introductory section: numerical analysts have developed a method called "Householder transformation" to bring a given vector in the direction of the first axis. The method presented here is based on a (generalized) rotation. It has several advantages over the Householder transformation, which uses a reflection instead of a rotation. It is interesting to compare the two approaches.

- Riccati equation: when one squares the square-root equation (8.22) (i.e., multiplies it to the right with its conjugate), one obtains, with $M := YY'$,

$$\begin{bmatrix} AMA' + BB' & AMC' + BD' \\ CMA' + DB' & CMC' + DD' \end{bmatrix} = \begin{bmatrix} M^{\langle -1 \rangle} + B_o B_o' & B_o D_o' \\ D_o B_o' & D_o D_o' \end{bmatrix}. \tag{8.57}$$

Since D_o has to be chosen minimal, it has a left inverse $D_o^{\dagger}$, so that $B_o = (AMC' + BD')(D_o^{\dagger})'$ as well as $(D_o D_o')^{\dagger} = (CMC' + DD')^{\dagger}$. It follows that M is a (semi)positive definite solution of the "Riccati" recursion (using the backward diagonal shift $M^{\langle -1 \rangle}$ on M):

$$M^{\langle -1 \rangle} = AMA' + BB' - (AMC' + BD')(CMC' + DD')^{\dagger}(CMA' + DB'). \tag{8.58}$$

In the matrix case, this forward recursion starts with "empty," and it should be recognized that it does not have to produce a strict positive definite solution, since the dimensions of M depend on the dimensions of Y, which may even disappear. One may argue that it is not wise to solve this recursion directly, as it produces the needed Y only indirectly and in a quadratic form, thereby losing numerical accuracy and end up not being positive definite due to the accumulation of errors, and requires the computation of a pseudo-inverse as well. The recursive outer–inner or inner–outer factorizations given in the theorem do not suffer from such problems.

- An LTI outer–inner factorization involves the separation of eigenvalues strictly outside the open unit disk ("outer" eigenvalues) from the other, "inner" ones (applied on the transition function $\Delta = A - BD^{-1}C$ of the inverse, or an adequately chosen partial inverse; see [30]). Such a separation of eigenvalues has been called a *dichotomy* in the literature. This is not surprising in the light of our discussion on the Lyapunov–Stein equation in Chapter 5, where we saw that the key difficulty lies in the implicit use of all the powers of the relevant transition matrix (A in that case). This is also the case here (now with Δ), but the added issue in the LTI inner–outer factorization is dichotomy, which can be done by computing the Schur eigen form (as was already advocated for the Lyapunov–Stein equation). This is certainly a clean way of doing things, since it works on the square-root versions in an as stable numerical way as possible. An interesting observation is that an outer–inner or inner–outer factorization amounts to an external factorization (left or right) on the inverse system, when it exists, and otherwise on a well chosen implicit partial subsystem. A further observation is that the outer–inner or inner–outer recursion *reveals the dichotomy*: the crucial Y matrix gets to be built up gradually by the recursion applied on a half infinitely indexed system.

8.8 Notes

- Although inner–outer and outer–inner factorizations are perhaps the most central operations in dynamical system theory, they have not been recognized as such in many treatments, because their far-reaching effects have often not been seen clearly, especially in the engineering community. However, already in the early times of Hardy space theory, their importance was recognized by the mathematicians working on complex function analysis, leading first to the Beurling theorem and then later, when matrix functions were considered, to the extension of the Beurling theorem known as the Beurling–Lax theorem. In a sweeping generalization of the basic ideas contained in Hardy space theory, Ringrose [58] and Arveson [7] set up a new algebraic category called *Nest Algebras*, for which the basic concepts behind inner–outer factorization hold, namely the properties of nested invariant subspaces defined by these authors (a nest algebra is an algebra for lower block triangular matrices). In more recent times, it has been realized that these concepts even extend usefully to nonlinear systems, especially the work of Willems [76], Ball and Helton [9] and van der Schaft [68] and their students have shown the way into that still not fully explored and very promising direction.
- A different approach (leading to the same effects) has come from estimation theory, and in particular the work of Thomas Kailath and his early students. When studying the Kalman filter and the somewhat pedestrian traditional way of computing state estimations, they realized that a more direct method would be based on the propagation of the square root of a covariance rather than the covariances themselves. This then led to the famous "square-root algorithm" for the Kalman filter [46], which, as we shall see in Chapter 10, is a direct implementation of inner–outer factorization on the assumed model.
- In the following chapters, we shall encounter many applications of inner–outer factorization theory: to estimation theory (the Kalman filter and the LU factorization), to system inversion theory and to control. In all these cases, what the factorization mainly achieves is what one could call a *dichotomy* on the inverse of the system. As we have seen, this is achieved mainly by an explicit eigenvalue computation in the LTI case or by other methods that involve implicit calculations of a potentially large number of powers of the transition matrix of the system, and hence dubious numerics. In the LTV case, the basic "square-root" equation is recursive, and it is not hard to see that the recursion is numerically stable, when the transition matrix of the original system is u.e.s., meaning that the effect of computation errors (or other disturbances) eventually disappear. (Continuous products play an important role in the disappearance, just like powers in the LTI case!) The LTI case can be obtained as a limiting case of an LTV recursion (e.g., a doubling procedure), but the computational cost may be high. Needless to say, many practical systems are essentially time-variant, and it pays to produce algorithms tailored to that case using instantaneous information on the system rather than a system model that is thought valid for all times (e.g., think about control methods for autonomous flying or driving).

- What happens with the outer–inner or inner–outer factorization of infinitely indexed systems? Just as with external factorization (see notes of Chapter 8), outer–inner or inner–outer factorization proceeds just as in the finitely indexed case, and the formulas derived in Section 8.3 apply on the local realization–provided some boundedness conditions hold, usually the boundedness of A, B, C, D and the u.e.s. property for A. However, there is an additional source of potential trouble. In contrast to the finitely indexed case, the outer factor T_o obtained in the factorization may still not be causally invertible *as a bounded operator*. It turns out that its unbounded inverse can be approximated by a sequence of bounded causal operators (in a sense to be defined precisely), but more cannot be claimed.

  This phenomenon already occurs with LTI systems. Take, for example, the LTI transfer function $T(z) = z(z-1)$; then z is the inner factor and $z-1$ the outer factor; $(z-1)^{-1}$ is unbounded but may be considered causal as the limit of $(z-(1+\epsilon))^{-1}$ when $\epsilon > 0$ goes to zero. In continuous-time LTI systems, the situation occurs when there are zeros on the imaginary axis in the original operator, a case that is very common with selective filters (e.g., band filters). Such systems are only invertible in a weak sense.

  How the theory of outer operators is made to work for infinitely indexed systems is beyond the scope of the present book; a full treatment is in [29].

9 The Kalman Filter as an Application

This chapter presents one of the most direct applications of outer–inter factorization: the Kalman filter. The purpose of the Kalman filter is the estimation of the state of a system from output measurements, where both the state and the measurements are affected by uncorrelated noisy disturbances. One assumes knowledge of (1) how the state evolution equations and the output equations are affected by noise and (2) the second-order statistical characteristics of the sources of noise (i.e., means and covariances). The goal is to estimate the state of the system while the system is evolving, using measured data up to index k, and this with minimal computational effort. This turns out to be an eminently recursive problem. We also consider a generalization, namely the problem of "smoothing," that is, refining past estimates using future measurement data.

Menu

Hors d'oeuvre
Linear State Estimation

Main Course
The Kalman Filter as Outer–Inner Factorization

Dessert
The Smoothing Problem

Dessert
Discussion Issues

9.1 Linear State Estimation Basics

The classical Kalman filter situation starts out with a given linear and time-variant stochastic model for the system whose state one wants to estimate recursively. We restrict ourselves here to discrete-time systems and assume them described by a minimal, linear, discrete-time realization:

$$\begin{cases} x_{k+1} &= A_k x_k + B_k u_k, \\ y_k &= C_k x_k + v_k. \end{cases} \tag{9.1}$$

As before, x_k is the state of the system at the index point k, but the inputs are now assumed to be stochastic vectors (i.e., unknown except for their statistical properties), and a vector of input noises and a vector of measurement or output noises are respectively represented by u_k and v_k. In the original formulation, these noise processes are assumed of zero mean and *uncorrelated* with each other at different time indices and with given covariances as a priori information.

Moreover, one assumes that the process starts at $k = 0$, with as initial input not only u_0 but also the initial state x_0, which is also assumed to be stochastic, zero mean and having a given, a priori known covariance P_0 uncorrelated with all other inputs.

The goal is the derivation of a new system, namely the *Kalman filter*, that computes an optimal estimate $\widehat{x}_k$ recursively for each new state x_k based on the measurement of past outputs from 0 to $k-1$, whereby "optimal" is understood as "minimizing the statistical quadratic error."

Let us make these assumptions more precise. We write the mean of a stochastic vector or matrix with the expectation operator $\mathbf{E}$ and assume zero means: $\mathbf{E}(x_0) = 0$, $\mathbf{E}(u_k) = 0$, $\mathbf{E}(v_k) = 0$. x_0, etc. The components of all these vectors (x_0, u_k, v_k) are stochastic variables, so $\mathbf{E}(x_k)$ is a vector of dimension η_k with components $[\mathbf{E}(x_k)]_{i=1:\eta_k} = 0$ and similarly for other stochastic vectors. All further means resulting from linear operations on those primary quantities will, of course, also be zero: $\mathbf{E}(y_k) = 0$ and $\mathbf{E}(x_k) = 0$ for all k, as they are all linearly dependent on the original stochastic variables.

Covariance statistics of the processes u_k and v_k are supposed known a priori as well; with $\delta_{k-\ell} := 1$ when $k = \ell$ and otherwise zero, the following covariances are assumed given: $\mathbf{E}(u_k u_\ell') := Q_k \delta_{k-\ell}$, $\mathbf{E}(v_k v_\ell') := R_k \delta_{k-\ell}$ and $\mathbf{E}(u_k v_\ell') = 0$ for all k and ℓ. (For example, for any k and ℓ, $\mathbf{E}(u_k u_\ell')$ is a matrix of covariances, which we assume zero for all $k \neq \ell$, while $\mathbf{E}(u_k u_k')$ is a positive definite matrix, which we assume to be nonsingular.) Also, x_0 is assumed uncorrelated with all the other noise sources, and the initial $P_0 := \mathbf{E}(x_0 x_0')$ is assumed known. (In a practical situation, these data would follow from an a priori statistical analysis.)

A Small Digression on Estimation

Suppose we wish to estimate a stochastic variable x from (measured) samples of n correlated stochastic variables $y_{i=1:n}$. We may try to determine a linear estimate $\widehat{x} := \sum_{i=1:n} a^i y_i$ for some coefficients a^i to be determined, chosen so as to minimize the quadratic estimation error $\mathbf{E}(\|x - \widehat{x}\|_2^2)$. Suppose moreover we know

1. that all the variables have zero mean value, and
2. all the covariances or correlations between these variables. (In other words, we know all the second-order statistics.)

The *orthogonality principle* then states *that the best possible linear estimate $\widehat{x}$ of x is such that the error $e := x - \widehat{x}$ is uncorrelated with the variables $y_{i=1:n}$*. This property (which is easy to prove, see further) suffices to determine the coefficients a^i. One step further assumes that both x and the y_i are *vectors* of stochastic variables, and the issue arrises on *how to use the orthogonality principle on vectors*. Each entry of $x = \mathrm{col}[x_k]$ can be estimated separately from the others using all the components of all the y_i. The question is how to assemble the individual variables into vectors and which norms to use on the vectors and matrices obtained, so that the results reduce to the original, individual estimates. We do this in the following steps.

Step 1: *A single variable x to be estimated using n other stochastic variables y_i.* We put the y_i into a single column stack $y = \text{col}[y_1\ y_2\ \cdots y_n]$ and write the estimate $\widehat{x} = ay$, in which a is a row vector: $a = [a^1\ a^2\ \cdots\ a^n]$. (For clarity and consistency, we annotate the components of a as columns.) The orthogonality principle now says $\mathbf{E}((x - ay)y') = 0$. With $P_{yy} := \mathbf{E}(yy')$ and $P_{xy} := \mathbf{E}(xy')$ (a row vector), the correlation matrix of the given data y and assuming this matrix to be nonsingular gives $a = P_{xy}P_{yy}^{-1}$. [Notice: if y is a stochastic vector, then $\mathbf{E}(y) := \text{col}[\mathbf{E}(y_k)]$, and similarly with stochastic matrices, the expectation of a vector is the vector of expectations by definition.]

Step 2: *Assume x is a vector itself.* Then each component x_k of x can be estimated separately as $\widehat{x}_k = \sum_i a_k^i y_i$ or, in the matrix notation $\widehat{x} = ay$, in which a is now a matrix. The orthogonality principle now reads $\mathbf{E}((x_k - \widehat{x}_k)y') = 0$ for individual $\widehat{x}_k$, and assembling this into the vector x, we obtain $\mathbf{E}((x - ay)y') = 0$ and again $a = \mathbf{E}(xy')P_{yy}^{-1} = P_{xy}P_{yy}^{-1}$, where $P_{xy} := \mathbf{E}(xy')$ and a are now matrices $((\mathbf{E}(xy'))_k^i = \mathbf{E}(x_k y_i'))$.

Step 3: *Suppose now the y_is are vectors themselves.* Then the next step is to stack the y_i further into one global column. In the sequel, we just keep track of all the (matrix) entities individually, while applying the orthogonality principle systematically (and recursively). This is done by remarking that for the component-wise optimized $\widehat{x}$, the orthogonality principle then simply states $\mathbf{E}((x - \widehat{x})y') = 0$, now with $y = \text{col}\ y_{1:n}$, in which each y_k is a vector itself and the a_k^i are now blocks. The estimation global error is again a matrix, and since each estimate $\widehat{x}_k$ is independently quadratically optimized given the global vector y, we have to show that the total sum $\mathbf{E}(\text{trace}(ee')) = \mathbf{E}\|e\|_F^2$ of all the entries of $e = x - \widehat{x}$ squared is also minimized when each entry is minimized separately. (*Note:* if e is a matrix of stochastic variables, then $\text{trace}(ee')$ equals the *Frobenius norm squared* $\|e\|_F^2$ of e, which is a quadratic norm on matrices viewed as stacked columns or stacked rows: it is just the sum of all the individual squared covariances.)

In the discussion so far, the variables x, y, etc. are stochastic, that is, probabilistic variables or vectors of probabilistic variables. Each entry could just as well be a sample vector from vectors in a vector space of "very large dimensions," whose natural inner products represent correlations (these may be normalized, e.g., by their number N). This "very large space" is often the space of data vectors, that is, sample measurements of the stochastic variables considered. Take, for example, the case of speech processing. Typically, one would sample speech at 8 kHz and consider segments of 20 ms, corresponding to 160 samples ($N = 160$). Such sample segments, represented as row vectors $\frac{1}{\sqrt{N}}u^{1:160}, \frac{1}{\sqrt{N}}v^{1:160}$, are then considered to live in a Euclidean space $\mathbf{R}^{160}$ with the inner product $u \cdot v = \frac{1}{N}\sum_{k=1}^{N} u^k v^k$. Sample spaces of even much larger dimension (like 3,000 samples) are not uncommon in signal processing. (For ease of writing, we put the row index at the bottom and the column index at the top. In tensor calculus, often the opposite convention is used, but we stick here to a more "natural" notation that indexes rows to the left of a matrix and columns on the top.) The important point is *that it does not matter whether x is a zero means stochastic variable or a data vector for the estimation or optimization theory to work, and the notation is adapted to have consistent matrix notation throughout.*

End of the Digression

Returning to the Kalman filter and summarizing: the linear least-squares estimate (llse) $\widehat{x}_k$ at stage k now has to be determined so that the trace of $P_{e_{x,k}e_{x,k}} := \mathbf{E}[(x_k - \widehat{x}_k)(x_k - \widehat{x}_k)']$ is minimized at each index k, *given the measured outputs* $y_{1:k-1} := \text{col}[y_0, \ldots, y_{k-1}]$. The vector $e_{x_k} := x_k - \widehat{x}_k$ is classically defined as the *innovations* at the index k. We define similarly the *output innovation* as $e_{y_k} := y_k - \widehat{y}_k$, where $\widehat{y}_k$ is the least-squares linear estimate of y_k given $y_{0:k-1}$ (shorthand for $e_{y_k}|_{y_{0:k-1}}$). We shall also use *normalized innovations* and indicate this with a bar, as, for example, in $\bar{e}_k = P_{e_k e_k}^{-1/2} e_k$. The Kalman filter is designed to be a recursive process that reads in the y_k's in sequence and determines from them the least-squares estimates $\widehat{x}_{k+1}$ at stage k as efficiently as possible, knowing $\widehat{x}_k$ from the previous calculation.

For convenience, we also introduce a shorthand inner product notation for zero mean stochastic variables or "large vectors," which takes care of calculations in which only second-order expectations play a role, and this in line with our previous discussion. Let u and v be zero-mean stochastic variables; then we write $u \cdot v' := \mathbf{E}(uv')$. Hence, the quadratic norm of a (zero-mean) variable u, $\|u\|^2 = \mathbf{E}(uu') = u \cdot u'$ equals its covariance (actually just $\mathbf{E}u^2$ in the real scalar case, or, in the complex case $\mathbf{E}|u|^2$), and two variables are *orthogonal* for this inner product when uncorrelated. (This is indeed a genuine inner product, since it satisfies all the requirements for an inner product, most importantly the fact that $u \cdot u' = 0 \Rightarrow u = 0$.)

Next, if we want to require that all components of u are orthogonal to *all* components of v, we should require

$$\mathbf{E}(uv') = \begin{bmatrix} \vdots \\ u_k \\ \vdots \end{bmatrix} \cdot \begin{bmatrix} \cdots & v'_\ell & \cdots \end{bmatrix} = u \cdot v' = 0 \tag{9.2}$$

as a matrix. As to the norm of a stochastic vector, say u, we use $\|u\|^2 = \text{trace}(u \cdot u')$. These definitions are consistent since they amount to a simplified annotation of the classical Frobenius norm.

Remark 1: In practice, each stochastic variable is often represented as a "data vector," that is, a vector of samples, and in the present formalism, we represent a data vector as a *row* vector. A (column) vector of stochastic variables is then represented as a "data matrix"; it is a column of rows, with each row corresponding to a single data vector, while each column is a coherent list of samples of the stochastic variables, taken jointly. In this case, the correlation matrices are called *scatter matrices*, but the algebra is the same as before. In particular, we may observe that if any data vector row$[x^{1:N}]$ (a row vector!) has a zero norm $x \cdot x' = 0$, then all components $x_k^2 = 0$.

Remark 2*: A stochastic variable lives in what is called a *Sigma Algebra* in Lebesgue–Borel measure theory (and related probability theory), that is, a collection of measurable subsets of events on which a (positive) *probability measure* is defined. We do not need the full power of measure theory to deal meaningfully with probabilistic variables, since the theory we need is only of *second order* and handles only means and correlations. The shortcut (which is most often taken) is to restrict the theory to the properties of stochastic variables rather than of sets of events. Corresponding to a (scalar) stochastic variable, say x, there is a *cumulative probability distribution* $P(\xi)$, with ξ a real variable, defined as the probability that $x \leq \xi$. $P(\xi)$ is an increasing function of ξ, which starts from 0 at $-\infty$ and increases gradually to 1 at $+\infty$. It may have jumps on the way, even an infinite numbers of them, or be continuous and even differentiable. However, as defined, it is continuous from the right,[1] since $P(\xi)$ contains the eventual jump at ξ. We adopt the standard assumption that x is *quadratically integrable*, by which it is meant that $\mathbf{E}(x^2)$ is finite. (It may also very well happen that $\mathbf{E}(x^2)$ is infinite, i.e., does not "exist," as is the case with Levy–Pareto distributions, a case we do not consider here.) In terms of the mass probability distribution, this means that $\int_{-\infty}^{\infty} \xi^2 dP(\xi) < \infty$. In this expression, the integral has to be interpreted as a Stieltjes integral (or, equivalently, a Lebesgue integral). On the space of zero mean, quadratically integrable variables, an inner product can then be formally defined as $(u, v) = \mathbf{E}(uv') = u \cdot v$. For further information on this matter, one may consult books on probability theory.

The Orthogonality Principle: Proof

Suppose x is a stochastic vector (i.e., a column stack of stochastic variables), and y is another stochastic vector, defined on the same stochastic space as x, and suppose that we have been able to measure y (e.g., using a collection of sensors). Suppose we wish to use this known data to estimate x in a linear fashion. Expressing linearity, let X_m be a matrix to be determined so that the linear estimate $\widehat{x}|y = X_m y$ is optimal, in the sense that $\text{trace}((x - \widehat{x}) \cdot (x - \widehat{x})') = \text{trace}((x - X_m y) \cdot (x - X_m y)')$ is minimal:

$$X_m = \text{argmin}_X (\text{trace}(x - Xy) \cdot (x - Xy)'). \tag{9.3}$$

[1] There is a simplified Lebesgue–Borel theory for functions on the real line called *Stieltjes integration theory*. This theory is equivalent and largely sufficient for almost all purposes. It is also older but less general in some of its aspects.

This is called *llse* or "linear least-squares estimation," perhaps the most common form of stochastic estimation.[2]

Standardly, we assume all necessary a priori statistics to be known. In this case, we shall need $P_{xx} := x \cdot x'$, $P_{xy} := x \cdot y'$ and $P_{yy} := y \cdot y'$. Moreover, and for definiteness of the solution, we may assume the stochastic variables in y to be linearly independent, that is, P_{yy} is nonsingular. The *orthogonality principle* says that *the estimate* $\widehat{x} = X_m y$ *is orthogonal to all the measured components of* y *individually*, that is, $(x - X_m y) \cdot y' = 0$, or, equivalently when P_{yy} is nonsingular, $X_m = P_{yx} P_{yy}^{-1}$.

Proof

X_m has to minimize trace$((x - Xy) \cdot (x - Xy)')$:

$$X_m = \text{argmin}_X \left(\text{trace}[P_{xx} - XP_{yx} - P_{xy}'X' + XP_{yy}X']\right). \tag{9.4}$$

As P_{yy} is assumed nonsingular, the expression between the square brackets can be rewritten as

$$(XP_{yy}^{1/2} - P_{xy}P_{yy}^{-1/2}) \cdot (P_{yy}^{1/2}X' - P_{yy}^{-1/2}P_{yx}) + (P_{xx} - P_{xy}P_{yy}^{-1}P_{yx}) \tag{9.5}$$

(using the standard trick of "quadratic completion"). Only the first (quadratic) term depends on X, and its trace will be minimal and hence zero when each factor is zero, that is, when X is chosen as $X_m = P_{yx}P_{yy}^{-1}$. The covariance of the error then becomes (note: $P_{xy} = P_{yx}'$)

$$P_{e_{x|y}, e_{x|y}} = e_{x|y} \cdot e_{x|y}' = P_{xx} - P_{xy}P_{yy}^{-1}P_{yx}. \tag{9.6}$$

□

9.2 The (Normalized) Model for the System

Let us write the given model equations in a normalized form, with $\bar{u}_k$ and $\bar{v}_k$ normalized (i.e., uncorrelated zero mean processes with unit covariance), and make the given covariances explicit in the square-root form:

$$\begin{cases} x_{k+1} = A_k x_k + \begin{bmatrix} B_k Q_k^{1/2} & 0 \end{bmatrix} \begin{bmatrix} \bar{u}_k \\ \bar{v}_k \end{bmatrix}, \\ y_k = C_k x_k + \begin{bmatrix} 0 & R_k^{1/2} \end{bmatrix} \begin{bmatrix} \bar{u}_k \\ \bar{v}_k \end{bmatrix}. \end{cases} \tag{9.7}$$

The complete model, starting at $k = 0$ with an initial state x_0 with the known covariance P_0, is shown in Fig. 9.1. To produce the initial state in a consistent fashion, a section at $k = -1$ has been added that produces x_0 from a unit variance white noise input vector $u_{-1} = \overline{x}_0$. This *dummy* section has $A_{-1} = \mathbf{I}$, $B_{-1} = P_0^{1/2}$, $C_{-1} = \cdot$ and $D_{-1} = \mathbf{-}$ (in this chapter we use fat fonts for empty entries, for added clarity).

[2] Only in the zero-mean Gaussian case, the conditional estimate $x|y$ is linear.

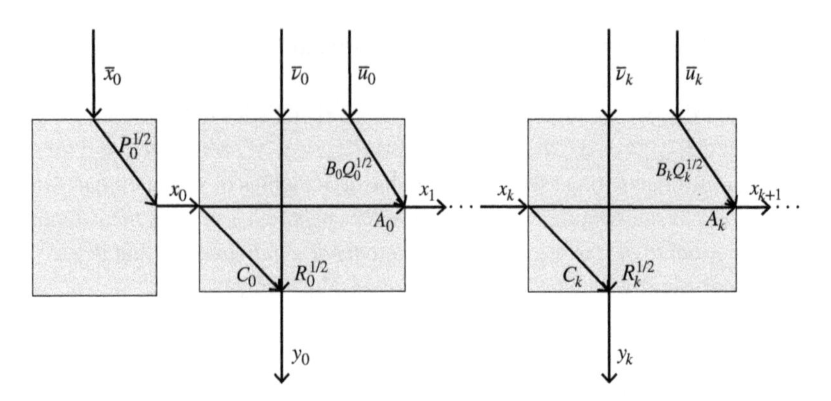

Figure 9.1 The model of the system (assumed known), whose state is to be estimated.

To be determined is the estimate $\widehat{x}_{k+1}$ with the smallest least-squares error, measured as a covariance, that is, such that the estimation error $e_{x,k+1}$ is orthogonal (component wise) on the known data, in this case $y_{0:k}$ (the orthogonality principle).

9.3 Outer–Inner Factorization

We show now that *an outer–inner factorization of the model given in Eqs. (9.7) and Fig. 9.1 solves the estimation problem directly.* The formal outer–inner factorization propagates an intermediary M_k according to the generic (upper) LQ factorization of Chapter 8,

$$\begin{bmatrix} A_k M_k & B_k \\ C_k M_k & D_k \end{bmatrix} = \begin{bmatrix} 0 & M_{k+1} & B_{o,k} \\ 0 & 0 & D_{o,k} \end{bmatrix} V_k, \tag{9.8}$$

to be specialized here for the values pertaining to the Kalman filter model. The first and preliminary section at $k = -1$ produces the initial condition M_0 for the first section at $k = 0$. It is a purely formal step, starting with $x_{-1} = \mathbf{I}$ and $M_{-1} = \cdot$

$$\begin{bmatrix} \mathbf{I} & P_0^{1/2} \\ \cdot & - \end{bmatrix} = \begin{bmatrix} M_0 & \mathbf{I} \\ - & \cdot \end{bmatrix} \begin{bmatrix} \mathbf{I} & I \\ \cdot & - \end{bmatrix} \tag{9.9}$$

producing $M_0 = P_0^{1/2}$. (This step would normally be skipped, and the initial condition on the state covariance just posited, but for completeness and to demonstrate consistency of the formalism, we mention it here in full.)

The outer–inner recursion starts then effectively at $k = 0$.

Step 0

$$\begin{bmatrix} A_0 P_0^{1/2} & B_0 Q_0^{1/2} & 0 \\ C_0 P_0^{1/2} & 0 & R_0^{1/2} \end{bmatrix} = \begin{bmatrix} 0 & M_1 & B_{o,0} \\ 0 & 0 & D_{o,0} \end{bmatrix} V_0, \tag{9.10}$$

(LQ factorization) in which V_0 is unitary. We show that it produces all the relevant quantities. Let Eq. (9.10) act on the inputs of the model, namely $\overline{x}_0$, $\overline{u}_0$ and $\overline{v}_0$, and determine what comes out: the factorization produces intermediate quantities that have to be identified. For simplifying purposes, R_0, P_0 and Q_0 are taken (square) nonsingular, and, for consistency, the system model is assumed to be reachable. As a consequence of reachability, $\begin{bmatrix} A_0 & B_0 \end{bmatrix}$ must be nonsingular (a necessary condition for reachability[3]). It follows that both $D_{o,0}$ and M_1 are square, nonsingular, because both the bottom block row and the whole matrix have full rank: $R_0^{1/2}, P_0^{1/2}$ and $Q_0^{1/2}$ by definition and the full matrix by reachability. Let now

$$\begin{bmatrix} \epsilon_1 \\ \epsilon_2 \\ \epsilon_3 \end{bmatrix} := V_0 \begin{bmatrix} \bar{x}_0 \\ \bar{u}_0 \\ \bar{v}_0 \end{bmatrix}. \tag{9.11}$$

Then the unitarity of V_0 together with the assumption that $\begin{bmatrix} \bar{x}_0 \\ \bar{u}_0 \\ \bar{v}_0 \end{bmatrix}$ are normalized uncorrelated makes all the epsilons normalized and uncorrelated as well, for $\epsilon \cdot \epsilon' = V_0 \begin{bmatrix} \bar{x}_0 \\ \bar{u}_0 \\ \bar{v}_0 \end{bmatrix} \cdot \begin{bmatrix} \bar{x}_0' & \bar{u}_0' & \bar{v}_0' \end{bmatrix} V_0' = V_0 I V_0' = I$. Filling in the epsilons in Eq. (9.10), one obtains

$$\begin{cases} x_1 &= M_1\epsilon_2 + B_{o,0}\epsilon_3, \\ y_0 &= C_0 x_0 + v_0 = D_{o,0}\epsilon_3. \end{cases} \tag{9.12}$$

These two equations show all the properties we need! The second equation shows that y_0 has zero mean (as we would expect) and covariance $D_{o,0}D_{o,0}'$, which equals $R_0 + C_0 P_0 C_0'$, so that $\epsilon_3 = \bar{e}_{y_0}$ is the *normalized output innovation* at this stage. To estimate x_1, the Kalman estimation filter uses the measured value of the actual output y_0, and since $D_{o,0}$ is invertible, also $\epsilon_3 = \bar{e}_{y_0} = D_{o,0}^{-1} y_0$ is known, and its value is available for the estimation of the new state x_1. Now, looking at the first equation, we see that it consists of two terms: one is known, namely $B_{o,0}\epsilon_3 = B_{o,0}D_{o,0}^{-1}y_0$, and the other, $M_1\epsilon_2$, is a zero-mean noise input with covariance $M_1 M_1'$, which is uncorrelated with ϵ_3. It follows immediately[4] that the estimate $\widehat{x}_1 = B_{o,0}\epsilon_3 = B_{o,0}\bar{e}_{y_0} = B_{o,0}D_{o,0}^{-1}y_0$, and hence the estimation error on x_1 is $x_1 - \widehat{x}_1 = M_1\epsilon_2$, with covariance $P_1 = M_1 M_1'$ and normalized innovation $\bar{e}_{x,1} = \epsilon_2$. This identifies $M_1 = P_1^{1/2}$, while $B_{o,0}$ is traditionally called the *normalized Kalman gain* at this stage.

The first stage leaves us with a cascade of a unitary section, an outer section and a propagation of the state as shown in Fig. 9.2. The factorization, together with the shown realizations for the individual factors, splits the state in two parts: one that is

[3] Let $\mathcal{R}_k$ be the kth reachability operator; then $\mathcal{R}_k \sim \begin{bmatrix} B_{k-1} & A_{k-1}\mathcal{R}_{k-1} \end{bmatrix}$. If $\mathcal{R}_k$ is to be nonsingular – that is, have a right inverse – then $\begin{bmatrix} B_{k-1} & A_{k-1} \end{bmatrix}$ cannot be singular, because, assuming that it were singular, then there would be a nonzero vector w, such that $w' \begin{bmatrix} B_{k-1} & A_{k-1} \end{bmatrix} = 0 \Rightarrow w'\mathcal{R}_k = 0$.

[4] Suppose a zero-mean stochastic variable $x = u_1 + u_2$, in which u_1 and u_2 are themselves zero mean stochastic variables and are uncorrelated, and suppose you measure the value of u_1. Then given that measurement, the best least-squares estimate $\widehat{x}$ for x will be $\widehat{x} = u_1$, and the covariance of the error will automatically be $(x - \widehat{x}) \cdot (x - \widehat{x})' = u_2 \cdot u_2'$. In the present case, as ϵ_2 and ϵ_3 are uncorrelated, so are $M_1\epsilon_2$ and $B_{o,0}\epsilon_3$, since $(M_1\epsilon_3 \cdot \epsilon_2' B_{o,0}') = M_1(\epsilon_3 \cdot \epsilon_2')B_{o,0}' = 0$, so the reasoning applies.

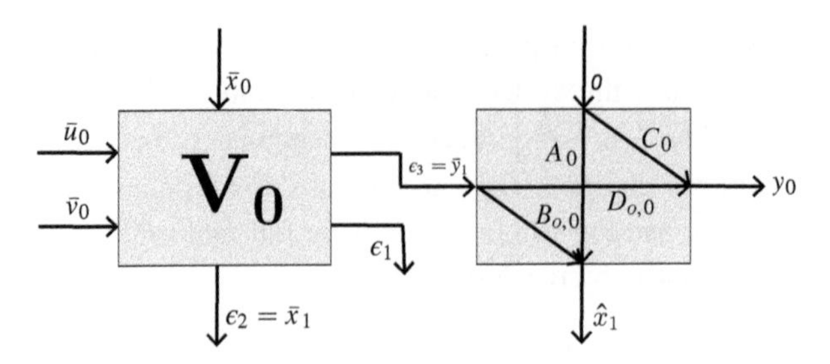

Figure 9.2 The 0th step in the outer–inner factorization of the system model, with relevant inputs and resulting outputs.

determined by the output y_0, namely the estimate $\widehat{x}_1$, and another that is orthogonal to it and normalized, namely the normalized innovation $\overline{x}_1$:

$$x_1 = M_1\overline{x}_1 + \widehat{x}_1. \tag{9.13}$$

Indeed, $\widehat{x}_1 \cdot \bar{x}_1' = B_{o,0}(\epsilon_3 \cdot \epsilon_2') = 0$. This property is propagated further in the recursion.

General step k

The outer–inner factorization looks now as follows in the general step, labeled k:

$$\begin{bmatrix} A_k M_k & B_k Q_k^{1/2} & 0 \\ C_k M_k & 0 & R_k^{1/2} \end{bmatrix} = \begin{bmatrix} 0 & M_{k+1} & B_{o,k} \\ 0 & 0 & D_{o,k} \end{bmatrix} V_k, \tag{9.14}$$

in which V_k is unitary and M_k is a connecting quantity that acquires the special meaning of "normalizing the innovation." The LQ factorization shown corresponds to a factorization of the model system $T = T_o V$, with $T_o \sim_c \{A, B_o, C, D_o\}$ and $V \sim_c \{A_V, B_V, C_V, D_V\}$. In the present case, all M_k and $D_{o,k}$ are square invertible, thanks to the assumed nonsingularity of R_k, Q_k and the minimality of T.

We claim as recursive hypothesis:

Proposition 9.1 *If the state input of the inner part (V_k) is the innovation $\bar{e}_k$ computed in the previous stage, and the state input of the outer filter is the kth estimate $\widehat{x}_k$ with the property $\widehat{x}_k \cdot \bar{e}_k' = 0$ so that $x_k = M_k\overline{e}_k + \widehat{x}_k$, then the new inner and outer parts update these quantities for $k+1$:*

$$x_{k+1} = M_{k+1}\overline{e}_{k+1} + \widehat{x}_{k+1}, \tag{9.15}$$

where

$$\left\{ \begin{array}{rcl} \begin{bmatrix} \epsilon_1 \\ \epsilon_2 \\ \epsilon_3 \end{bmatrix} & = & V_k \begin{bmatrix} \overline{e}_k \\ \overline{u}_k \\ \overline{v}_k \end{bmatrix}, \\ \widehat{x}_{k+1} & = & B_{o,k}\epsilon_3 + A_k\widehat{x}_k \\ & = & B_{o,k}D_{o,k}^{-1}y_k + (A_k - B_{o,k}D_{o,k}^{-1}C_k)\widehat{x}_k, \\ \overline{e}_{k+1} & = & \epsilon_2. \end{array} \right. \tag{9.16}$$

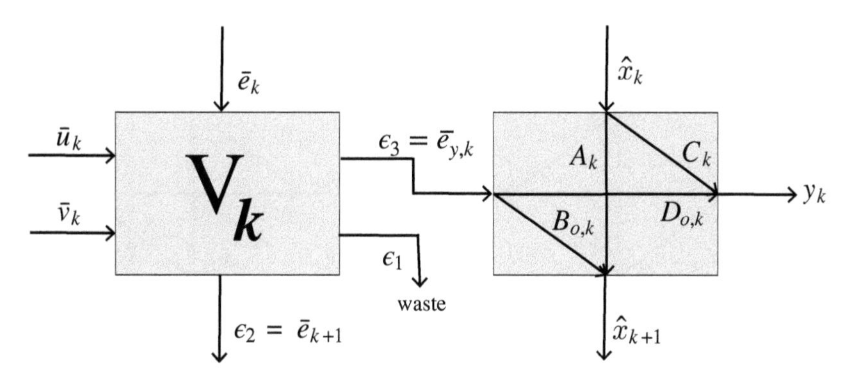

Figure 9.3 The kth step in the outer–inner factorization of the model system.

Assuming the covariance of the input innovation to be $M_k M_k' = (x_k - \widehat{x}_k)\cdot(x_k - \widehat{x}_k)'$, *this property is updated for the next stage:* $M_{k+1}M_{k+1}' = (x_{k+1} - \widehat{x}_{k+1})\cdot(x_{k+1} - \widehat{x}_{k+1})'$ *(Fig. 9.3).*

Proof

The proof follows the same pattern as in Step 0, and is only slightly more complicated due to the appearance of $\widehat{x}_k$. Using the kth stage of the inner–outer factorization (Eq. (9.14)), let

$$\begin{bmatrix} \epsilon_1 \\ \epsilon_2 \\ \epsilon_3 \end{bmatrix} := V_k \begin{bmatrix} \bar{e}_k \\ \bar{u}_k \\ \bar{v}_k \end{bmatrix}. \tag{9.17}$$

We have again that the ϵ's are orthonormal (with the stochastic inner product, this means: zero-mean, uncorrelated and of covariance one, properties that get preserved in the product with the unitary matrix V_k), and they are also orthogonal to $\widehat{x}_k$, because $\bar{e}_k' \cdot \widehat{x}_k = 0$, as well as $\bar{u}_k \cdot \widehat{x}_k' = 0$. Since $x_{k+1} = A_k x_k + B_k Q_k^{1/2}\bar{u}_k = (A_k M_k \bar{e}_k + B_k Q_k^{1/2}\bar{u}_k) + A\widehat{x}_k$, and $A_k M_k \bar{e}_k + B_k Q_k^{1/2}\bar{u}_k = M_{k+1}\epsilon_2 + B_{o,k}\epsilon_3$, we have by the outer–inner factorization (9.14) (see Fig. 9.3),

$$\begin{cases} x_{k+1} &= M_{k+1}\epsilon_2 + B_{o,k}\epsilon_3 + A_k\widehat{x}_k, \\ y_k &= C_k\widehat{x}_k + D_{o,k}\epsilon_3, \end{cases} \tag{9.18}$$

in which the terms in each of the two expressions are mutually orthogonal, $D_{o,k}$ and M_{k+1} are square nonsingular, thanks to (1) the nonsingularity of R_k and (2) the presumed minimality of the state-space model (i.e., the reachability of $\{A, B\}$ – see the previous footnote – which all together make the system matrix right invertible). Remark now first that $D_{o,k}\epsilon_3$ is just zero mean "noise" (i.e., orthogonal on everything that is not yet related to it) added to the known quantity $C_k\widehat{x}_k$, so that $\widehat{y}_k = C_k\widehat{x}_k$, where $\widehat{y}_k$ is the estimate of y_k given $y_{0:k-1}$. After measuring y_k, also $\epsilon_3 = D_{o,k}^{-1}(y_k - \widehat{y}_k) = \bar{e}_{y_k}$ is known, and it follows from the first equation that $\widehat{x}_{k+1} = B_{o,k}\bar{e}_{y_k} + A_k\widehat{x}_k$ and that the new innovation $e_{k+1} = x_{k+1} - \widehat{x}_{k+1} = M_{k+1}\epsilon_2$,

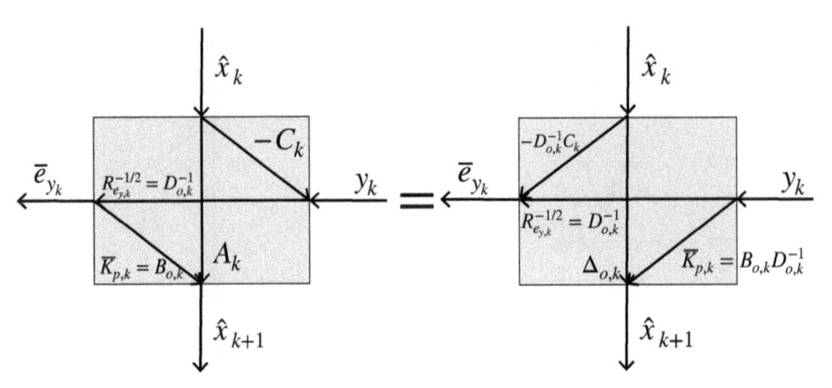

Figure 9.4 The Kalman estimation filter is the inverse of the outer factor. Notice: $\widehat{x}_k \cdot \bar{e}'_{y_k} = 0$.

in which ϵ_2 is zero-mean, uncorrelated with ϵ_3 and $\widehat{x}_k$. The output of the inner part V_k is hence $\bar{e}_{k+1}$, and the covariance of the estimation error $x_{k+1} - \widehat{x}_{k+1} = M_{k+1}\epsilon_2$ is $(x_{k+1} - \widehat{x}_{k+1}) \cdot (x_{k+1} - \widehat{x}_{k+1})' = M_{k+1}M'_{k+1}$. □

The square root of the covariance of the output innovation is $D_{o,k}$, commonly written as $R^{1/2}_{e_y,k}$, while $B_{o,k}$ is commonly known as the *normalized Kalman gain* $\bar{K}_k$. The factorization and the resulting identifications in the more traditional notation – also with $M_k = P_k^{1/2}$, with P_k being the covariance of the innovation in x_k – are shown in Fig. 9.3. The resulting Kalman estimation filter is then the inverse of the outer factor, shown in Fig. 9.4.

In preparation for Section 9.4, and using the notation of Chapter 8, here is the inner factor explicitly written as

$$\left[\begin{array}{c|cc} C_W & B_{W,1} & B_{W,2} \\ \hline A_V & B_{V,1} & B_{V,2} \\ C_V & D_{V,1} & D_{V,2} \end{array}\right]_k \begin{bmatrix} \bar{e}_x \\ \bar{u} \\ \bar{v} \end{bmatrix}_k = \begin{bmatrix} \epsilon_1 \\ \epsilon_2 \\ \epsilon_3 \end{bmatrix}_k, \tag{9.19}$$

where, with $\Delta_o := A - B_o D_o^{-1} C$,

$$\left[\begin{array}{c|cc} C_W & B_{W,1} & B_{W,2} \\ \hline A_V & B_{V,1} & B_{V,2} \\ C_V & D_{V,1} & D_{V,2} \end{array}\right]_k = \left[\begin{array}{c|cc} * & * & * \\ \hline M_{k+1}^{-1}\Delta_{o,k}M_k & M_{k+1}^{-1}Q_k^{1/2} & -M_{k+1}^{-1}B_{o,k}D_{o,k}^{-1}R_k^{1/2} \\ D_{o,k}^{-1}C_k M_k & 0 & D_{o,k}^{-1}R_k^{1/2} \end{array}\right], \tag{9.20}$$

in which the * block row is the orthogonal complement of the bottom two rows, providing an orthonormal basis for the kernel of $\left[\begin{array}{c|cc} A_k M_k & Q_k^{1/2} & 0 \\ C_k M_k & 0 & R_k^{1/2} \end{array}\right]$ – it represents the part of the input noise that is discarded (information lost!) and that does not play a role further on in the recursion (nor in the remaining treatment). Remark also that the subblock (2,3),(1,3) is an equivalent state-space realization of the Kalman filter. (We will use this property in Section 9.4 on smoothing.)

The Affine Case

There is, of course, interest in extending the linear case to the affine case (and then further to nonlinear). The extension to affine does not add much difficulty; let us develop some notations for it. Now, we would have as an initial state $X_0 = \overline{X}_0 + x_0$, in which $\overline{X}_0$ is the initial mean value by definition, and x_0, is zero mean with covariance $x_0 \cdot x_0' := P_0$. For stage k, we have now

$$\begin{cases} X_{k+1} &= A_k X_k + Q_k^{1/2}\overline{u}_k + E_k, \\ Y_k &= C_k X_k + R_k^{1/2}\overline{v}_k + F_k, \end{cases} \tag{9.21}$$

in which E_k and F_k are the fixed (nonstochastic) model parameters, and $\overline{u}_k$ and $\overline{v}_k$ are the (zero-mean) independent white noise as before. This set of equations splits linearly into two sets: one deterministic that propagates the mean values, and the other stochastic, just as before: $X_k = \overline{X}_k + x_k$, $Y_k = \overline{Y}_k + y_k$,

$$\begin{cases} \overline{X}_{k+1} &= A_k \overline{X}_k + E_k, \\ \overline{Y}_k &= C_k \overline{X}_k + F_k \end{cases} \tag{9.22}$$

and

$$\begin{cases} x_{k+1} &= A_k x_k + Q_k^{1/2}\overline{u}_k, \\ y_k &= C_k x_k + R_k^{1/2}\overline{v}_k, \end{cases} \tag{9.23}$$

and the previous treatment on the stochastic part (outer–inner factorization) remains valid and produces estimates that then must be "corrected" by the affine parts involved. It may be observed that a similar reasoning applies in the nonlinear case, when one considers variational models along a trajectory. In fact, "affine" may be considered a first step toward variational calculus.

9.4 Smoothing*

Smoothing consists in improving on an estimate by using future information (and hence delaying the decision). The simplest case is to improve on $\widehat{x}_k$ using the new measurement y_k. We call this "instantaneous smoothing" because it involves a quantity that is immediately available and was used to compute $\widehat{x}_{k+1}$, namely y_k. Let us denote this new estimate $\widehat{x}_{k|k} = \mathbf{E}[x_k | y_{0:k}]$ (while $\widehat{x}_k = \mathbf{E}[x_k | y_{0:k-1}]$). We have $y_k = C_k \widehat{x}_k + D_{o,k}\overline{e}_{y_k}$ (an orthogonal decomposition of y_k): $\overline{e}_{y_k}$ depends only linearly on variables that are orthogonal to $\widehat{x}_k$, namely e_{x_k} and $\overline{v}_k$. $\widehat{x}_k$ is already the projection of x_k on $\bigvee y_{0:k-1}$; it must now be augmented by the projection on $\bigvee \overline{e}_{y_k}$: $\widehat{x}_{k|k} = \widehat{x}_k + \phi_k \overline{e}_{y_k}$, where by the orthogonality principle ($(x_k - \widehat{x}_{k|k}) \perp \bigvee(\widehat{x}, \overline{e}_{y_k})$), the matrix of coefficients has to satisfy

$$(x_k - \widehat{x}_{k|k}) \cdot \overline{e}_{y_k}' = 0. \tag{9.24}$$

Hence $\phi_k = e_{x_k} \cdot \overline{e}'_{y_k}$, since $x_k = e_{x_k} + \widehat{x}_k$. Now, by the Kalman filter (Fig. 9.4) we have $\overline{e}_{y_k} = -D_{o,k}^{-1} C_k \widehat{x}_k + D_{o,k}^{-1} y_k$; hence,

$$\begin{aligned}\phi_k &= e_{x_k} \cdot \overline{e}'_{y_k} \\ &= e_{x_k} \cdot (x'_k C'_k + \overline{v}'_k R_k^{1/2} - \widehat{x}'_k C'_k) D_{o,k}^{-\prime} \\ &= e_{x_k} \cdot e'_{x_k} C'_k D_{o,k}^{-\prime} \\ &= P_k C'_k D_{o,k}^{-\prime} \end{aligned} \tag{9.25}$$

since $e_{x_k} \cdot \overline{v}'_k = 0$ and $e_{x_k} \cdot e'_{x_k} = P_k (= M_k M'_k)$. It follows that the new estimation error is given by

$$x_k - \widehat{x}_{k|k} = e_{x_k} - P_k C'_k D_{o,k}^{-\prime} \overline{e}_{y_k}. \tag{9.26}$$

This amounts to an improvement of $\phi_k \phi'_k$ on P_k, the variance of the previous error (since $\widehat{x}_k \perp \overline{e}_{y_k}$). An alternative characterization of ϕ_k is $\phi_k = M_k C'_{V,k}$ since $C_{V,k} = D_{o,k}^{-1} C_k M_k$, or, to put this differently, one obtains ϕ_k by inverting the inner factor, inputting it with col $\begin{bmatrix} I & 0 & 0 \end{bmatrix}$, and then denormalizing the result by premultiplying with M_k. This idea generalizes.

Further Smoothing

Further smoothing can be done using the information obtained down the line and utilizing the same principles. The outer–inner factorization builds up an orthonormal basis for the output space $\bigvee y_{0:n} = \bigvee \overline{e}_{y_{0:n}}$. We may continue decomposing e_{x_k} according to that future basis:

$$\widehat{x}_{k|n} = \widehat{x}_k + \phi_{k,k} \overline{e}_{y_k} + \cdots + \phi_{k,n} \overline{e}_{y_n} \tag{9.27}$$

for $n \geq k$, with

$$\phi_{k,n} = e_{x_k} \cdot \overline{e}'_{y_n} \tag{9.28}$$

since the orthogonality principle requires $(x_k - \widehat{x}_{k|n}) \cdot \overline{e}'_{y,n} = 0$, as before (and $\phi_{k,k} = \phi_k$ from the previous instantaneous smoothing). Let us set up a recursion to compute $\phi_{k,n+1}$ from the previous data. To do this comfortably, we shall need the state transition matrix of the inverse of the outer filter (which is the Kalman filter), namely,

$$\Delta_{o,k} := A_k - B_{o,k} D_{o,k}^{-1} C_k. \tag{9.29}$$

Then the following holds:

Proposition 9.2 *For $n > k$,*

$$\phi_{k,n+1} = P_k \Delta'_{o,k} \cdots \Delta'_{o,n} C'_{n+1} D_{o,n+1}^{-\prime}. \tag{9.30}$$

Proof

We have, from the model (with NIN standing for "new independent noise"),

$$\begin{aligned}\overline{e}_{y,n+1} &= D_{o,n+1}^{-1}(y_{n+1} - C_{n+1}\widehat{x}_{n+1}) \\ &= D_{o,n+1}^{-1}(C_{n+1}x_{n+1} + \text{NIN} - C_{n+1}\widehat{x}_{n+1}) \\ &= D_{o,n+1}^{-1}(C_{n+1}e_{x,n+1} + \text{NIN}).\end{aligned} \tag{9.31}$$

It follows that

$$\phi_{k,n+1} = e_{x,k} \cdot \overline{e}'_{y,n+1} = e_{x,k} \cdot e'_{x,n+1} C'_{n+1} D^{-\prime}_{o,n+1}. \tag{9.32}$$

A recursion on $E_{k,n+1} := e_{x,k} \cdot e'_{x,n+1}$ is what is needed:

$$\begin{aligned}e_{x,n+1} &= x_{n+1} - \widehat{x}_{n+1} \\ &= A_n x_n + \text{NIN} - A_n\widehat{x}_n - B_{o,n}\overline{e}_{y,n} \\ &= A_n e_{x,n} - B_{o,n}\overline{e}_{y,n} + \text{NIN}.\end{aligned} \tag{9.33}$$

Hence,

$$e_{x,k} \cdot e'_{x,n+1} = e_{x,k} \cdot (e'_{x,n} A'_n - \overline{e}'_{y,n} B'_{o,n}), \tag{9.34}$$

yielding

$$E_{k,n+1} = E_{k,n} A'_n - \phi_{k,n} B'_{o,n}. \tag{9.35}$$

The result is

$$\begin{cases} E_{k,n+1} &= E_{k,n} A'_n - \phi_{k,n} B'_{o,n}, \\ \phi_{k,n+1} &= E_{k,n+1} C'_{n+1} D_{o,n+1}, ^{-\prime} \end{cases} \tag{9.36}$$

with starting values $E_{k,k} = P_k$ and $\phi_{k,k} = P_k C'_k D^{-\prime}_{o,k}$, while $\overline{e}_{y,n} = D_{o,n}^{-1}(y_n - C_n\widehat{x}_n)$, obtained by inverting the outer filter. Filling in $\phi_{k,n} = E_{k,n} C'_n D^{-\prime}_{o,n}$, the recursion on $E_{k,n+1}$ then gives

$$E_{k,n+1} = E_{k,n}(A'_n - C'_n D^{-\prime}_{o,n} B'_{o,n}) = E_{k,n}\Delta'_{o,n}, \tag{9.37}$$

with starting value (see Section 9.3) $E_{k,k} = P_k$, so that $E_{k,n+1} = P_k \Delta'_{o,k+1} \cdots \Delta'_{o,n}$. □

As in the previous instantaneous case, it appears that

$$D_{o,n}^{-1} C_n \Delta_{o,n-1} \cdots \Delta_{o,k+1} M_k \tag{9.38}$$

is the last block in the (1,3) entry in the, now further globalized, inner factor $V^{n\geq k} := V_n V_{n-1} \cdots V_k$. Its C-entry is the block-column vector

$$\text{col}\left[\, D_{o,k}^{-1} C_k \quad D_{o,k+1}^{-1} C_{k+1} \Delta_{o,k} \quad \cdots \quad D_{o,n}^{-1} C_n \Delta_{o,n-1} \cdots \Delta_{o,k+1} \,\right] M_k \tag{9.39}$$

and $\phi_{k,n} = M_k C'_{V^{n\geq k}}$. This means that the increasing smoothing of x_k can be computed recursively, as stated in Proposition 9.3.

Proposition 9.3 *The order $n - k$ smoothing filter at the index k is given by*

$$\widehat{x}_{k|n} = \widehat{x}_k + M_k [C_{V^{n\geq k}}]' e_{y_{k:n}}, \tag{9.40}$$

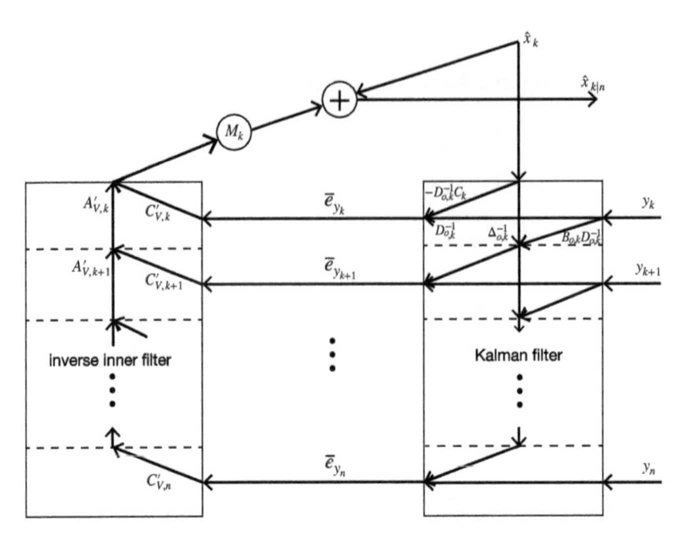

Figure 9.5 The recursive smoothing filter at the index k using data up to and including n.

which may be implemented recursively by reversing the inner factor $V^{n\geq k}$ as shown in Fig. 9.5.

Recursive update: One may observe that two subsequent smoothings of the same order $n-k$ share a large component, namely $C'_{V,k+1}\overline{e}_{y_{k+1}} + \cdots + [A^{n>k}_{V,i}]'C'_{V,n}\overline{e}_{y_n}$. If one wishes to perform systematical smoothing, one has to compute a schema of the type (assuming $k \gg \eta$ and all entries block matrices)

$$\begin{bmatrix} I & a_1 & a_1a_2 & \cdots & a_i^{0<\eta} & 0 & 0 & \cdots & 0 \\ & I & a_2 & a_2a_3 & \cdots & a_i^{1<\eta+1} & 0 & \cdots & 0 \\ & & \ddots & \ddots & \ddots & \ddots & \ddots & \ddots & \vdots \\ & & & & & & & & a_i^{k<(\eta+k-1)} \end{bmatrix} \begin{bmatrix} f_0 \\ f_1 \\ \vdots \\ f_k \end{bmatrix}. \tag{9.41}$$

Without use of the recursive update, the number of floating point matrix computations would be $k\frac{(\eta-1)\eta}{2}$ versus $k + \frac{\eta(\eta+1)}{2}$ when the recursive update can indeed be used, which presumes the availability of the full set of Kalman filtered data.

9.5 Discussion Issues

- We have required minimality of the Kalman model filter. It is interesting to see how this assumption is used. Let us first check the time-variant case.

  In stage 0, minimality amounts, perhaps surprisingly, to $\bigvee(B_0, A_0)$ spanning the full dimension of x_1 (or, $w'[B_0\ A_0] = 0 \Rightarrow w' = 0$). This is because the almost trivial -1 stage is still lacking in the treatment. In that stage, the incoming state, which is needed for a strictly orthodox time-variant treatment, has to be provided. This (zero-mean stochastic) state x_0 with (nonsingular) covariance $P_0 = M_0M_0'$ is generated simply by inputting it at index -1. The model and outer–inner factorization for stage -1 is then simply

$$\begin{bmatrix} | & P_0^{1/2} \\ \cdot & - \end{bmatrix} = \begin{bmatrix} P_0^{1/2} & | \\ - & \cdot \end{bmatrix} \begin{bmatrix} | & I \\ \cdot & - \end{bmatrix}. \tag{9.42}$$

This makes $B_{-1} = P_0^{1/2}$, the other entries in the system model empty and $M_0 = P_0^{1/2}$. The reachability matrix at index 0, $\mathcal{R}_1 = \begin{bmatrix} A_0 B_{-1} & B_0 \end{bmatrix}$ is hence to be supposed nonsingular for the system to be reachable. (It is easy to see that if $\begin{bmatrix} A_0 & B_0 \end{bmatrix}$ is singular (independent rows), then there are states in x_1 that cannot be reached, namely the states in the (nonzero) co-kernel of this first reachability matrix.)

In the general stage k, the inductive hypothesis makes M_k square nonsingular, and one has to show that M_{k+1} will be square nonsingular as well. For that, it is necessary and sufficient that $\begin{bmatrix} A_k M_k & B_k \end{bmatrix}$ has independent rows, so that it compresses to a square nonsingular matrix by orthogonal column transformation. Since M_k is square invertible, $\begin{bmatrix} A_k M_k & B_k \end{bmatrix}$ will have independent rows if and only if $\begin{bmatrix} A_k & B_k \end{bmatrix}$ has independent rows as well, which is a consequence of the assumed reachability, since $\mathcal{R}_{k+1} = \begin{bmatrix} A_k \mathcal{R}_k & B_k \end{bmatrix}$. As a further observation, from the inner–outer theory, we have the minimal factorization $M_k = \mathcal{R}_{k-1}\mathcal{R}'_{V,k-1}$ for all k, which implies that both $\mathcal{R}_{k-1}$ and $\mathcal{R}_{V,k-1}$ are nonsingular.

A further useful observation is that the "stage k" case can be brought back to the "stage 0" case, by integrating all the stages from 0 to k into one global stage. With the original minimal filter realization at the index i given by $\begin{bmatrix} A_i & B_i \\ C_i & D_i \end{bmatrix}$, the filter section combining the steps $0 \cdots k$ has the (cascaded) realization $\begin{bmatrix} A_{[k]} & B_{[k]} \\ C_{[k]} & D_{[k]} \end{bmatrix}$, in which

$$\begin{cases} A_{[k]} & := & A_k \cdots A_0 \\ B_{[k]} & := & \begin{bmatrix} (A_k \cdots A_1 B_0) & \cdots & B_k \end{bmatrix} \left(= \begin{bmatrix} A_k B_{[k-1]} & B_k \end{bmatrix} \right) \\ C_{[k]} & := & \mathrm{col}\begin{bmatrix} C_0 & \cdots & (C_k A_{k-1} \cdots A_0) \end{bmatrix} = \left(\mathrm{col}\begin{bmatrix} C_{[k-1]} & C_k A_{[k-1]} \end{bmatrix} \right) \\ D_{[k]} & := & T_{0:k,0:k}. \end{cases} \tag{9.43}$$

When outer–inner factored, this filter produces, as before in the initial step, both the estimate $\widehat{x}_{k+1}$ and the normalized innovation $P_{k+1}^{-1/2} e_{x,k+1}$, for exactly the same reasons as before. (We do not repeat the argument, because this cascaded filter can just be considered to be the initial step in its own right.) This brings the proof back to the proof given for "stage 0," now with a complete reachability matrix $[B_{[k]} \;\; A_{[k]} B_{[k-1]}]$ reaching up to x_{k+1}.

In the *time-invariant case*, two initialization strategies are common: either one reduces the problem to the time-variant case, where one starts out measuring at the index point 0 and assumes an initial state x_0 whose covariance is known. This actually reduces the problem to the previous case, and one may then study how the estimation evolves with increasing indices, with the only difference being that now the subsequent A_k, C_k, Q_k and R_k are all the same. In case the model is internally stable, in the sense that $\lim_{k\to\infty} A^k = 0$, it is not too hard to show that M_k eventually reaches a nonsingular fixed point (still assuming the minimality of the model representation, of course) when $k \to \infty$, and the prediction filter becomes gradually independent of the initial state. Alternatively, one may study time-invariant outer–inner factorizations of the original transfer function directly, which would also involve some stability conditions on the original model. How this can be done is indicated in the appendix of Chapter 8.

- *Riccati equation*. We have derived the Kalman filter using outer–inner factorization. Traditionally, an opposite road is followed: one derives the Kalman estimation by "brute force," just solving the equations that follow from the orthogonality principle mentioned at the start of this chapter and so deriving an equation for the covariance P_{k+1} of the innovation recursively from P_k for each k. On the basis of these derivations, Kailath and his coworkers [46] derived what has been called the *square-root algorithm* for updating the square root $P_k^{1/2}$ of the covariance rather than the covariance itself. This square-root algorithm is identical to the outer–inner factorization, and, as we have done, the direct derivation of the Kalman filter from outer–inner appears to be simpler and more insightful than the original brute force derivation. The original (quadratic) equation is called a *Riccati equation*. In our TV case, it is simply a recursive update equation (the original term Riccati equation was given in honor of the mathematician who studied differential equations with a quadratic term), while in the LTI case it becomes a fixed-point equation, which is not easy to solve. However, it is easy to derive this Riccati equation directly from the outer–inner or, equivalently, square-root equation conversely. From Eq. (9.14), which we can write shorthand $\mathbf{T}_k = \mathbf{T}_{o,k} V_k$, we find, after postmultiplication with the transpose and using $V_k V_k' = I$, $\mathbf{T}_k \mathbf{T}_k' = \mathbf{T}_{o,k} \mathbf{T}_{o,k}'$, which produces

$$\begin{cases} P_{k+1} + B_{o,k} B_{o,k}' &= A_k P_k A_k' + B_k Q_k B_k', \\ B_{o,k} D_{o,k}' &= A_k P_k C_k', \\ D_{o,k} D_{o,k}' &= C_k P_k C_k' + R_k. \end{cases} \tag{9.44}$$

From these equations, $B_{o,k}$ and $D_{o,k}$ can be eliminated and introduced in the equation for P_{k+1}. This produces in sequence $D_{o,k} = (C_k P_k C_k' + R_k)^{1/2}$ (which is invertible, thanks to the nonsingularity of R_k); then $B_{o,k} = A_k P_k C_k' (C_k P_k C_k' + R_k)^{-1/2}$ and, finally, the recursive matrix Riccati equation

$$P_{k+1} = A_k P_k A_k' + B_k Q_k B_k' - A_k P_k C_k' (R_k + C_k P_k C_k')^{-1} C_k P_k A_k'. \tag{9.45}$$

Quite a bit of effort in the literature is devoted to studying this equation and deriving properties that can often easily be obtained just from the outer–inner factorization (such as the existence of a guaranteed positive definite solution).

9.6 Notes

The Kalman filter was conceived and derived by a few people in parallel in the period 1958–1961, to name: Stratonovitch, Kalman, Bucy and somewhat earlier by Thorvald Thiele and Swerling. It played a key role in the development of the Apollo navigation computer, which was devised by Schmidt of the NASA navigation research group at

Ames Laboratories in Mountain View, following a visit by Kalman. The great advantage of the new approach was its recursive character: it allowed, given the available data at a certain point in time, to make the best possible incremental choices for the next step [48]. This meant that first, one would have to estimate as accurately as possible the state of the rocket, given noisy position and velocity measurements and, next, derive from the estimates the necessary controls to move that state forward in the desired way (the intended trajectory, or a new updated desirable one). "Reachability" and "observability" obviously had to play a central role there.

From that point on, Kalman started to develop the "state-space theory" for dynamical systems in a systematic way, focusing on these most essential concepts (as we are also doing in this book, in the wake of the approach proposed by Kalman). Although the first derivations of the Kalman filter were for time-variant systems, it soon seemed that the time-invariant case would lead to a richer algebraic content, and, moreover, most of the community was geared toward LTI systems and input–output rather than state-space descriptions. The connection between the matrix algebra for state-space descriptions (the A, B, C, D formalism) and the traditional transfer function approach was soon firmly established, and a host of new algebraic results followed which strengthened both sides: it provided the state-space people with the firm algebraic foundation of module theory (matrix functions and series calculus) and the classical Laplace and z-transform calculus with new ways of characterizing the "degree" of a system instead of the cumbersome Smith–McMillan canonical forms. It all seemed like an ideal symbiosis, despite the fact that it could be generalized neither to time-variant nor to nonlinear systems.

With the advent of numerical calculus, the situation changed dramatically, and the emphasis returned from transfer function calculus to matrix algebra. The method of choice in numerical analysis is the use of orthogonal (or unitary) transformations, and it is no wonder that pretty soon after the discovery of the Kalman filter and the rather ad hoc computations connected to it came the idea of using the notion of "innovation" instead, which, inductively, led to a new type of algorithm to compute the Kalman estimation filter based on orthogonal computations: the "square-root algorithm," first proposed by Kailath and Morf [47, 52]. It was later found out that this algorithm is actually a special case of inner–outer factorization. Turning the tables around, one can use it as a basis for the development of the necessary Bayesian innovation theory needed for the Kalman filter. This has been the approach that we have followed in this book.

10 Polynomial Representations

Representations of a system as ratios of simpler objects play a major role in the development of system theory: they are an essential ingredient in solving many problems in control theory and numerical algebra, as we have already shown. In this chapter, we treat a new type of external coprime factorization, which works with unilateral series expansions, and therefore can be used to treat unstable systems directly, using external representations and the one-sided shift algebra. Instead of using an inner function in the (left or right) denominator of the representation (as discussed in Chapter 7), we now use polynomials in the shift Z, that is, causal staircase matrices. This representation generalizes the fractional coprime factorizations of classical matrix fractional calculus to time-variant calculus. The polynomial factorization is obtained via an operation called *deadbeat control*, which is perhaps the simplest possible and hence most fundamental control action one might imagine, and is therefore important already just from that point of view. Interestingly, the classical Bezout identities generalize to the LTV case, although there is no Euclidean algorithm available in that case (it is replaced by a careful consideration of pre-images.)

Menu

Hors d'oeuvre
Z-Polynomials

Main Course
Ratios of Z-Polynomials and Bezout Identities

An Addendum
The LTI Case Revisited

Dessert
Notes

10.1 Fractional Z-Polynomial Representations

In this chapter, we develop an alternative external factorization, representing a causal operator T as a ratio of two minimal polynomials in the shift Z (or dually a ratio of minimal polynomials in Z'). Such polynomials correspond to lower triangular matrices with a staircase form, that is, matrices with zero entries whose support forms a staircase. (Matrices with nonzero support in a band around the main diagonal form a special case.) We shall, of course, aim at minimal, that is, coprime representations. It is a remarkable fact that such representations do exist in the time-variant and quasi-separable case and that they can easily be derived as well, providing for an alternative external factorization theory, much in the same spirit as the celebrated polynomial representations for the LTI case (which is usually derived using the Euclidean greatest common divisor algorithm, a technique that is not available in the time-variant situation).

Deadbeat Control

The key to generating polynomial representations is a simple method called "deadbeat control" [33], which we review in this subsection. Let us assume that we dispose of a minimal time-variant system realization $\{A,B,C,D\}$ (with block diagonal matrices), and let us position ourselves at some index k and consider the state space, call it $\mathcal{X}_k$, at that point. Let us try to generate, from the index point k on, a minimal set of inputs $u_k, u_{k+1}, \ldots$ aiming at bringing the state x_k to zero in as few steps as possible. If successful, this will produce a control law that "beats the state to death" as fast as possible, hence the term "deadbeat control." Before deriving the deadbeat control law, let us make a couple of observations.

First of all, any state x_k that belongs to the kernel of A_k does not have to be beaten to zero; A_k already does that in the present stage k. Next, suppose that the state x_k belongs to the *pre-image for A_k of the range of B_k*[1]; then there will be inputs u_k such that $x_{k+1} = A_k x_k + B_k u_k = 0$, because we have, in that case, $B_k u_k \in \bigvee A_k$, the span of the columns of A_k (alternatively also named the range of A_k), and hence also $u_k := -B_k^+ A_k x_k$, where B_k^+ is any pseudo-inverse of B_k; hence, $u_k = -F_{k,1} x_k$ with $F_{k,1} = B_k^+ A_k$.

This idea (from [33, 34]) can be made recursive. Let us concentrate first on the recursive generation of an adequate basis for the pre-images by A_k. Suppose that in the next stage $k+1$, x_{k+1} can be "beaten to death" by some u_{k+1} in at most one step, by which is meant: *there exists a u_{k+1} such that $A_{k+1}x_{k+1} + B_{k+1}u_{k+1} = 0$*; then x_{k+1} belongs to the pre-image by A_{k+1} of the range of B_{k+1}, and all such x_{k+1} form a subspace; let η be a basis for it (notice: $\bigvee \eta$ contains the kernel of A_{k+1}). Now consider the pre-image by A_k (call it $\mathcal{S}$) of the sum of the spaces "range of B_k" and $\bigvee \eta$ (which we denote as $\mathcal{S} := \text{pre}_{A_k} \bigvee\{B_k, \eta\}$); then for any vector $x_k \in \mathcal{S}$, there shall be an input u_k so that $A_k x_k + B_k u_k \in \eta$. In stage $k+1$, this vector can then be "beaten to death" by the procedure given in the previous paragraph, for states x_{k+1} that can be beaten to death in at most two steps, and so on. See Fig. 10.1 for an illustration.

[1] Suppose $a: \mathcal{X} \to \mathcal{Y}$ is a map from $\mathcal{X}$ to $\mathcal{Y}$; then a pre-image of any element $y \in \mathcal{Y}$ for a is an element $x \in \mathcal{X}$ such that $y = ax$. In the case of linear maps, the notion extends to spaces: if $\mathcal{S}_+$ is a subspace of $\mathcal{Y}$, then its pre-image for a is the largest subspace $\mathcal{S}$ such that $ax \in \mathcal{S}_+$ for all $x \in \mathcal{S}$. A common mathematical notation is $\mathcal{S} = a^{-1}\mathcal{S}_+$, although typically a is not invertible – not even square.

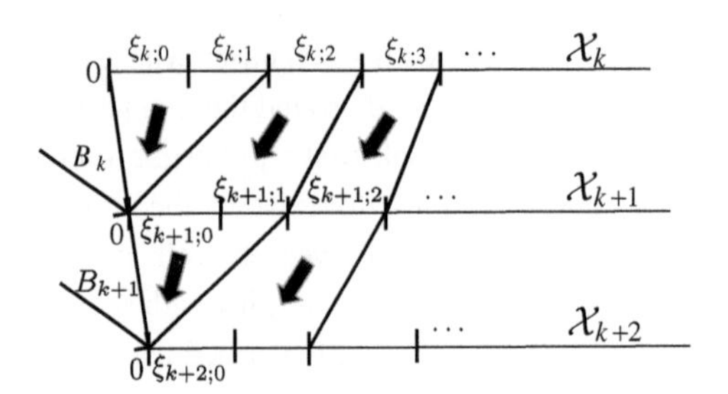

Figure 10.1 States that can be beaten to death in two stages.

The idea can be formalized as follows, using the following lemmas.

Lemma 10.1 *Let* $A\colon \mathbf{R}^k \to \mathbf{R}^\ell$ *and* $B\colon \mathbf{R}^m \to \mathbf{R}^\ell$ *(matrices of dimensions* $\ell \times k$ *and* $\ell \times m$ *respectively) and let* $\overline{B}$ *be a basis for the joint space* $\bigvee A \cap \bigvee B$*; then*

1. *the pre-image of* $\bigvee B$ *by* A *(to be denoted as* $\mathrm{pre}_A B$*) is* $\bigvee[\xi_0|A^+\overline{B}]$*, in which* ξ_0 *is a basis for the kernel of* A *and* A^+ *is any pseudo-inverse of* A*;*
2. *there exists a matrix* K_B *such that* $\overline{B} = BK_B$*;*
3. $\xi_0 \perp \bigvee[A^\dagger \overline{B}]$*, where* $A^\dagger$ *is the Moore–Penrose pseudo-inverse of* A*.*

Proof

Most of the statements follow almost directly from the definitions. Since $\overline{B}$ is a basis for $\bigvee A \cap \bigvee B$, there is a matrix R such that $\overline{B} = AR$. Now,

On 1:

a. $\bigvee[\xi_0|A^+\overline{B}] \subset \mathrm{pre}_A B$, because (1) $A\xi_0 = 0$ and (2) $AA^+\overline{B} = \overline{B}$ since $\overline{B} = AR$ and $AA^+A = A$ by definition of pseudo-inverse, so that $AA^+\overline{B} = AA^+AR = AR = \overline{B}$.

b. Conversely, $\mathrm{pre}_A B \subset \bigvee[\xi_0|A^+\overline{B}]$: suppose $x \in \mathrm{pre}_A B$; then there is u such that $Ax = Bu$ and hence u_1 such that $Ax = \overline{B}u_1$, since Bu is in $\bigvee A \cap \bigvee B$; let now $x_1 := A^+\overline{B}u_1$ and consider $A(x - x_1) = Ax - AA^+\overline{B}u_1 = Ax - AA^+ARu_1 = Ax - Ax = 0$; hence, $x \in \bigvee[\xi_0|A^+\overline{B}]$.

On 2: this is immediate from the definition of $\overline{B}$;

On 3: because the range of $A^\dagger$ is the range of A' and is therefore orthogonal to the kernel of A. □

Remark: If we denote by A^{-1} the map from $y \in \mathbf{R}^n$ to subsets in $\mathbf{R}^m$ that maps y to its pre-image ($A^{-1}y = \mathrm{pre}_A\, y$), then we can write $A^{-1}y = (A^\dagger y) \oplus \mathcal{K} = (A^+y) + \mathcal{K}$, where $\mathcal{K} = \bigvee \xi_0$ is the kernel of A. Each $A^{-1}y$ defines an equivalence class of vectors that are mapped by A to the same y.

Lemma 10.2 *A vector* x *belongs to the pre-image of* B *by a matrix* A *if and only if for some vector* y*,* $\begin{bmatrix} x \\ -y \end{bmatrix}$ *belongs to the kernel of* $\begin{bmatrix} A & B \end{bmatrix}$*.*

Proof

Equivalent to the definition! □

The lemma can be used as follows. Let us denote by ξ_0 an orthonormal basis for the subspace of $\mathcal{X}_k$ that can be beaten to death in one step, using B, ordered so that $\xi_{0,0}$ is a basis for the kernel of A_k and $[\xi_{0,0}|\xi_{0,1}]$ forms a basis for the pre-image by A_k of B_k. Consider a full LQ factorization of the pair

$$\begin{bmatrix} A_k & B_k \end{bmatrix} = \begin{bmatrix} 0 & R \end{bmatrix} \begin{bmatrix} Q_{1,1} & Q_{1,2} \\ Q_{2,1} & Q_{2,2} \end{bmatrix}, \tag{10.1}$$

then all $x_k \in \bigvee Q'_{1,1}$ will be in $\bigvee[\xi_0|\xi_1]$ and conversely, since $\begin{bmatrix} Q'_{1,1} \\ Q'_{1,2} \end{bmatrix}$ spans the kernel of $\begin{bmatrix} A & B \end{bmatrix}$. Moreover, $\bigvee\{\xi_0, Q'_{1,1}\} = \bigvee[\xi_0|\xi_1]$ and, for any $x_k \in \bigvee[\xi_0|\xi_1]$, the corresponding $u_k = -B_k^+ A_k x_k$ for any B_k^+.

Remark: An accurate way to compute the joint space of some $\bigvee A$ and $\bigvee B$ would be to compute *principal vectors*. Given an orthonormal basis U for $\bigvee A$ and V for $\bigvee B$, this method consists in computing the SVD of $V'U$. The singular vectors corresponding to the singular value 1 form an orthonormal basis for the intersection $(\bigvee A) \cap (\bigvee B)$.

To obtain the global result recursively, a streamlined way to proceed is as follows, using a backward recursion (there are more sophisticated methods numerically, but the result presented gives a closed form expression). To start, we may assume, for each k, that

$$\bigvee (A_k|B_k) = \mathcal{X}_{k+1}$$

for, if it were not, there would be vectors in either $\mathcal{X}_k$ and/or $\mathcal{X}_{k+1}$ that are unreachable. Equivalently, one may assume $R_k := A_k A'_k + B_k B'_k$ nonsingular.

Let us position ourselves at the index k assuming that all x_k can be beaten to death in at most $c_k + 1$ steps and that we dispose of an orthonormal basis $[\eta_0|\eta_1|\cdots|\eta_{c_k-1}|\cdots]$ for the state x_{k+1} such that for each ℓ any state in $\bigvee \eta_{0:\ell}$ can be beaten to death in at most $\ell + 1$ steps, with $\bigvee \eta_0$ containing the kernel of A_{k+1}.

By hypothesis, the range of A_k is contained in (but not necessarily equal to) $\bigvee[B_k|\eta_{0:c_k-1}]$. Our goal now is to propagate the (orthonormal) basis η in $\mathcal{X}_{k+1}$ to a similar basis for $\mathcal{X}_k$ backward, in order to produce an orthonormal basis $\xi_{0:c_k}$ with the property that $\xi_{0:\ell}$ maps to $\eta_{0:\ell-1}$ for any $0 \le \ell \le c_k$, using A_k and B_k. To simplify the notation, let us write $A_k \leftarrow A$ and $B_k \leftarrow B$ and let $\overline{B}$ be a basis for $\bigvee A \cap \bigvee B$. In the sequel, we shall need the following lemmas concerning the pre-images that play a role in the map from $\mathcal{X}_k$ plus the input space U_k to $\mathcal{X}_{k+1}$:

Lemma 10.3 $\begin{bmatrix} \xi \\ y \end{bmatrix} \in \mathrm{pre}_{[A|B]}[B|\eta]$ *if and only if* $\xi \in \mathrm{pre}_A[B|\eta]$.

Proof

If $\begin{bmatrix} \xi \\ y \end{bmatrix} \in \mathrm{pre}_{[A|B]}[B|\eta]$, then there exist u and a such that $A\xi + By = Bu + \eta a$; hence $A\xi = B(u-y) + \eta a$; Conversely, if $\xi \in \mathrm{pre}_A[B|\eta]$, then $A\xi + B0 = B\overline{u} + \eta\overline{a}$ for some conformal $\overline{u}$ and $\overline{a}$. □

Lemma 10.4 *Assume $AA' + BB' = R$ nonsingular, and let $\bigvee \xi_{0,0} = \text{kernel}(A)$ and $\bigvee \xi_{0,1}$ such that $\bigvee[\xi_{0,0}\ \xi_{0,1}]$ spans the subspace of $\mathcal{X}_k$ that can be brought to zero by A and B; then*

$$\bigvee[\xi_{0,0}|\xi_{0,1}] = \bigvee[\xi_{0,0}|A^\dagger\overline{B}] = \bigvee[\xi_{0,0}|A'R^{-1}B]. \tag{10.2}$$

Proof

The first equality is by the definition of $\overline{B}$ (see also Lemma 10.1). Concerning the second, let $x \in \bigvee[\xi_{0,0}|A'R^{-1}B]$; then there are vectors u and v such that $x = \xi_{0,0}v + A'R^{-1}Bu$, and hence $Ax = AA'R^{-1}Bu = (I - BB'R^{-1})Bu = B(I - B'R^{-1}B)u$; hence $x \in \bigvee[\xi_{0,0}|\xi_{0,1}]$. Conversely, if x can be brought to zero by $[A|B]$, then there is u such that $Ax - Bu = 0$. Since $\mathcal{X}_k = \bigvee(A') \oplus \text{kernel}(A)$, there exist y and v such that $x = A'R^{-1}y + \xi_0 v$; hence $Ax = AA'R^{-1}y = Bu$. Hence, $AA'R^{-1}y = y - BB'R^{-1}y = Bu$ and $y = B(u + B'R^{-1}y)$, so that $y \in \bigvee B$. Hence, $x \in \bigvee[\xi_{0,0}|A'R^{-1}B]$. □

From the previous discussion, we now have

Proposition 10.5 *Let*

- *$\xi_{0,0}$ be an orthonormal basis for the kernel of A,*
- *$R = AA' + BB'$ nonsingular,*
- *B nonsingular (with no loss of generality),*
- *$[\xi_{0,0}|\xi_{0,1}]$ an orthonormal basis for $\bigvee[\xi_0|A'R^{-1}B]$,*
- *the matrices $r := r_{1:c_k}$, $f := f_{1:c_k}$ and $a_{0:c_k-1,2:c_k}$ be defined by left-to-right orthonormalization of*

$$\left[\,\xi_{0,0} \mid A'R^{-1}B \mid A'R^{-1}\eta_{0:c_k-1}\right] \tag{10.3}$$

as follows (empty entries may occur!):

$$\xi_{0:c_k} = \left[\,\xi_{0,0} \mid A'R^{-1}B \mid A'R^{-1}\eta_{0:c_k-1}\right]\left[\begin{array}{cc|cccc} I & r_0 & r_1 & \cdots & & r_{c_k} \\ \hline 0 & f_0 & f_1 & \cdots & & f_{c_k} \\ 0 & 0 & a_{0,1} & \cdots & & a_{0,c_k} \\ \vdots & \vdots & & \ddots & & \vdots \\ 0 & 0 & 0 & \cdots & & a_{c_k-1,c_k} \end{array}\right], \tag{10.4}$$

with $\xi_{0:c_{k-1}}$ being an orthonormal basis for $\mathcal{X}_k$.
Then
$A_f := A - BF$ maps $\xi_{0:c_k}$ to $\eta_{0:c_k-1}$, where

$$\begin{cases} A_f &= \eta_{0:c_{k-1}} a\xi', \\ F &= (I - B'R^{-1}B)f\xi' - B'R^{-1}A_f = B^\dagger(A - A_f). \end{cases} \tag{10.5}$$

Proof

Expanding Eq. (10.4), we obtain $\xi = \xi_{0,0}r + A'R^{-1}Bf + A'R^{-1}[0\ 0\ \eta_{0:c_k-1}a]$ (with $r_{-1} := I$ if it exists and $f_{-1} := 0$). Applying A, noticing that $AA'R^{-1}Bf = B(I - B'R^{-1}B)f$ and $AA'R^{-1}\eta a = \eta a - BB'R^{-1}\eta a$, one finds (if η has components η_ℓ with

$\ell > c_{k-1}$, then a has to be enlarged with zero: these components play no role)

$$A\xi = \eta a + B[(I - B'R^{-1}B)f - B'R^{-1}\eta a]. \tag{10.6}$$

From the controllability hypothesis, ξ is a full orthonormal basis for $\mathcal{X}_k$. Applying ξ' from the right gives the result. □

Example 10.6 Both the procedure and the notation may be illustrated by a couple of simple examples. Given the data

$$A = \begin{bmatrix} 0 & 1 & 0 \\ 1 & 0 & 0 \\ 1 & 1 & 1 \end{bmatrix}, B = \begin{bmatrix} 1 \\ 0 \\ 1 \end{bmatrix}, \left[\eta_0 \mid \eta_1 \mid \eta_2\right] = \left[\begin{array}{c|c|c} 1 & 0 & 0 \\ 0 & 1 & 0 \\ 0 & 0 & 1 \end{array}\right], \tag{10.7}$$

we find $\xi_{0,0} = \begin{bmatrix} \mathbf{I} \\ \mathbf{I} \\ \mathbf{I} \end{bmatrix}$,

$$R = \begin{bmatrix} 2 & 0 & 2 \\ 0 & 1 & 1 \\ 2 & 1 & 4 \end{bmatrix}, R^{-1} = \begin{bmatrix} 3/2 & 1 & -1 \\ 1 & 2 & -1 \\ -1 & -1 & 1 \end{bmatrix}, A'R^{-1} = \begin{bmatrix} 0 & 1 & 0 \\ 1/2 & 0 & 0 \\ -1 & -1 & 1 \end{bmatrix}. \tag{10.8}$$

To be orthogonalized is

$$\left[\xi_{0,0} \mid A'R^{-1}B \mid A'R^{-1}\eta_0 \mid A'R^{-1}\eta_1 \mid A'R^{-1}\eta_2\right]. \tag{10.9}$$

Hence,

$$\left[\begin{array}{c|c|c|c|c} \mathbf{I} & 0 & 0 & 1 & 0 \\ \mathbf{I} & 1/2 & 1/2 & 0 & 0 \\ \mathbf{I} & 0 & -1 & -1 & 1 \end{array}\right] \left[\begin{array}{c|c|c|c} \cdot & \mathbf{-} & \mathbf{-} & \mathbf{-} \\ \mathbf{I} & 2 & 1 & 1 \\ \mathbf{I} & 0 & -1 & -1 \\ \mathbf{I} & 0 & 0 & 1 \\ \mathbf{I} & 0 & 0 & 0 \end{array}\right] = \left[\begin{array}{c|c|c|c} \mathbf{I} & 0 & 0 & 1 \\ \mathbf{I} & 1 & 0 & 0 \\ \mathbf{I} & 0 & 1 & 0 \end{array}\right] \tag{10.10}$$
$$:= \left[\xi_{0,0} \mid \xi_{0,1} \mid \xi_1 \mid \xi_2\right],$$

and the result is

$$A_f = \eta a \xi' = \left[\begin{array}{c|c|c} 1 & 0 & 0 \\ 0 & 1 & 0 \\ 0 & 0 & 1 \end{array}\right] \left[\begin{array}{c|c|c|c} \mathbf{I} & 0 & -1 & -1 \\ \mathbf{I} & 0 & 0 & 1 \\ \mathbf{I} & 0 & 0 & 0 \end{array}\right] \left[\begin{array}{ccc} \mathbf{-} & \mathbf{-} & \mathbf{-} \\ \hline 0 & 1 & 0 \\ \hline 0 & 0 & 1 \\ \hline 1 & 0 & 0 \end{array}\right] = \begin{bmatrix} -1 & 0 & -1 \\ 1 & 0 & 0 \\ 0 & 0 & 0 \end{bmatrix},$$
$$F = \begin{bmatrix} 1 & 1 & 1 \end{bmatrix}, \tag{10.11}$$

and the feedback F is effective in bringing the state in $\mathcal{X}_k$ down to zero in three steps instead of four.

Example 10.7 Given the data

$$A = \begin{bmatrix} 0 & 0 & 0 \\ 1 & 0 & 0 \\ 0 & 1 & 0 \end{bmatrix}, B = \begin{bmatrix} 1 \\ 1 \\ 0 \end{bmatrix}, \begin{bmatrix} \eta_0 \mid \eta_1 \mid \eta_2 \end{bmatrix} = \left[\begin{array}{c|c|c} 1 & 0 & 0 \\ 0 & 1 & 0 \\ 0 & 0 & 1 \end{array}\right], \tag{10.12}$$

we find $\xi_{0,0} = \begin{bmatrix} 0 \\ 0 \\ 1 \end{bmatrix}$,

$$R = \begin{bmatrix} 1 & 1 & 0 \\ 1 & 2 & 0 \\ 0 & 0 & 1 \end{bmatrix}, R^{-1} = \begin{bmatrix} 2 & -1 & 0 \\ -1 & 1 & 0 \\ 0 & 0 & 1 \end{bmatrix}, A'R^{-1} = \begin{bmatrix} -1 & 1 & 0 \\ 0 & 0 & 1 \\ 0 & 0 & 0 \end{bmatrix}. \tag{10.13}$$

To be orthogonalized is

$$\begin{aligned} &\begin{bmatrix} \xi_{0,0} \mid A'R^{-1}B \mid A'R^{-1}\eta_0 \mid A'R^{-1}\eta_1 \mid A'R^{-1}\eta_2 \end{bmatrix}: \\ &\left[\begin{array}{c|c|c|c|c} 0 & 0 & -1 & 1 & 0 \\ 0 & 0 & 0 & 0 & 1 \\ 1 & 0 & 0 & 0 & 0 \end{array}\right] \left[\begin{array}{c|c|c} 1 & 0 & 0 \\ 0 & 0 & 0 \\ 0 & -1 & 0 \\ 0 & 0 & 0 \\ 0 & 0 & 1 \end{array}\right] = \left[\begin{array}{c|c|c} 0 & 1 & 0 \\ 0 & 0 & 1 \\ 1 & 0 & 0 \end{array}\right] \\ &:= \begin{bmatrix} \xi_{0,0} \mid \xi_{0,1} \mid \xi_2 \mid \xi_3 \mid \xi_4 \end{bmatrix}, \end{aligned} \tag{10.14}$$

and the result is

$$\begin{aligned} A_f = \eta a \xi' &= \left[\begin{array}{c|c|c} 1 & 0 & 0 \\ 0 & 1 & 0 \\ 0 & 0 & 1 \end{array}\right] \left[\begin{array}{c|c|c} 0 & -1 & 0 \\ 0 & 0 & 0 \\ 0 & 0 & 1 \end{array}\right] \left[\begin{array}{ccc} 0 & 0 & 1 \\ \hline - & - & - \\ \hline 1 & 0 & 0 \\ \hline - & - & - \\ \hline 0 & 1 & 0 \end{array}\right] = \begin{bmatrix} -1 & 0 & 0 \\ 0 & 0 & 0 \\ 0 & 1 & 0 \end{bmatrix}, \\ F &= \begin{bmatrix} 1 & 0 & 0 \end{bmatrix}. \end{aligned} \tag{10.15}$$

The Global Result

The point at issue is now whether *any* state at *all* indices k can be beaten to death in a finite number of steps or, in other words, whether for each k, there exists a c_k such that $\xi_k := \bigvee \begin{bmatrix} \xi_{k;0} & \xi_{k;1} & \cdots & \xi_{k;c_k} \end{bmatrix}$ spans the whole state space $\mathcal{X}_k$ at each index point k, and, in the case of infinite indices, whether there is an upper bound to the c_k. The answer to this question follows from the input-controlled state evolution, and we state this as a separate proposition. To state the proposition comfortably, let us use the *continuing product* notation: $A^{k+c>k} := A_{k+c-1} \cdots A_{k+1}$ for an integer $c > 1$.

Proposition 10.8 *Any state x_k at any index k of the system described by $x_{k+1} = A_k x_k + B_k u_k$ can be brought to zero in c_k steps, if and only if for each k,*

$$\bigvee \left[A^{k+c_k-1>k} B_k \mid \cdots \mid B_{k+c_k-1} \right] \subset \bigvee A^{k+c_k>k-1}. \tag{10.16}$$

A sufficient condition for this is that the system is reachable at any index k.

Proof

The proof is immediate from the global input–output relation, starting from a state x_k and using inputs $u_k, \ldots, u_{k+c-1}$ for some integer $c \geq 1$:

$$x_{k+c} = A^{k+c>k-1} x_k + \left[A^{k+c-1>k} B_k \mid \cdots \mid B_{k+c-1} \right] u_{k:k+c-1} \tag{10.17}$$

and a reasoning the same as before. The sufficiency claim follows from the fact that if the system is fully reachable (i.e., each state is reachable), then the partial reachability matrices $\left[A^{k+c-1>k} B_k \mid \cdots \mid B_{k+c-1} \right]$ have to reach full rank for $c := c_k$, with c_k being the minimal value at each k. (In the case of an infinite number of indices, one might require the "local horizon" c_k to be uniformly bounded.) □

Definition 10.9 A system is called controllable of order c if and only if condition (10.16) is satisfied for some $c_k \leq c$ at all k.

We see that the controllability condition is weaker than the reachability condition. (This is typical for discrete-time systems; in continuous-time systems driven by an ordinary differential equation, the two conditions coincide.) In the sequel, we shall impose minimality to keep things simple, and reachability will hence be assumed.

The computation of the ξ_k as well as $A_{f,k}$ and F_k goes by the backward determination of pre-images starting at the last non-empty state space. Assume knowledge of a basis $\xi_{k+1;0:c_{k+1}}$ for $\mathcal{X}_{k+1}$ such that states in $\bigvee \xi_{k+1;0:\ell}$ ($\ell \leq c_{k+1}$) can be "beaten to death" in at most ℓ steps, and given the incoming data A_k and B_k, one computes a similar basis for $\mathcal{X}_k$ as $\xi_{k;0:c_k}$ by orthonormalization, using Proposition 10.5. The proposition produces the resulting feedback law F_k at each index point k as well. The net result is a (block) diagonal matrix $F = \mathrm{diag} F_k$, which is such that ZA_f with $A_f := A - BF$ is nilpotent of order $c = \mathrm{minsup}_k c_k$.

10.2 Ratios of Z-Polynomials

Assuming controllability, the deadbeat construction for a system $\{A, B, C, D\}$ produces, as shown in Section 10.1, a diagonal matrix F such that ZA_f, with $A_f := A - BF$, is nilpotent of some minimal order c given by the deadbeat construction of Propositions 10.5 and 10.8. Now, assume the original representation $\{A, B, C, D\}$ minimal (which has controllability as a consequence and can always be achieved in a preliminary step) and consider the matrix P_r with (again minimal) realization

$$P_r \sim_c \begin{bmatrix} A - BF & B \\ -F & I \end{bmatrix}. \tag{10.18}$$

It is polynomial in Z of order c, where $c := \text{minsup}(c_k)$. And its inverse

$$P_r^{-1} \sim_c \begin{bmatrix} A & B \\ F & I \end{bmatrix} \tag{10.19}$$

is automatically a block lower matrix. Let us now define $\Delta_r := TP_r$; then

$$\begin{aligned} \Delta_r &= \left[D + C(I - ZA)^{-1}ZB\right]\left[I - F(I - ZA_f)^{-1}ZB\right] \\ &= D + (C - DF)(I - ZA_f)^{-1}ZB, \end{aligned} \tag{10.20}$$

and hence

$$\Delta_r \sim_c \begin{bmatrix} A - BF & B \\ C - DF & D \end{bmatrix} \tag{10.21}$$

is itself polynomial in Z of order at most c, and we have obtained a representation of the original operator T as a ratio of two polynomials of order c in Z: $T = \Delta_r P_r^{-1}$. This representation is minimal, assuming the original to be minimal as well, because P_r^{-1} determines its state dimensions as the dimensions of A.

For a complementary, dual factorization, consider the dual system, noticing that $Z'(I - A'Z')^{-1} = (I - Z'A')^{-1}Z'$, $T' = D' + B'(I - Z'A')^{-1}Z'C'$, and applying the "deadbeat control" running time backward, we obtain an anti-causal realization

$$P'_\ell \sim_a \begin{bmatrix} A' - C'G' & C' \\ -G' & I \end{bmatrix} \tag{10.22}$$

with G' now replacing F as "deadbeat controller." The reachability map for the dual system is the adjoint of the observability map of the original system. Let d_k be the deadbeat control horizon at the index $k + 1$ (because stage k in the dual system has x_{k+1} an input), and let $d = \text{minsup}(d_k)$; then P_ℓ will have order d. Following the same treatment as before, we find Δ'_ℓ, and hence

$$T = P_\ell^{-1}\Delta_\ell, \tag{10.23}$$

with

$$P_\ell \sim_c \begin{bmatrix} A_g & -G \\ C & I \end{bmatrix}, \; A_g := A - GC, \; \Delta_\ell \sim_c \begin{bmatrix} A_g & B - GD \\ C & D \end{bmatrix}. \tag{10.24}$$

Both P_ℓ and P_r have minimal realizations, and therefore have the same local degrees (given by the dimensions of A).

10.3 Bezout Identities*

Bezout identities play an important role in determining the properties of the factors in the external factorizations derived in Section 10.2. In particular, they are helpful in proving coprime properties of the factors, by which it is meant that the factors do not have common, nontrivial divisors. These properties are then further exploited in deriving various control laws.

Let $T = \Delta_r P_r^{-1} = P_\ell^{-1}\Delta_\ell$, with P_r, P_ℓ, Δ_r and Δ_ℓ as defined in Section 10.2, and consider the joint polynomial matrix, thereby defining two new polynomials M and N:

$$\begin{bmatrix} M & -\Delta_r \\ N & P_r \end{bmatrix} :\sim_c \left[\begin{array}{c|cc} A_f & -G & B \\ \hline -(C-DF) & I & -D \\ -F & 0 & I \end{array}\right]. \tag{10.25}$$

The state transition matrix of the inverse of this operator is now

$$A_f - \begin{bmatrix} -G & B \end{bmatrix} \begin{bmatrix} I & D \\ 0 & I \end{bmatrix} \begin{bmatrix} -(C-DF) \\ -F \end{bmatrix} = A_g, \tag{10.26}$$

hence also nilpotent, and we find as realization, for the inverse,

$$\left[\begin{array}{c|cc} A_g & -G & B-GD \\ \hline C & I & D \\ F & 0 & I \end{array}\right] \sim_c \begin{bmatrix} P_\ell & \Delta_\ell \\ R & S \end{bmatrix}. \tag{10.27}$$

The LTV *Bezout relations* follow:

$$\begin{bmatrix} P_\ell & \Delta_\ell \\ R & S \end{bmatrix} \begin{bmatrix} M & -\Delta_r \\ N & P_r \end{bmatrix} = \begin{bmatrix} I & \\ & I \end{bmatrix} \tag{10.28}$$

or, in a detailed form

$$\begin{cases} P_\ell M + \Delta_\ell N &= I, \\ -R\Delta_r + SP_r &= I. \end{cases} \tag{10.29}$$

From these, it follows that P_ℓ and Δ_ℓ are *left-coprime* in the sense that any common invertible left polynomial factor in Z has to be *unimodular*, that is, it must have a polynomial inverse as well, and dually for P_r and Δ_r, which must be *right-coprime.* To put it another way, there cannot be a meaningful cancelation in the factorization $P_\ell^{-1}\Delta_\ell$ nor in $\Delta_r P_r^{-1}$, reflecting the minimality of the factorizations.

10.4 Polynomial Representations for LTI Systems*

LTI systems are not fundamentally different from the matrix case just treated; some simplifications *and* some complications occur as usual. In a nutshell, the Hankel operators at each index point are equal, but they have infinite indices, so that orthonormal base vectors have infinite indices as well, but finite dimension when the system has a finite-dimensional state space. Nonetheless, we know that we can obtain an $\{A, B, C, D\}$ realization from a restricted version, after which the realization can be converted to an input normal form (or an output normal form) by solving $P = BB' + APA'$ for the reachability Gramian P (or for the observability Gramian $Q = C'C + A'QA$) – preferably in a square-root form; see Chapter 5. These fixed-point Lyapunov–Stein equations are solvable under broad conditions, but in order to obtain converging bases for the reachability and observability spaces, we have to ask A to be strictly stable. Here, A is just a constant matrix, and the notion of strict stability (i.e., all eigenvalues of A are strictly inside the unit disc of the complex plane) and u.e.s. coincide.

Therefore, the question arises: *What can be done when the stability condition is not satisfied?* In that case, one cannot reasonably speak of a decent Hankel operator without making further assumptions. Previously, our strategy was to restrict the Hankel operator to finitely indexed submatrices, but that is what one could call an "ad hoc" solution. It may be that the fixed-point Lyapunov–Stein equation is still solvable, so that either an operator P or Q is well defined, although not obtainable through a series development, and, moreover, they may (or will) no longer be positive definite. For a scalar example, let $B = 1$ and $A = 2$; then we would have $P = 1 + 4P$ and hence $P = -1/3$; there is no positive definite solution. From the theory of solving fixed-point Lyapunov–Stein equations (see the appendix in Chapter 5), we know that the Lyapunov–Stein system of equations will be nonsingular if and only if the eigenvalues λ_k satisfy the condition $1 - \lambda_i \overline{\lambda_k} \neq 0$ for all relevant i and k. If they do not, there is likely no solution, as can already be seen by the simple example $A = 1$ and $B = 1$.

Hence, a different approach is called for to handle unstable systems, generally defined as systems for which the input–output (behavioral), or equivalently, Hankel map is unbounded. We already know how to derive realizations for such systems, but we now want to develop a "system theory" for them that, like in the stable u.e.s. case, characterizes reachability and observability spaces. It turns out that this can be elegantly done with polynomial representations. In the remainder of this section, we give an account of the LTI theory, using the same method as for the LTV case. The approach we present is therefore different from the classical approach based on module theory, because the latter does not extend to LTV systems (these do not generate modules in a straightforward way). The "deadbeat control" method has the advantage to produce the desired forms directly, using numerical algebra, rather than indirectly, using algebraic properties of rings and modules. The use of polynomial representations is "natural" in the unstable context, because it only handles one-sided series or operators (series or operators whose support does not extend to $-\infty$), so that they can multiply each other meaningfully.

Let us then consider *causal* discrete-time LTI systems and assume that they have a well-defined response for every input with a finite time support. Let the input dimension (which is now constant over all time) be m, output dimension n and e^i be the ith natural vector in $\mathcal{R}^m$, that is, $e^i_k = \delta(i - k)$ for $k = 1 \cdots m$. Then $T^i := Te^i$ will be the ith impulse response, which, because of causality, will only be nonzero from the index $k = 0$ on. Writing this as $T^i(z) := \sum_{k=0}^{\infty} z^k T^i_k$ and stacking inputs and outputs, we obtain the $n \times m$ transfer function $T(z) := \begin{bmatrix} T^1(z) & \cdots & T^m(z) \end{bmatrix}$. All this is well defined, whether or not the system is stable – in the case of an unstable system, the magnitude of the T^i_k will keep on increasing with k, often exponentially. With some abuse of notation, we can also write $T(z) = \sum_{k=0}^{\infty} z^k T_k$, with the T_k now constant $n \times m$ matrices.

The easiest way to construct an external factorization for this type of LTI transfer functions, assuming they possess a finite-dimensional state space, is to use "deadbeat control," based on a preliminary realization: we already know how to obtain a minimal $\{A,B,C,D\}$ realization from a finite version of a sufficiently large partial Hankel matrix (see Chapter 4)

$$H_k := \text{Han}(T_1,T_2,\cdots,T_{2k-1}) := \begin{bmatrix} T_1 & T_2 & \cdots & T_k \\ T_2 & T_3 & .\cdot^{\cdot} & T_{k+1} \\ \vdots & .\cdot^{\cdot} & .\cdot^{\cdot} & \vdots \\ T_k & T_{k+1} & \cdots & T_{2k-1} \end{bmatrix}, \tag{10.30}$$

so let us assume that we have this realization available. We may then *define* $(I - zA)^{-1} := I + zA + z^2A^2 + \cdots$, and we shall have $T(z) = D + C(I - zA)^{-1}zB$ as a one-sided formal series in z, in which A^k may grow exponentially, depending on the location of the eigenvalues[2] of A.

From a given and reachable pair $\{A,B\}$, the deadbeat analysis detailed in the next subsection produces a matrix F such that $A-BF$ is nilpotent (i.e., such that there is an integer k for which $(A - BF)^k = 0$), and hence $P \sim_c \begin{bmatrix} A-BF & B \\ -F & I \end{bmatrix}$ is polynomial. Hence, $P(z) = I - F(I - z(A - BF))^{-1}zB$ (as a matrix polynomial in z), and we may compute, as a formal series,

$$\begin{aligned} \Delta(z) &:= T(z)P(z) = \left[D + C(I - zA)^{-1}zB\right]\left[I - F(I - z(A - BF))^{-1}zB\right] \\ &= D + (C - DF)(I - z(A - BF))^{-1}zB, \end{aligned} \tag{10.31}$$

so that $\Delta(z)$ is also polynomial and, again formally in one-sided series calculus, $T(z) = \Delta(z)P^{-1}(z)$. One can check directly that $P(z)^{-1} \sim_c \begin{bmatrix} A & B \\ F & I \end{bmatrix} = I + zFB + z^2FAB + \cdots := Q(z)$ formally, since $P(z)Q(z) = Q(z)P(z) = I$, a product of a finite series in z with a formal series in z, for which the computation of individual terms is finite and hence well defined. The same is also true for the formal product $\Delta(z)Q(z) = \Delta(z)P(z)^{-1}$. Hence, we have a consistent z-series algebraic theory and $T(z)$ has a fractional representation as $\Delta(z)P(z)^{-1}$, which we would call a right factorization, to be denoted henceforth as $\Delta_r(z)P_r(z)^{-1}$. Likewise, one may define a left factorization $T(z) = P_\ell(z)^{-1}\Delta_\ell(z)$. This factorization can also be expressed as an "external" product, for example, $T(z) = (z^{-\kappa}\Delta_r(z))(z^{-\kappa}P_r(z))^{-1}$, where κ is chosen as the largest of the orders (largest exponent in z) occurring in either $P_r(z)$ or $\Delta_r(z)$. Note that the realizations given for the P's and the Δ's do not have to be minimal. For example, if $T(z)$ is already polynomial, the P's would disappear, and likewise if $T(z)$ happens to be the formal inverse of a polynomial.

[2] One-sided series, for example, the series in z^k with $k \geq K$ for some K, can be multiplied with each other, even when their coefficients become unbounded when $k \to \infty$, because the multiplication of two such series only involves finite computations of the convolution type.

10.5 Discussion Items

- **Elementary examples:** External factorizations are somewhat peculiar in our setup, in particular with semi-infinite or infinite indices, which are worth considering in an exploratory way. Look at related examples to see the peculiarities. For example, try the left and right external factorizations of the following causal matrices:

$$\begin{bmatrix} \boxed{1} & & & \\ 1/2 & 1 & & \\ & 1/2 & 1 & \\ & & 1/2 & 1 \end{bmatrix}, \begin{bmatrix} \boxed{1} & & & \\ 1/2 & 1 & & \\ & 1/2 & 1 & \\ & & \ddots & \ddots \end{bmatrix}, \\ \begin{bmatrix} \boxed{1} & & & \\ 2 & 1 & & \\ & 2 & 1 & \\ & & 2 & 1 \end{bmatrix}, \begin{bmatrix} \boxed{1} & & & \\ 2 & 1 & & \\ & 2 & 1 & \\ & & \ddots & \ddots \end{bmatrix}, \tag{10.32}$$

  where the second and last matrices are half infinite.
- **Nerode equivalences with polynomials:** Consider the external factorization with polynomial matrices. How can such factorizations be interpreted in terms of "Hankel-type" maps, that is, maps for strict past to future or strict future to past? What would be appropriate spaces on which to apply such maps?
- **An interesting issue is how to control the state optimally.** The deadbeat control is a minimum time control, but it seems to have major drawbacks as a control strategy. Which would those be? Much "trajectory control" (e.g., the Apollo mission or keeping an airplane to its planned trajectory) uses a "differential model," that is, a model where the state is actually the deviation from the nominal trajectory in the global state space. "Keeping to the trajectory" is then trying to keep the deviation small. Often this has to be done with a limited energy budget. How can this be done?

  To begin, we should be clear about our model, in particular how it deals with "energy," since the various components of both the input and the state may have different physical dimensions (e.g., the state may be of the form $x = \begin{bmatrix} r \\ v \end{bmatrix}$, with r a position and v a velocity). It pays to normalize variables so that we, as engineers, know where we stand energy-wise. (In the example, we may want to characterize the kinetic energy of the system as $\|v\|^2$, where v is then a normalized velocity $v := \sqrt{\frac{m}{2}} v_r$, with v_r the actual velocity.) Similar considerations may be made about the inputs.

  So to keep the discussion simple, let us assume an LTI situation and that the goal of the control is to reduce $\|x\|^2$ using an input on which there is a limit at each step: $\|u_k\|^2 \leq L$ for each k – henceforth, we just put $u \leftarrow u_k$ for the control at a specific step k. Let us also assume that (1) B is nonsingular (otherwise one can reduce the input space) and (2) we have already established that the one-step unconstrained minimal norm optimal control is $u = -B^\dagger Ax$, with $B^\dagger$ the Moore–Penrose inverse of B, is too large with $\|u\|^2 > L$. What is then the one-step minimal norm optimal control with $||u||^2 = L$? It is given by solving a constrained optimization problem on the control u with Lagrangian

$$\mathcal{L} = (x'A' + u'B')(Ax + Bu) + \lambda(u'u - L) \tag{10.33}$$

and λ as a Lagrange multiplier on the constraint.

Requiring $\nabla_u \mathcal{L} = 2B'Ax + 2B'Bu + 2\lambda u = 0$ at the optimal point, we find

$$u_{\text{opt}} = -(\lambda + B'B)^{-1}B'Ax, \tag{10.34}$$

and the control law is now given by $-Fx$, with $F = (\lambda + B'B)^{-1}B'A$ instead of $B^\dagger A = (B'B)^{-1}B'A$. λ parametrizes the norm reduction, and it obviously reduces the input norm

$$||u||^2 = x'A'B(\lambda + B'B)^{-2}B'Ax = L. \tag{10.35}$$

At this point, one can play various control games. We see that the last expression is dependent on x, for example, when $Ax = 0$, no control is needed, which we know already from the deadbeat control, and if $||B^\dagger Ax||^2 \leq L$, then $u = -B^\dagger Ax$ would do as well.

However, often one wishes a control that works for all states within a certain range, say $||x|| \leq M$ (using Euclidean norms throughout). This one obtains by requiring $||(\lambda + B'B)^{-1}B'A|| \leq \sqrt{\frac{L}{M}}$, taking the smallest possible λ that satisfies this equation. (The bigger λ, the smaller the norm. If B is just a vector and $\beta = B'B$, we could take $\lambda = \sqrt{\frac{M}{L}}||B'A|| - \beta$ when positive.) How good is the solution then? One extra criterion that is often used is to require the stability of the system under the feedback law, so whatever F is chosen, one would require $A - BF$ to have all its eigenvalues inside the unit disk. This requirement is automatically satisfied for the deadbeat control (where all eigenvalues are zero) but needs extra attention in the more general case. (It may even be that there is no solution, of course.) Let us explore this issue further for the LTI case.

- **Companion form:** In classical or traditional control methods for single input–single output systems, the companion form for A and B in the "controller canonical form" corresponding to a transfer function

$$T(z) = d + \frac{c_{\delta-1}z^{-\delta+1} + \cdots + c_1 z^{-1} + c_0}{z^{-\delta} + a_{-\delta+1}z^{-\delta+1} + \cdots a_1 z^{-1} + a_0} \tag{10.36}$$

is given by

$$T(z) \sim_c \left[\begin{array}{ccccc|c} 0 & 1 & 0 & \cdots & 0 & 0 \\ & 0 & 1 & \ddots & 0 & 0 \\ & & \ddots & \ddots & \vdots & \vdots \\ & & & 0 & 1 & 0 \\ -a_0 & -a_1 & \cdots & \cdots & -a_{\delta-1} & 1 \\ \hline c_0 & c_1 & \cdots & \cdots & c_{\delta-1} & d \end{array}\right]. \tag{10.37}$$

It is pretty straightforward (and a good exercise) to do a deadbeat-control analysis on this form! The form is very popular with control engineers, because it allows easy *pole placement*. Using $F = \mathrm{col}[F_0, F_1, \ldots, F_{\delta-1}]$, we find

$$A_f = A - BF = \left[\begin{array}{cccc|c} 0 & 1 & 0 & \cdots & 0 \\ & 0 & 1 & \ddots & 0 \\ & & \ddots & \ddots & \vdots \\ & & & 0 & 1 \\ \hline -a_0 - F_0 & -a_1 - F_1 & \cdots & \cdots & -a_{\delta-1} - F_{\delta-1} \end{array}\right]. \tag{10.38}$$

For example, putting all $F_i = -a_i$ produces deadbeat control, and an arbitrary denominator for the controlled system is achieved by setting $F_i = -a_i + p_i$ for a desired characteristic polynomial $\chi_{A_f}(\lambda) = \lambda^\delta + p_{\delta-1}\lambda^{\delta-1} + \cdots + p_0$ of A_f. The approach can be generalized to LTI multiport systems, thanks to the Heymann–Hautus lemma; see the literature on this matter [41]. Needless to say, working on the characteristic polynomial has its numerical problems (ill-conditioning of the roots) and is only suitable for low-dimensional problems.

11 Quasi-separable Moore–Penrose Inversion

A quasi-separable matrix description is the specification of a general matrix as the sum of a (block) lower part and a block upper part, whereby either part is given by a state-space realization. Many systems are not invertible; some have left or right inverses, but all have a Moore–Penrose inverse, that is, a linear operator that provides the best possible match for inversion, using a quadratic criterium. This chapter shows the remarkable property that the Moore–Penrose inverse of a system with quasi-separable description has itself a quasi-separable description, which can be derived efficiently and recursively from the original description (i.e., with computational complexity linear in the dimensions of the matrix), using orthogonal transformations exclusively.

Menu

Hors d'oeuvre
The Causal Matrix Case

Main Course
The Full Glory of Quasi-separable Moore–Penrose Inversion

Cheese Platter
A 4×4 Exemplary Case

Dessert
Infinite Indices

11.1 Introduction

A central problem in many applications is the inversion of a system, be it a system of equations, a dynamical system or just a matrix. Here we consider this issue for the general case of quasi-separable systems, that is, systems that decompose additively in a lower (causal) part and an upper (anti-causal) part, each one with a given realization. Often there is no exact inverse at all, or there are just many possible inverses. A causal system may not have a causal inverse but may very well have a fully quasi-separable one. When an inverse does not exist, then the more general question is: Given a (non-invertible) system equation $Tu = y$, in which y is given and the value, or at least an estimate of u is desired, what would be a best estimate u_a for u, so that the error $\|y - y_a\|$, with $y_a = Tu_a$, is minimal (for a well-chosen norm)? When a (Euclidean) quadratic norm is used to qualify the "closeness" of an approximation (i.e., $\|y - y_a\|_2$), then the result, given by the Moore–Penrose inverse is remarkable: there exists a linear operator $T^\dagger$ defining $u_a = T^\dagger y$ for which $y_a = Tu_a$, $\|y - y_a\|_2$ is minimal *and* of all possible u for which $Tu = y_a$, $\|u_a\|_2$ is minimal itself in the quadratic norm as well.

In the case of quasi-separable systems, it turns out that, in addition, the Moore–Penrose inverse $T^\dagger$ has the same overall separable complexity as the original system, although the division of state dimensions between the causal and anti-causal parts is likely to be different in the inverse as compared to the original. It is perhaps this fact that makes quasi-separability theory so powerful.

In Chapter 8, we studied the outer–inner factorization of a causal (lower) transfer function T, which leads to a factorization

$$T = \begin{bmatrix} 0 & T_{o\ell} \end{bmatrix} \begin{bmatrix} V_1 \\ V_2 \end{bmatrix} = T_{o,\ell} V_2, \tag{11.1}$$

in which $T_{o\ell}$ is causally left invertible and V is causal unitary; see especially the prototype example in Section 8.2. $T_{o\ell}$ is a tall, block lower matrix whose columns form a (nonorthonormal) basis of the range of T (typically in the echelon form!). Although $T_{o\ell}$ is left invertible (it is left outer), a further step is needed to compute its Moore–Penrose inverse. Because $T_{o\ell}$ is a block lower matrix, this can be done by a simple, recursive QR factorization, which amounts to (is the same as) an inner–outer factorization on $T_{o\ell}$:

$$T_{o\ell} = \begin{bmatrix} U_1 & U_2 \end{bmatrix} \begin{bmatrix} 0 \\ T_o \end{bmatrix} = U_2 T_o, \tag{11.2}$$

in which U is now unitary and T_o is causally right invertible. Since $T_{o\ell}$ is also left invertible, the same is true for $T_o = U_2' T_{o\ell}$, and T_o therefore has a causal inverse T_o^{-1} – it is a block lower operator with square invertible blocks on the main diagonal.

Let us remark that the first outer–inner recursion goes forward, while the second, inner–outer recursion uses the result of the first and goes *backwards*; see the example in Section 8.2. This perhaps unfortunate circumstance is unavoidable: U_2 forms a block lower orthonormal basis for the range of both T and $T_{o,\ell}$, and the inner–outer algorithm computes a realization for that basis starting from the rightmost column and moving upward in the northwest direction, dual but otherwise identical to the outer–inner algorithm that works forward, starting from the top diagonal block and moving downward in the southeast direction. Efficiency results from the systematic use of state-space realizations.

We have already used the Moore–Penrose inverse in the more restricted setting of the Bellman filter (Chapter 1). We shall see that the computation of the Moore–Penrose inverse of a full quasi-separable system requires one external factorization and two factorizations of inner–outer type. Nonetheless, the algorithm proposed achieves linear computational complexity in the full matrix dimension.

Hence, we see that the inner–outer factorization plays a major role in determining the Moore–Penrose inverse, because it produces the necessary information on constructing the contributions to causal and anti-causal parts in the inverse, as well as additional information on kernels and co-kernels useful in producing more general approximate inverses, all called "pseudo-inverses."

11.2 The Basic Theorem

In Section 11.1, we indicated that we can convert any causal transfer operator T recursively to the form $T = V_\ell T_o V_r$, in which all three factors are causal, V_ℓ is isometric, T_o has a causal inverse T_o^{-1} and V_r is co-isometric. We show now that $T^\dagger = V_r' T_o^{-1} V_\ell'$ is the Moore–Penrose pseudo-inverse. In the quasi-separable (full matrix) case, to be derived in Section 11.3, we shall obtain a more general but similar form, which leads to the more general result of the following theorem, in which T is not assumed causal.

Theorem 11.1 *The Moore–Penrose pseudo-inverse $T^\dagger$ of $T = V_\ell T_o V_r$, in which V_ℓ is isometric, V_r is co-isometric and T_o has a bounded inverse, is given by $T^\dagger = V_r' T_o^{-1} V_\ell'$ and provides the unique minimum norm solution $u = T^\dagger y$ to the optimization problem* $\min_v(\|Tv - y\|_2)$, *for any given y.*

Proof
Let $y = y_1 + y_2$, with $y_1 = V_\ell V_\ell' y$ the orthogonal projection of y on the range of V_ℓ. For the Euclidean norm, we then have $\|Tu - y\|_2^2 = \|V_\ell(T_o V_r u - V_\ell' y) - y_2\|_2^2 = \|T_o V_r u - V_\ell' y\|_2^2 + \|y_2\|_2^2$ because $y_2 \perp V_\ell(\cdots)$ by definition of y_1. Therefore, the quadratic error is at least $\|y_2\|_2^2 = \|(I - V_\ell V_\ell')y\|_2^2$, which is independent of u. All inputs u that minimize $\|T_o V_r u - V_\ell' y\|_2$ make it zero: these all satisfy $V_r u = T_o^{-1} V_\ell' y$, thanks to the invertibility of T_o. All errors minimizing u hence have the form $u = u_1 + u_2$, with $u_1 = V_r' V_r u$ (the projection of u on the range of V_r'), and $u_2 \perp u_1$; hence, $\|u\|_2^2 = \|u_1\|_2^2 + \|u_2\|_2^2$ and $V_r u = V_r u_1$. u_1 is therefore the unique solution with the minimum norm in the collection of all possible solutions that achieve the minimal error: when $u = V_r' T_o^{-1} V_\ell' y$, then $u = u_1$, $\|T_o V_r u - V_\ell' y\|_2 = 0$. □

Hence:

- all minimal error solutions have the form $u = V_r' T_o^{-1} V_\ell' y + (I - V_r' V_r)x$, in which x is a compatible but otherwise arbitrary vector, and the second term is orthogonal on the first;
- the only smallest and hence minimal norm solution is $u := V_r' T_o^{-1} V_\ell' y$;
- the Moore–Penrose operator that provides the solution is unique as well, as it is the well-defined map $y \mapsto \min\|\operatorname{argmin}_v(\|Tv - y\|_2)\|_2$. Although $T^\dagger$ is unique, the decomposition $T = V_\ell T_0 V_r$ is not. In the case T is causal, then all three factors can be made causal as well, but even so, they are not defined uniquely. For example, one can execute the inner–outer decomposition first, followed by outer–inner on the remaining right-outer factor, or conversely.

11.3 The Full Quasi-separable Case

Now, we assume our matrix T to have a realization for its causal part and another for its anti-causal part. To connect with the literature in numerical linear algebra, we change notation, keeping the general conventions we have adopted so far – actually only the symbols used will change. Hopefully, the reader will easily recognize the various components of the state-space realizations in the alternative notation. The added benefit of this approach turns out to be the discovery of the hidden structure in what may appear purely numerical manipulations at first. To help the imagination, we develop the general theory in parallel with a 4×4 quasi-separable (block-)matrix T in terms of its realizations as

$$T = \begin{bmatrix} d_1 & g_1h_2 & g_1b_2h_3 & g_1b_2b_3h_4 \\ p_2q_1 & d_2 & g_2h_3 & g_2b_3h_4 \\ p_3a_2q_1 & p_3q_2 & d_3 & g_3h_4 \\ p_4a_3a_2q_1 & p_4a_3q_2 & p_4q_3 & d_4 \end{bmatrix}, \tag{11.3}$$

in which all the entries are block matrices of compatible dimensions. (One should easily recognize the realization of the lower – causal – part as $\{a, q, p, d\}$ and of the strictly upper – strict anti-causal – part as $\{b, h, g, 0\}$; we shall make this explicit soon.) We do not have to assume minimality at this point, although that may help keeping the dimensions of all the entries as low as possible. The corresponding general expression for quasi-separable T in terms of our standard formalism is then

$$T = p(I - Za)^{-1}Zq + d + g(I - Z'b)^{-1}Z'h. \tag{11.4}$$

The goal of the algorithm is to convert T to the form $T = V_\ell R V_r$, where V_ℓ is isometric, R is upper triangular and square invertible and V_r is co-isometric. The Moore–Penrose inverse is then $T^\dagger = V_2 R^{-1} U_2'$ (see Theorem 11.1). We want to do this by working exclusively on the given realizations, not on the direct entries of T.

The added complication with respect to the purely causal case is that the lower and the upper parts in T contribute both to the upper and lower parts of the resulting Moore–Penrose inverse. This necessitates the mixup of the lower and upper parts in a meaningful way. Several choices are possible to achieve this. The choice we make here consists of the following phases (to be motivated by the order of computations):

Phase 1: convert the lower part to upper using postmultiplication with an anti-causal unitary matrix and combine the result with the upper part;
Phase 2: perform an anti-causal inner–outer factorization on the combined result;
Phase 3: reduce the result further so that the resulting outer factor is invertible.

(Phases 2 and 3 are as treated as in the causal case but now with anti-causal matrices. The extra phase one is needed to get the combined result.)

In the version presented here, the algorithm works by forward recursion first on the columns of the lower part to convert the full matrix to an upper matrix (phase 1), followed by and combined with another forward recursion that produces an anti-causal right outer factor (phase 2, of course, together with unitary transformations), which, in phase 3, is then reduced to an anti-causal invertible factor, this time using a backward recursion. In the process, more information on the operator and its Moore–Penrose inverse is, of course, produced, such as various ranges and kernels. Let us work out details of the three phases, trying to keep notation under control and using the 4×4 example as an illustration.

Phase 1: Conversion to the Upper Part

The first step in the algorithm converts the lower part to the upper part, using the lower realization exclusively. The resulting combined upper part is converted in parallel with the transformations of the first step, and this is combined with a full determination of the kernel of T. The procedure determines the matrix V and an intermediate matrix, the upper matrix T_u, which is such that $T = T_u V$, where T_u is upper block triangular and V is causal unitary (i.e., the inner part).

Conversion of the Lower Part

V is exclusively determined by the lower part, which we consider first (we shall look at the combination thereafter). Let

$$T_\ell := \begin{bmatrix} d_1 & 0 & 0 & 0 \\ p_2 q_1 & d_2 & 0 & 0 \\ p_3 a_2 q_1 & p_3 q_2 & d_3 & 0 \\ p_4 a_3 a_2 q_1 & p_4 a_3 q_2 & p_4 q_3 & d_4 \end{bmatrix}. \tag{11.5}$$

We see that the first column is "dominated" by $\begin{bmatrix} d_1 \\ q_1 \end{bmatrix}$, so let us find a unitary matrix $\begin{bmatrix} d'_{V1} & q'_{V1} \end{bmatrix}$ (the notation anticipates on the interpretation to be given soon) such that

$$\begin{bmatrix} d_1 \\ \hline q_1 \end{bmatrix} \begin{bmatrix} d'_{V1} & q'_{V1} \end{bmatrix} = \begin{bmatrix} D_1 & G_1 \\ \hline 0 & Y_2 \end{bmatrix} \tag{11.6}$$

is an upper LQ factorization, with Y_2 left invertible (i.e., compressing to the right, starting with the last row and yielding Y_2 in the upper column echelon form). D_1 may be thought compressed further as $\begin{bmatrix} 0 & D_1^c \end{bmatrix}$, with D_1^c left invertible, and resulting in a full column of zeros that will be canceled eventually. (What has been made upper must still be combined with the original upper part!) Applying $\begin{bmatrix} d'_{V1} & q'_{V1} \end{bmatrix}$ to the right on the first column produces a first transformation

$$\begin{aligned} &\begin{bmatrix} d_1 & 0 & 0 & 0 \\ p_2 q_1 & d_2 & 0 & 0 \\ p_3 a_2 q_1 & p_3 q_2 & d_3 & 0 \\ p_4 a_3 a_2 q_1 & p_4 a_3 q_2 & p_4 q_3 & d_4 \end{bmatrix} \begin{bmatrix} d'_{V1} & q'_{V1} \end{bmatrix} \\ &= \left[\begin{array}{c|cc|cc} D_1 & G_1 & 0 & 0 & 0 \\ \hline 0 & p_2 Y_2 & d_2 & 0 & 0 \\ 0 & p_3 a_2 Y_2 & p_3 q_2 & d_3 & 0 \\ 0 & p_4 a_3 a_2 Y_2 & p_4 a_3 q_2 & p_4 q_3 & d_4 \end{array}\right]. \end{aligned} \tag{11.7}$$

The procedure now moves to the second block column of the result with $\begin{bmatrix} p_2Y_2 & d_2 \end{bmatrix}$ as a diagonal element in the second position and computes a new (upper) LQ factorization:

$$\left[\begin{array}{cc} p_2Y_2 & d_2 \\ \hline a_2Y_2 & q_2 \end{array}\right] \begin{bmatrix} p'_{V2} & a'_{V2} \\ d'_{V2} & q'_{V2} \end{bmatrix} = \left[\begin{array}{cc} D_2 & G_2 \\ \hline 0 & Y_3 \end{array}\right], \tag{11.8}$$

with Y_3 left invertible. (Notice: the row subdivision is forced by the data – as indicated by the horizontal bar – while the column subdivision is a consequence of the algorithm, by which Y_3 is restricted to being left invertible, i.e., its columns are linearly independent.) Applying $\begin{bmatrix} p'_{V2} & a'_{V2} \\ d'_{V2} & q'_{V2} \end{bmatrix}$ on the second block column from the right now produces

$$\left[\begin{array}{c|c} G_1p'_{V2} & G_1a'_{V2} \\ \hline D_2 & G_2 \\ 0 & p_3Y_3 \\ 0 & p_4a_3Y_3 \end{array}\right], \tag{11.9}$$

so the next submatrix to be converted to upper is the remainder lower part

$$\left[\begin{array}{cc|c} p_3Y_3 & d_3 & 0 \\ \hline p_4a_3Y_3 & p_4q_3 & D_4 \end{array}\right], \tag{11.10}$$

which is achieved by postmultiplication with $\begin{bmatrix} p'_{V3} & a'_{V3} \\ d'_{V3} & q'_{V3} \end{bmatrix}$, and then one step more produces a resulting upper matrix

$$\begin{bmatrix} D_1 & G_1p'_{V2} & G_1a'_{V2}p'_{V3} & G_1a'_{V2}a'_{V3}p'_{V4} \\ 0 & D_2 & G_2p'_{V3} & G_2a'_{V3}p'_{V4} \\ 0 & 0 & D_3 & G_3p'_{V4} \\ 0 & 0 & 0 & D_4 \end{bmatrix}, \tag{11.11}$$

to which should be added the upper part, column transformed and repartitioned by the unitary transformation matrices just derived. For example, after the second step, one shall have as a transformed full matrix

$$\left[\begin{array}{c|cc|cc} D_1 & G_1p'_{V2} + g_1h_2d'_{V2} & G_1a'_{V2} + g_1h_2q'_{V2} & g_1b_2h_3 & g_1b_2b_3h_4 \\ \hline 0 & D_2 & G_2 & g_2h_3 & g_2b_3h_4 \\ 0 & 0 & p_3Y_3 & d_3 & g_3h_4 \\ 0 & 0 & p_4a_3Y_3 & p_4q_3 & d_4 \end{array}\right]. \tag{11.12}$$

The column partitioning moves one notch to the left, etc. and the detailed formulas become messy. Much clearer notation and insight are obtained by moving to the global notation, as we do next.

The Combined Upper System

Using our globalized diagonal notation (e.g., $a := \text{diag}[a_i]$ and dummies where needed), we have

$$T = p(I - Za)^{-1}Zq + d + g(I - Z'b)^{-1}Z'h. \tag{11.13}$$

Let $V = d_V + p_V(I - Za_V)^{-1}Zq_V$ be a representation of a (required minimal) inner matrix that will make $T_u = TV'$ upper; then we must have

$$\begin{aligned}&(p(I - Za)^{-1}Zq + d)(d'_V + q'_V Z'(I - a'_V Z')^{-1} p'_V)\\&:= D + GZ'(I - a'_V Z')^{-1} p'_V\end{aligned} \tag{11.14}$$

anti-causal for $\{G, D\}$ to be determined, that is, making the causal part anti-causal. (See the treatment of external factorization in Chapter 7.) The mixed term splits as usual:

$$\begin{aligned}&(I - Za)^{-1}Zqq'_V Z'(I - a'_V Z')^{-1}\\&= (I - Za)^{-1}ZaY + Y + Ya'_V Z'(I - a'_V Z')^{-1},\end{aligned} \tag{11.15}$$

with Y satisfying the forward recursion: $Y^{\langle -1\rangle} = Z'YZ = qq'_V + aYa'_V$, yielding

$$\begin{cases} pYp'_V + dd'_V &= D,\\ aYp'_V + qd'_V &= 0,\\ pYa'_V + dq'_V &= G,\\ aYa'_V + qq'_V &= Y^{\langle -1\rangle},\end{cases} \tag{11.16}$$

in which Y is requested to be as small as possible. Putting this in matrix form (following the "natural" ordering used in the example) gives the forward recursion

$$\begin{bmatrix} d_k & p_k Y_k \\ q_k & a_k Y_k \end{bmatrix} \begin{bmatrix} d'_V & q'_V \\ p'_V & a'_V \end{bmatrix} = \begin{bmatrix} D_k & G_k \\ 0 & Y_{k+1} \end{bmatrix}. \tag{11.17}$$

This is recognized as a (block lower) LQ factorization, with Y left invertible and hence as small as possible (and D may be thought of the form $\begin{bmatrix} 0 & D^c \end{bmatrix}$ producing intermittent columns of zeros that may be dropped later).

It should now be obvious that the operations we did in the beginning of this section consist exactly of the recursions we just derived. The formula also shows that all quantities are derived directly from the given realizations, at least so far. We still have to take care of the transformation on the strictly upper part:

$$g(I - Z'b)^{-1}Z'h(d'_V + q'_V(I - Z'a'_V)^{-1}Z'p'_V). \tag{11.18}$$

This term appears to be irreducible in general and has to be added to the previous, yielding as the result of this first stage:

$$\begin{aligned}&g(I - Z'b)^{-1}Z'hd'_V + g(I - Z'b)^{-1}Z'hq'_V(I - Z'a'_V)^{-1}Z'p'_V\\&\qquad + D + G(I - Z'a'_V)^{-1}p'_V\\&= D + \begin{bmatrix} g & G \end{bmatrix}\left(I - Z'\begin{bmatrix} b & hq'_V \\ 0 & a'_V \end{bmatrix}\right)^{-1} Z'\begin{bmatrix} hd'_V \\ p'_V \end{bmatrix},\end{aligned} \tag{11.19}$$

in which, as usual, $Z' = \begin{bmatrix} Z' & 0 \\ 0 & Z' \end{bmatrix}$, where Z and Z' are just shift operators. In other words, the calculation so far just consists in computing D, G and the realization of V, with y an intermediate quantity.

The quasi-separable realization of T has now been converted to a regular realization of the upper (anti-causal) matrix $T_u = TV'$. Let us rename it, updating the upper realization as $T_u \sim_a \begin{bmatrix} b & h \\ g & d \end{bmatrix}$ with

$$\begin{cases} b & \leftarrow & \begin{bmatrix} b & hq'_V \\ 0 & a'_V \end{bmatrix}, \\ h & \leftarrow & \begin{bmatrix} hd'_V \\ p'_V \end{bmatrix}, \\ g & \leftarrow & \begin{bmatrix} g & G \end{bmatrix}, \\ d & \leftarrow & D. \end{cases} \tag{11.20}$$

Phase 2: Inner–Outer on the Upper Part

The next two phases work on the upper matrix and are intended to determine a complete inner-outer-inner decomposition of T_u, together with ranges and co-ranges:

$$T_u = \begin{bmatrix} v'_1 & v'_2 \end{bmatrix} \begin{bmatrix} 0 & 0 \\ 0 & R \end{bmatrix} \begin{bmatrix} U'_1 \\ U'_2 \end{bmatrix}, \tag{11.21}$$

in which v' and U' are upper co-inner and R is upper and upper invertible. This then goes in two steps: (1) determine a maximal left-co-inner (hence upper) v' such that $\begin{bmatrix} 0 \\ T_o \end{bmatrix} := vT_u$ is upper and right invertible, and then (2) determine a maximal inner (lower) U such that $T_oU = \begin{bmatrix} 0 & R \end{bmatrix}$ with R being upper invertible. Again, all operations are done on the realizations. The operations are very similar to what happened in the causal case as described in Chapter 8, except that now the operator is upper and the operations are all dual to the lower case. In fact, they would be identical if they were executed on the transpose T'_u.

The operation with v on T_u in the example starts on $\begin{bmatrix} h_1 \\ d_1 \end{bmatrix}$ and runs forward, just as in phase 1. We do not repeat the sequence of steps in detail; it is simply easier to work direct on the system representations:

$$vT_u = (d_v + p_v(I - Za_v)^{-1}Zq_v)(d + g(I - Z'b)^{-1}Z'h)? =?d_o + g_o(I - Z'b)^{-1}Z'h, \tag{11.22}$$

which, much as before (and as detailed before), works out to

$$\begin{bmatrix} p_{wk} & d_{wk} \\ a_{vk} & q_{vk} \\ p_{vk} & d_{vk} \end{bmatrix} \begin{bmatrix} y_k b_k & y_k h_k \\ g_k & d_k \end{bmatrix} = \begin{bmatrix} 0 & 0 \\ y_{k+1} & 0 \\ g_{ok} & d_{ok} \end{bmatrix}, \tag{11.23}$$

which is again a QR factorization in which d_{ok} is right invertible, and the intermediate y_{k+1} takes care of the transfer of a reduced row from stage k to $k+1$: it carries information on the kernel to the previous column, ending (in the example at stage 4) with a column of zeros and producing the desired $\begin{bmatrix} 0 \\ T_o \end{bmatrix} := \begin{bmatrix} v_1 \\ v_2 \end{bmatrix} T_u$, in which T_o is upper, with diagonal elements that are right invertible by construction, so that T_o itself is right causally invertible, and v partitions as indicated. The co-kernel of T_u (and of T) is clearly given by all $(xv_1')' = v_1x'$ with x an arbitrary dimension-compatible vector.

Phase 3: Computing the Range, the Co-kernel and R

This phase is very similar to the previous, except that the recursion will now move backward and the factorization will be an anti-causal outer–inner factorization, this time on T_o obtained in the previous step. So, we look for a maximal causal unitary U such that UT_o is still upper. Although the operations are much as before, here are the formulas. The goal is

$$(d_o + g(I - Z'b)^{-1}Z'h_o)(d_U + p_U(I - Za_U)^{-1}Zq_U)? =?d_R + g_R(I - Z'b)^{-1}Z'h_0, \tag{11.24}$$

leading to

$$\begin{bmatrix} b\widehat{y} & h_o \\ g\widehat{y} & d_o \end{bmatrix} \left[\begin{array}{c|cc} q_{U_1} & a_U & q_{U_2} \\ d_{U_1} & p_U & d_{U_2} \end{array}\right] = \left[\begin{array}{c|cc} 0 & \widehat{y}^{<+1>} & g_R \\ 0 & 0 & d_R \end{array}\right], \tag{11.25}$$

where $\hat{y}$ is again an intermediate quantity – of course different from before – and we have transposed rows and columns as in Step 1 to accommodate the order, in which the quantities arise in the matrix representation. This is now a backward QR factorization, in which d_R is made square invertible and $\hat{y}$ takes care of transmitting data to the next stage. This step then finally results in

$$\begin{bmatrix} 0 & 0 \\ 0 & R \end{bmatrix} = \begin{bmatrix} 0 \\ T_o \end{bmatrix} \begin{bmatrix} U_1 & U_2 \end{bmatrix} \tag{11.26}$$

and

$$T = \begin{bmatrix} v_1' & v_2' \end{bmatrix} \begin{bmatrix} 0 & 0 \\ 0 & R \end{bmatrix} \begin{bmatrix} U_1' \\ U_2' \end{bmatrix} V = v_2'RU_2'V, \tag{11.27}$$

where $\bigvee v_1'$ spans the co-kernel of T, $\bigvee(V'U_1)$ spans the kernel, $\bigvee v_2'$ spans the range, $\bigvee V'U_2$ spans the co-range and R is a square, upper and invertible matrix. Finally, the Moore–Penrose inverse is hence found to be

$$T^\dagger = V'U_2R^{-1}v_2 \tag{11.28}$$

by Theorem 11.1, obtained as a product of four unilateral transfer operators, each characterized by its (causal or anti-causal) realization.

The resulting algorithm hence boils down to three recursions: two forward (one "external" and the other "inner–outer" and one backward "outer–inner"). The first recursion establishes the full mixup of the dynamics of the inverse, while the last two reestablish the division between causal and anti-causal of the inverse.

The Algorithm

The first two forward recursions can be (and will be) executed by the so-called *lazy evaluation*, very much in the style of the classical Gaussian elimination. That is, in order and intertwined. For both, we have a forward progression of the state-space model ($1 \to 2 \to 3 \cdots$) with an intermediate matrix (which we have termed Y and y respectively) The model for the second recursion is only dependent on the original model at the current index level, and the model obtained in the first recursion at the same index. Indicating the first recursion by the roman digit "I" and the second by the roman digit "II," the full forward recursion, combining Phases 1 and 2 is hence I.1 $\to$ II.1 $\to$ I.2 $\to$ II.2 $\cdots$, whereby it might be remarked that there are only dependences between (for all relevant k) I.k and II.k, and then further I.k and I.$(k + 1)$ as well as II.k and II.$(k + 1)$, but not between II.k and I.$(k + 1)$, allowing for considerable pipelining in the algorithm. Only Phase 3 of the algorithm necessitates postprocessing of the resulting (nonsingular but not yet square) T_o via a theoretically unavoidable backward recursion. However, the latter can often be avoided if further information on the system is available, for example, that it is originally nonsingular, but that may require recourse to a different variant than the one presented here. Be all this as it may, the algorithm given is (1) numerically efficient as it is linear in the size of the original matrix and (2) numerically stable as it uses only orthogonal transformation. For further information on the numerics, see [16].

11.4 Doubly Infinitely Indexed Matrices*

In the case of doubly infinitely indexed matrices (hence also in the LTI case), the mechanics of finding the Moore–Penrose inverse is very much the same, but there are important differences in the interpretation of the results. For starters, some assumptions are needed concerning the "regularity" of the original state-space representation for T. Let us first look at the semi-infinitely indexed case, say starting at $k = 1$ and running to $+\infty$, a very common case in practice, and let us use the notation of Section 11.3. a, q, p, d, b, h, g have become the operators on an infinitely dimensional space, and boundedness must be considered. A first issue is the stability of the inverses $(I - Za)^{-1}$ and $(I - Z'b)^{-1}$. We already saw that we can define *one-sided* inverses, just by one-sided series expansions,

$$\left\{ \begin{array}{lcl} (I - Za)^{-1} & = & I + Za + Z^2 a^{<-1>} a + Z^3 a^{<-2>} a^{<-1>} a + \cdots , \\ (I - Z'b)^{-1} & = & I + Z'b + Z'^2 b^{<+1>} b + Z'^3 b^{<+2>} b^{<+1>} b + \cdots , \end{array} \right. \tag{11.29}$$

in which the one-sidedness of the lower part runs forward with causal shifts and that of the upper part runs backward with anti-causal shifts. The sums all contain finitely computable terms (provided one disposes of a way of computing the realizations all the way to ∞), but such series cannot, for example, be multiplied with each other, because in that case infinite sums would arise whose convergence would be doubtful. Even the multiplication of the upper part with the input is doubtful, as the input runs forward and the upper part's formal representation runs backward. One could think that such problems occur exclusively in the infinite indexing case, but that is wishful thinking: already finite matrices have to deal with potentially very large entries, and the study of stability in the infinitely indexed case can shed considerable light on the situation. Take, for example, $a = 2$ constant for $k \geq 2$: the kth term in the series expansion is then 2^k, which quickly diverges. Applying this to an input series that starts only at $k = 1$ is *in principle* no problem, as the sums remain finite, but the result gets exponentially out of bounds and very quickly looses meaning. In the case of the upper part, the situation is even worse, as the input series now has to have finite support for the sum to converge, but also then, there is a problem of significance.

So, in infinitely indexed cases, one has to somehow restrict the context. This is done, in the first place, by restricting the space to which possible inputs and eventually also possible outputs may belong, and hence requesting some kind of boundedness on the operators as well. Realization theory plays an important role in dealing with this issue. In this book, we mostly require the inputs to belong to an ℓ_2-space, namely a space that we denote $\ell_2^{\mathbf{m}}$, in which $\mathbf{m} = \{\ldots, m_{-1}, m_0, m_1, \ldots\}$ is a sequence of indices; the space is viewed as the direct sum of Euclidean spaces $\mathbf{R}^{m_k}$ or $\mathbf{C}^{m_k}$, with the overall norm given by $\|u\| = \sqrt{\sum_{k=-\infty}^{+\infty} \|u_k\|^2}$ and whereby only those infinitely indexed series belong to the space that have a convergent norm. Next, one usually requires the original operator T to be a bounded map from an input space $\mathbf{R}^{\mathbf{m}}$ to an output space $\mathbf{R}^{\mathbf{n}}$, but there is no guarantee that resulting operators, obtained through, for example, external or inner–outer factorization are bounded as well. To decide about this, one has to analyze further.

Using realization theory, assuming ℓ_2-boundedness and finite-state realizations at each index point k, one can investigate the Hankel map H_k at each position k, which will be a finite-dimensional operator, that is, an operator whose range and co-range have an equal, finite dimension, which we typically call η_k. H_k maps bounded strict past input sequences to future output sequences, which, according to the assumptions of boundedness of the operator T, will also all belong to an appropriate ℓ_2 space. It follows that one can choose a finite-dimensional orthonormal basis for either the reachability space or observability space at each index point k. (The proof is standard but requires some elementary Hilbert space theory; here we only make the result plausible.) Hence, at each k, $H_k = \mathbf{O}_k \mathbf{R}_k$, in which, for example, $\mathbf{R}_k$ is co-isometric and $\mathbf{O}_k$ consists of a finite basis of ℓ_2 columns, or, vice versa, $\mathbf{O}_k$ is isometric and $\mathbf{R}_k$ consists of a finite basis of ℓ_2 rows. Let us consider the first case: an orthonormal reachability basis as represented in the notation of Section 11.3. Realization theory results then in diagonal operators a and q for which $\begin{bmatrix} a & q \end{bmatrix}$ is isometric and $\mathbf{R} = (I - Za)^{-1} Zq$. In particular, a and q are bounded (again: elementary Hilbert space proof). Also, the central diagonal term d in T will automatically be bounded, as it is a partial operator of the bounded input–output map. What about the resulting p? For p, more or less a similar argument as for d holds: it can be viewed (with some work) as a suboperator of the original map. So, at this point, there is not much doubt about the boundedness of the operators in the input (or equally output) normal form.

However, this does not yet guarantee the convergence of the inverse $(I - Za)^{-1}$. Consider, for example, an input normal form with a co-isometric input pair

$\begin{bmatrix} a_k & q_k \end{bmatrix} = \begin{bmatrix} \sqrt{1-2^{-2|k|}} & 2^{-|k|} \end{bmatrix}$; then $[a^{\langle \pm \ell \rangle}]_k \sim 1$ for $0 < \ell \ll |k|$ and large $|k|$, and the lower diagonal, unilateral matrix $(I - Za)^{-1}$ fills up with an infinite number of entries almost equal to 1 when the indices become large: the Neumann series will not converge, at least not in norm (it would be somewhat of the type $I+Z+Z^2+\cdots$)! Still, the operator T may be bounded, in particular when p_k goes to zero fast enough. The global reachability operator becomes gradually singular in this case, making the state harder and harder to reach when k progresses (its norms tend to $\sqrt{k}2^{-|k|}$). To continue with the example, the pair $\begin{bmatrix} \sqrt{1-2^{-2|k|}} & 2^{-|k|} \end{bmatrix}$ can be completed to unitary:

$$V \sim_c \begin{bmatrix} \sqrt{1-2^{-2|k|}} & 2^{-|k|} \\ -2^{-|k|} & \sqrt{1-2^{-2|k|}} \end{bmatrix} \tag{11.30}$$

and defines a bounded operator V with unitary realization. *However, V as a global operator is not unitary; it is merely co-isometric*. This is a situation that cannot arise in the finite matrix case, namely a system with a unitary realization, but which is not unitary itself.

The situation can be remedied by requesting the state transition operator of the original operator to be uniformly exponentially stable, that is, requesting the spectral radius of Za to be strictly less than 1. This property is independent of the basis chosen, as long as the realization is minimal and allows the external factorization of Step 1 to go through. However, when reaching the inner–outer or outer–inner part (Steps 2 and 3), new convergence problems may come up. First, inner functions may turn up that have unitary realizations but are not inner – indicating the presence of special kernels – or R may turn out not be boundedly invertible, although outer. In that case, it has only a *dense causal range*. This situation may already arise with LTI systems, for example, the transfer function $T(z) = 1 - z$ would produce it: it suffices to have a zero on the unit circle. (People then talk about "lack of dichotomy.") All these delicate questions are explored in detail in [29].

11.5 Discussion Items

- *The difference between external and inner–outer*: The external recursion in Phase 1 (Eq. (11.17)) looks very much like the (transpose of the) inner–outer recursion in Phase 2 (Eq. (11.23)), although the effect these recursions have is very different. In Phase I, the algorithm converts lower to upper, while in Phase II, the algorithm separates the causal from the anti-causal part of the (Moore–Penrose) inverse, at least partially (because Phase 3 also contributes to that separation). In the first (external) case, the first quantity computed is the connecting matrix Y_{k+1}, while in the second case (inner–outer), it is the output matrix d_{ok}. Perhaps not surprisingly all together: in the first phase, there is an arrow reversal in the state propagation, while in the second, the reversal is in the input–output connection.

11.6 Notes

1. Original proof that the Moore–Penrose inverse of a quasi-separable system involves inner and outer functions of the same degree as the original was given in Alle-Jan van der Veen's thesis [69], reproduced in [29]. Further developments and proofs can be found in [37] and [16], from which this example is taken. The term "quasi-separable" was, to the best of our knowledge, introduced by the late Israel Gohberg to mark the (substantial) generalization with respect to semi-separable matrices; see [36].
2. It is a remarkable fact that in this Euclidean-norm case, the minimum norm, minimum error is linearly dependent on the input: it is produced by the linear operator $T^{\dagger}$. This fact has made the Moore–Penrose inverse extremely useful and common. When another norm needs to be minimized (like the ℓ_1 norm in many control applications, or the ℓ_∞ norm in design applications), a very different approach has to be followed. In view of the very different geometric appearance of the unit ball in an $\mathcal{R}^n$ space, the ℓ_2 minimum may be far away from the ℓ_1 or the ℓ_∞ minimum, because the latter two favor certain directions, while the ℓ_2 norm has a uniform unit ball. In particular, the ℓ_1 norm will favor directions along base lines or base planes and therefore yield sparse solutions, which can be very useful in a number of applications.
3. An alternative method to obtain the Moore–Penrose inverse and full information on ranges, kernels and co-kernels is the SVD. Inner–outer or outer–inner factorizations are recursive methods of the QR type (depending on the ordering of the matrices and whether one works on columns or rows) and hence do not compute eigenvalues nor singular values. They also do not have the same level of accuracy as the SVD, although they are "backward stable," since only orthonormal transformations are used at each stage of the computation, assuming that a realization is given. If a realization still has to be determined from input–output data, then further numerical errors may ensue, but in many practical problems, the realization is actually given rather than the matrix entries themselves. An extensive discussion on realization algorithms can be found in Chapter 4. Although the SVD gives higher accuracy, it does not provide the result in a quasi-separable form. However, in most cases, there is an in-between possibility, whereby the global operations are recursive, while the local orthogonalizations use the SVD. Whether such an approach is numerically valuable depends on the convergence properties of the realization and in particular the spectral radius of the transition matrices Za and $Z'b$, as discussed in the text.
4. The SVD is often used to provide for low-rank approximations to the original system, whereby "small" singular values are put to zero, thereby reducing the rank of the representation. As stated in the previous remark, the SVD looses the quasi-separable form. An in-between and often attractive solution is to use "model reduction": reducing the local degrees (dimensions of the state space) in the quasi-separable representation before performing inner–outer factorizations for Moore–Penrose inversion.

Part II

Further Contributions to Matrix Theory

12 LU (Spectral) Factorization

The factorization of a matrix into a lower and an upper part, also called LU factorization, is one of the oldest and most basic problems in matrix algebra. It is closely related to Gaussian elimination and can, with some difficulty, be used to solve linear systems of equations. However, in system theory, LU factorization acquires special meaning: it is identical to the ubiquitous problem of spectral factorization, used to solve a variety of problems, such as a unitary embedding of a contractive transfer function, or optimal stochastic filtering (sometimes called "Wiener filtering"). The traditional LU-factorization method using Gaussian elimination suffers from serious numerical instabilities due to the fact that the factorization may not exist. In this chapter, we give necessary and sufficient conditions for existence plus a novel, numerically stable algorithm to compute the LU factorization of a mixed causal–anti-causal system with given realizations, using exclusively stable numerical operations (inner–outer factorization and scaling).

Menu

Hors d'oeuvre
Gaussian Elimination and LU Factorization

Main Course
LU Factorization for Quasi-separable Operators Using Outer–Inner Factorization

Dessert
A Tridiagonal Example

12.1 Introduction

Gaussian elimination is perhaps the "mother" of all numerical matrix operations. You know it from your first course in algebra; for example, when you want to solve a system of equations, say

$$\begin{cases} x + 3y &= 1, \\ 5x - y &= 3, \end{cases} \tag{12.1}$$

you start out by multiplying the first equation with 5 and subtracting it from the second, to produce a new equivalent system

$$\begin{cases} x + 3y &= 1, \\ -16y &= -2, \end{cases} \tag{12.2}$$

so that the second equation is solved as $y = 1/8$. Backsubstituting y in the first produces the solution $x = 5/8$. It is, of course, clear that the procedure cannot be executed in the order shown when the first coefficient is zero, and in that case, a reshuffling of the equations is necessary. When such an anomaly does not occur and the original matrix is invertible, then a streamlined (in order) way of performing the Gaussian elimination leads to a recursive factorization of the original matrix, let us call it T, into two factors L and U as $T = LU$, in which L is lower triangular and invertible, while U is upper triangular and invertible. We briefly discuss this classical method.

Gaussian elimination and LU factorization

Let T be a matrix with scalar entries of dimension $n \times n$ indexed starting with 1.

The first step in the LU factorization.
Assume T_1^1 is nonzero. It is called the first *pivot*. Let

$$L1 := \begin{bmatrix} T_1^1 & 0 \\ T_{2:n}^1 & I \end{bmatrix}, \; U1 := \begin{bmatrix} 1 & [T_1^1]^{-1} T_1^{2:n} \\ 0 & I \end{bmatrix} \tag{12.3}$$

and

$$\text{Tr2} := \begin{bmatrix} 1 & 0 \\ 0 & T2 \end{bmatrix}, \text{ with } T2 := \begin{bmatrix} T_{2:n}^{2:n} - [L1]_{2:n}^1 [U1]_1^{2:n} \end{bmatrix}. \tag{12.4}$$

Then

$$T = L1 \; \text{Tr2} \; U1 = \begin{bmatrix} T_1^1 & 0 \\ T_{2:n}^1 & I \end{bmatrix} \begin{bmatrix} 1 & 0 \\ 0 & T2 \end{bmatrix} \begin{bmatrix} 1 & [T_1^1]^{-1} T_1^{2:n} \\ 0 & I \end{bmatrix} \tag{12.5}$$

produces a product of a lower triangular matrix $L1$, a matrix with a reduced dimension $T2$ and an upper triangular matrix $U1$. It should be remarked that, in the traditional algorithm, $L1$ and $U1$ are produced at this point, while the production of entries in $T2$ is postponed to the next steps: the product $[L1]_{2:n}^1 [U1]_1^{2:n}$ just consists of one multiplication for each entry in $T2$: $[T2]_i^j = [L1]_i^1 [U1]_1^j$ and will be done at stage $\min(i, j)$. (This is called *lazy evaluation*: evaluate when needed!) $T2$ itself does not have to be computed explicitly, as it gets reduced in further steps. To keep with our indexing conventions, we index $T2$ starting with 2; notice also that Tr2 and $T2$ are just logical names for matrices occurring in step 2 of the recursion.

The general step.
Assume now we have computed k steps and are moving now to step $k + 1$. At the start of this next step, we have reached

$$T = \text{Lk Trk Uk} = \begin{bmatrix} L_{1:k}^{1:k} & 0 \\ L_{k+1:n}^{1:k} & I \end{bmatrix} \begin{bmatrix} I & 0 \\ 0 & T(k+1) \end{bmatrix} \begin{bmatrix} I & U_{1:k}^{k+1:n} \\ 0 & I \end{bmatrix}, \tag{12.6}$$

in which $T(k + 1) = T_{k+1:n}^{k+1:n} - [\text{Lk}]_{k+1:n}^{1:k} [\text{Uk}]_{1:k}^{k+1:n}$, with the later product consisting, element-wise, of the inner product of a row of Lk with a column of Uk. Remark that the stage reached so far actually already defines a whole collection of final elements in L (namely its first k columns) and U (its first k rows). In this step, both the $(k+1)$th column of L and the $(k+1)$th row of U will be defined, as well as all the data needed for the next step.

The computation of the $(k+1)$th step now starts out with the calculation of the $(k+1)$th *pivot*, which we give a special symbol s_{k+1}:

$$s_{k+1} = T_{k+1}^{k+1} - [\mathrm{Lk}]_{k+1}^{1:k}[\mathrm{Uk}]_{1:k}^{k+1}. \tag{12.7}$$

It will also be the top-left element of $T(k+1)$: $[T(k+1)]_{k+1}^{k+1} = s_{k+1}$. The LU factorization will be able to continue (without shuffling of further rows) if and only if $s_{k+1} \neq 0$.

The $(k+1)$th column of L is now computed as $L_{1:k}^{k+1} = 0$ and $L_{k+1:n}^{k+1} = T_{k+1:n}^{k+1} - [\mathrm{Lk}]_{k+1:n}^{1:k}[\mathrm{Uk}]_{1:k}^{k+1}$, that is, the original matrix element minus an inner product of an already existing subrow of L and subcolumn of U. Similarly, the $(k+1)$th row of U will be computed from the existing matrix and an inner product of already computed subrows and subcolumns, but now a division with the pivot is necessary:

$$\begin{aligned} &U_{k+1}^{1:k} = 0,\ U_{k+1}^{k+1} := 1, \\ &U_{k+1}^{k+1:n} = s_{k+1}^{-1}\left(T_{k+1}^{k+1:n} - [\mathrm{Lk}]_{k+1}^{1:k}[\mathrm{Uk}]_{1:k}^{k+1:n}\right). \end{aligned} \tag{12.8}$$

The result of this step is then

$$T = L(k+1)\mathrm{Tr}(k+1)U(k+1), \tag{12.9}$$

in which

$$\mathrm{L(k+1)} = \left[\begin{array}{c|cc} L_{1:k}^{1:k} & 0 & 0 \\ \hline L_{k+1}^{1:k} & s_{k+1} & 0 \\ L_{k+2:n}^{1:k} & L_{k+2:n}^{k+1} & I \end{array}\right] \tag{12.10}$$

(only the right under block is actually computed, i.e., the $(k+1)$th column),

$$\mathrm{U(k+1)} = \left[\begin{array}{c|cc} U_{1:k}^{1:k} & U_{1:k}^{k+1} & U_{1:k}^{k+1:n} \\ \hline 0 & 1 & U_{k+1}^{k+1:n} \\ 0 & 0 & I \end{array}\right] \tag{12.11}$$

(and again only the bottom-right block is computed, i.e., the $(k+1)$th row), and finally

$$T(k+2) = T_{k+2:n}^{k+2:n} - L_{k+2:n}^{1:k+1}U_{1:k+1}^{k+2:n}, \tag{12.12}$$

where the entries of L and U have already been computed in this and previous steps.

Pivots and Schur Complements

The key quantity in the LU factorization is hence the pivot. The pivot is actually the Schur complement of $T_{1:k}^{1:k}$ in $T_{1:k+1}^{1:k+1}$. A quick reminder: let T be divided into submatrices $T = \begin{bmatrix} A & B \\ C & D \end{bmatrix}$, with A square and invertible; then T factors as

$$\begin{bmatrix} A & B \\ C & D \end{bmatrix} = \begin{bmatrix} A & 0 \\ C & I \end{bmatrix}\begin{bmatrix} I & 0 \\ 0 & S \end{bmatrix}\begin{bmatrix} I & A^{-1}B \\ 0 & I \end{bmatrix}, \tag{12.13}$$

with $S = D - CA^{-1}B$ being the Schur complement of A in T. T is invertible when both A and S are, and we have $\det T = \det A \det S$, but these conditions are, of course, not necessary, since A may not be invertible, in which case S does not exist. In the case of a straight LU factorization, we shall have that all the principal submatrices $T_{1:k}^{1:k}$ are invertible. Clearly, also $\det T = s_1 \cdots s_n$ (with $s_1 := T_{1,1}$).

Further Considerations

When T is *strictly positive definite*, that is, when T is symmetric (in the complex case, Hermitian) – $T = T'$ – and such that for all vectors $x \neq 0$, $x'Tx > 0$, then the factorization procedure can be modified slightly to produce $T = LL'$, called a *Cholesky factorization*. All pivots are positive in this case, and instead of leaving the new column of L in the $(k+1)$th undisturbed, one divides it with $\sqrt{s_{k+1}}$, and also the $(k+1)$th row gets divided by $\sqrt{s_{k+1}}$. L is in a sense a generalized square root of T – the exact square root of the positive definite T exists also: $T = (T^{1/2})^2$, with $T^{1/2}$ square symmetric. In most applications, an LL' factorization suffices.

In case T is only positive definite but not strictly, then the procedure can be adapted to yield a nonsquare factor L, thanks to the fact that in the positive definite case, a nonzero pivot leads to the corresponding column and row being zero, so that the factorization can be reduced as that point.

Instead of putting the pivot as diagonal in the L factor and dividing the rows of U by it, one can in general also leave the pivot in the middle and divide also the columns of L by it to produce a so-called LDU factorization: $T = LDU$, with unit main diagonals in L and U, and $D = \text{diag}(s_k)$. This new L is the old L multiplied to the right with D^{-1}. Needless to say, the extra division may induce more calculations and may hence be uneconomical, but that may depend on subsequent calculations.

Finally, the LU factorization also works for block matrices, in which case, the pivots become matrices themselves and must all be nonsingular for the method to proceed. In the classical case, this may not be all that interesting, but in our quasi-separable case, this will be how we proceed, because often we are actually *given* block representations to start with.

12.2 LU Factorization for Quasi-separable Matrices

All variants of classical LU factorizations suffer from the need to compute partial Schur complements (pivots) explicitly and then to use potentially inaccurate values further on in the recursion. This is the main cause of numerical instability. The canonical factorizations of Chapters 7 and 8 offer a very different approach, which only uses numerically stable factorizations in the recursion. Local Schur complements are then easily and stably derived from them but not used in the recursion.

We assume that we are given a quasi-separable system representation consisting of a lower (causal) and an upper (anti causal) part:

$$T = C_c(I - ZA_c)^{-1}ZB_c + D + C_a(I - Z'A_a)^{-1}Z'B_a. \tag{12.14}$$

The goal is to determine an LU factorization for T in the classical matrix sense, when it exists, or give necessary and sufficient conditions of existence in general.

To ensure uniqueness and meaningfulness, some regularity conditions on the factors must be imposed. An LU factorization of a system matrix only makes sense when each of the factors has a proper (and bounded) inverse of the same type, causal for the L-factor and anti-causal for the U-factor – the original purpose of LU factorization was the system inversion of a system that was already presumed to be invertible, but many more recent applications also require multiplicative causal–anti-causal decompositions of the inverse as well as of the original. Therefore, the following (classical) definitions:

Definition 12.1 We say that an operator T possesses an LU factorization (or, equivalently, a spectral factorization) if and only if there exist an outer operator L and a conjugate-outer operator U such that $T = LU$.

This definition holds for both finitely or infinitely indexed systems. To avoid borderline cases in the infinitely indexed case, we strengthen this definition and require the outer factors to have bounded inverses. (Since outer only requires dense range, the boundedness of the inverse is not guaranteed in general. In the finitely indexed case, the existence of the bounded inverse is automatic.)

Definition 12.2 We say that an operator T possesses a strict LU factorization if and only if there exist an outer system L with bounded inverse and a conjugate-outer system U with bounded inverse such that $T = LU$.

It is not too hard to show, but outside the scope of the treatment here, that such a strict LU factorization is necessarily minimal (i.e., no nonminimal states are introduced in the product when minimal realizations for each of the factors are used).

The strict LU-factorization problem is solved when necessary and sufficient existence conditions for the factors L and U are obtained, and an efficient and numerically stable algorithm to find them is produced. That is what we now proceed to do. We concentrate our efforts on U, for L a similar but dual strategy will be valid. (We shall see that U can be determined independently from L.) As we require U to be outer invertible, we can also and without loss of generality normalize the instantaneous term D_U to be unity ($D_U = I$).

Preliminaries

The first observation is that if T has a strict LU factorization and W is a minimal inner function such that TW is lower, then also UW is lower, for $UW = L^{-1}(TW)$, a product of lower operators by assumption. (In the LTI theory, W' characterizes the poles of U.) Let us assume that T is such that its anti-causal part is uniformly reachable; then because of soon-to-be-proven Proposition 12.3, W will be bi-inner (i.e., causal and unitary). W is easily determined when the original $\begin{bmatrix} A_a & B_a \end{bmatrix}$ of T is in the input normal form (which we just assume) so that $W = D_W + C_W(I - ZA_W)^{-1}ZB_W$, with $A'_W = A_a$, $C'_W = B_a$ and $\begin{bmatrix} A_W & B_W \\ C_W & D_W \end{bmatrix}$ a unitary realization for W with appropriately chosen B_W and D_W (for more explanation, see Chapter 7). Next we define $V := D_V + C_V(I - ZA_V)^{-1}ZB_V$ as the inner factor in an inner–outer factorization of $TW = T_oV$, with T_o outer, to be obtained by an outer–inner factorization algorithm on TW. Because of the assumed properties of strict invertibility, V can also be characterized by the relation $V(TW)^{-1} = T_o^{-1}$ or, to put it differently, V is the smallest inner function that pushes $(TW)^{-1}$ back to causality. Because of the independent interest of this property, we state it as a separate proposition.

Proposition 12.3 *[Only for infinitely indexed systems, in the finite matrix case, the property is automatic.]*

Let T be a boundedly invertible operator such that its anti-causal part is uniformly reachable (the latter can be assumed to be in the input normal form), and let $T_u = TW$, where W is the minimal inner factor that makes T_u causal. Then W is bi-inner. In addition, in the outer–inner factorization $T_u = T_oV$ of T_u, V is bi-inner as well.

Proof

In the previous section, we already indicated that W is inner by construction,[1] with realization (assuming the input normal form for the anti-causal part T_a of T):

$$W \sim_c \begin{bmatrix} A_a' & B_W \\ B_a' & D_W \end{bmatrix}. \tag{12.15}$$

From the outer–inner factorization theory, we also know that V will be bi-inner if and only if $\ker(T_u\cdot) = 0$ (see, e.g., [29, p. 150]). But T_u is invertible with T and hence the condition is satisfied. □

As a consequence A_V will be u.e.s. (see [29, p. 125]). A further property needed in the sequel is that of *maximal phase systems*.

Definition 12.4 We say that a causal system T given by a minimal realization $T = D + C(I - ZA)^{-1}ZB$ is maximal phase if and only if it has a bounded anti-causal inverse (the boundedness condition being only needed for infinitely indexed systems.)

The following characterization of a maximum phase system holds:

Proposition 12.5 *The causal system T with (bounded and u.e.s.) minimal realization $T = D+C(I-ZA)^{-1}ZB$ is maximal phase system if and only if (1) $\begin{bmatrix} A & B \\ C & D \end{bmatrix}$ is square and invertible, and (2)[only for infinitely indexed system] $\widehat{A}$ in $\begin{bmatrix} \widehat{A} & \widehat{B} \\ \widehat{C} & \widehat{D} \end{bmatrix} = \begin{bmatrix} A & B \\ C & D \end{bmatrix}^{-1}$ is u.e.s.*

Proof

The "if" part is easily checked by computing $TT^{-1} = T^{-1}T = I$ with $T^{-1} := \widehat{D}+\widehat{C}(I - Z'\widehat{A})^{-1}Z'\widehat{B}$ together [in the case of infinitely indexed systems] with the fact that the conditions stated for the realization of T^{-1} makes it bounded and u.e.s.

[1] Because it has a unitary realization!

For the converse: Suppose that T has a bounded anti-causal inverse T^{-1}, then we have to show that the realization of T is square and invertible. We do this by first deriving isomorphic minimal realizations for T and T^{-1} on the basis of their observability spaces, and then constructing invertible transition maps at each index k. Focus on any index k and consider the decomposition in past and future spaces with respect to the index k:

$$T := \begin{bmatrix} E_k & 0 \\ H_k & F_k \end{bmatrix}, \tag{12.16}$$

with corresponding decomposition of the input and output spaces as $u := \begin{bmatrix} u_p \\ u_f \end{bmatrix}$ and $y := \begin{bmatrix} y_p \\ y_f \end{bmatrix}$, and corresponding projection operators Π_p and Π_f with respect to k. We can now establish the isomorphism between the observability spaces of T and T^{-1} at each arbitrary stage k as follows. Let u_p be a strictly past input of T with respect to k and let (still with respect to k) $y = \begin{bmatrix} y_p \\ y_f \end{bmatrix} = T \begin{bmatrix} u_p \\ 0 \end{bmatrix}$. y_f is then a "natural response" of T at k. Next, consider $\begin{bmatrix} u_p \\ 0 \end{bmatrix} - T^{-1} \begin{bmatrix} y_p \\ 0 \end{bmatrix}$; then because T^{-1} is anti-causal, this input will belong to the strict past and be of the form $\begin{bmatrix} \widetilde{u}_p \\ 0 \end{bmatrix}$. The isomorphism sought is $\widetilde{u}_p \leftrightarrow y_f$, which is an isomorphism because $T \begin{bmatrix} \widetilde{u}_p \\ 0 \end{bmatrix} = \begin{bmatrix} 0 \\ y_f \end{bmatrix}$ and T has a bounded inverse. Hence $\widetilde{u}_p$ is a natural response to y_f for T^{-1} (at the present index point k for the direct system, and $k-1$ for the inverse system). Supposing that an orthonormal basis is chosen for the space $\{\widetilde{u}_p\}$, we shall have $\|\widetilde{u}_p\|_2 = \|x_k\|_2$. Let now $y_f := \begin{bmatrix} y_{f;k} \\ y_{f;k+1,:} \end{bmatrix}$, the natural response to x_k. Applying a single impulse $u_k \delta_k$ to T produces an additional future response, which we denote as $\begin{bmatrix} h_k \\ h_{k+1,:} \end{bmatrix}$, so that the total response to the input $\begin{bmatrix} \widetilde{u}_p \\ u_k \\ 0 \end{bmatrix}$ is $\begin{bmatrix} 0 \\ y_{f;k} + h_k \\ y_{f;k+1,:} + h_{k+1,:} \end{bmatrix}$. From the definition of the observability operator $\mathcal{O}_{k+1}$ and the assumption that we deal with a minimal realization, so that $\mathcal{O}_{k+1}$ has a left inverse, we now have $x_{k+1} = \mathcal{O}_{k+1}^{\dagger}(y_{f;k+1,:} + h_{k+1,:})$, again an isomorphism, because $y_{f;k+1,:} + h_{k+1,:}$ is, again, a natural response and belongs to the range of $\mathcal{O}_{k+1}$. From the construction, we have

$$\begin{bmatrix} \widetilde{u}_p \\ u_k \\ 0 \end{bmatrix} = T^{-1} \begin{bmatrix} 0 \\ y_{f;k} + h_k \\ y_{f;k+1,:} + h_{k+1,:} \end{bmatrix}, \tag{12.17}$$

and because of the isomorphism $\widetilde{u}_p \sim x_k$ and $y_{f;k+1,:} + h_{k+1,:} \sim x_{k+1}$, this produces the nonsingular map

$$\begin{bmatrix} x_k \\ u_k \end{bmatrix} = \begin{bmatrix} \widehat{A}_k & \widehat{B}_k \\ \widehat{C}_k & \widehat{D}_k \end{bmatrix} \begin{bmatrix} x_{k+1} \\ y_k \end{bmatrix}, \tag{12.18}$$

which clearly is a right inverse of the original. The result then follows either from the observed isomorphisms in the construction just given, or from the same (but dual) reasoning, but now applied on the reachability part. (Remark: the state x_{k+1} in the original system would normally be denoted as x_k in the inverse system, according to our conventions for causal and anti-causal systems.) □

The simplest maximum phase systems are the inners. They all have anti-causal inverses, but there are many others, of course.

The Method

We are now ready to formulate the factorization algorithm. It consists of the following three steps:

Step 1: Determination of W

W is defined as the minimal bi-inner factor that makes $T_u := TW$ causal: $W \sim_c \begin{bmatrix} A_a' & B_W \\ B_a' & D_W \end{bmatrix}$ when the anti-causal part T_a of T is in the input normal form.

Step 2: Determine V

V is defined as the inner factor in the outer–inner factorization of $T_u = T_o V$. (It is the crucial step, which, in the LTI situation, results in the matching of zeros to poles. It has a corresponding matching effect in the more general case considered here.) A minimal realization for T_u is found from the product TW and easily computed as

$$\begin{bmatrix} A_u & B_u \\ C_u & D_u \end{bmatrix} := \left[\begin{array}{cc|c} A_c & B_c B_a' & B_c D_W \\ 0 & A_a' & B_W \\ \hline C_c & DB_a' + C_a A_a' & DD_W + C_a B_W \end{array}\right]. \tag{12.19}$$

(Similarly, one easily checks that $\begin{bmatrix} M_c & \\ & I \end{bmatrix}$ is the reachability Gramian of T_u, with M_c being the reachability Gramian of T_c.) A square-root algorithm will now produce the desired outer–inner factorization $T_u = T_o V$:

$$\begin{bmatrix} A_u Y & B_u \\ C_u Y & D_u \end{bmatrix} \cdot \left[\begin{array}{c|cc} C_n' & A_V' & C_V' \\ D_n' & B_V' & D_V' \end{array}\right] = \left[\begin{array}{c|cc} 0 & Y^{\langle -1 \rangle} & B_o \\ 0 & 0 & D_o \end{array}\right], \tag{12.20}$$

where the echelon form on the right-hand side is obtained by compressing the columns to the right, starting with the last row. The algorithm shows that no factorization exists, when any of the following situations occur:

1. There is an index at which D_{ok} is nonsquare. T_o is then not strictly outer, and an invertible U does not exist.
2. C_n and D_n do not vanish (there is a kernel, T is not invertible).
   D_o has to be square invertible, and $Y^{\langle -1 \rangle}$ is full rank by definition. The computation is a forward recursion, with Y_{k+1} being computed from Y_k.
   A third, more technical condition that only applies to infinitely indexed systems could force an abortion as well, and that is:
3. [Only for infinitely indexed systems:] A_V is not u.e.s. As A_V is contractive by construction, this needs at least $\lim\sup_{k\to\infty} \|A_{Vk}\| = 1$, but if T is regular, then T_u will also be regular, and V then has to be bi-inner, so if it is not, then T is not boundedly invertible.

We can proceed (in a numerical stable way) to the next step whether or not these terminating conditions occur. (In the latter case, the recursion remains valid and yields further information on the operator T, which, however, is guaranteed not to have a regular LU factorization.)

Step 3: Determine the Result

We start by remarking that Y, being defined by

$$Y^{\langle -1 \rangle} = A_u Y A_V' + B_u B_V', \tag{12.21}$$

splits in two components $Y = \begin{bmatrix} Y_1 \\ Y_2 \end{bmatrix}$, in which $R := Y_2$ satisfies the recursion $R^{\langle -1 \rangle} = A_W R A_V' + B_W B_V'$. The result now hinges on properties shown in Proposition 12.6 and Theorem 12.7.

Proposition 12.6 *Suppose that $T = LU$ is a strict LU factorization with U normalized; then there will exist a (conformal) diagonal operator F such that*

$$U \sim_a \begin{bmatrix} A_a & B_a \\ F & I \end{bmatrix}. \tag{12.22}$$

Proof

The proof is a direct consequence of $U = L^{-1}T$, in which L^{-1} is causal. It follows that anti-causal U has the same reachability space as $T_a := C_a(I - Z'A_a)^{-1}Z'B_a$. Hence, $F := C_U$ is the only parameter that still remains to be determined in a realization for U, which borrows the reachability data A_a and B_a from T_a. □

The Main Result

We now have all the ingredients for the main result.

Theorem 12.7 *Suppose that T is a regular quasi-separable operator, W is the minimal bi-inner such that $T_u = TW$ is causal (lower), $T_u = T_o V$ is an outer–inner factorization, with V being a bi-inner right factor, and suppose that R defined by the Lyapunov–Stein recursion*

$$R^{\langle -1\rangle} = A_W R A_V' + B_W B_V' \tag{12.23}$$

is square, bounded and boundedly invertible, then a strict LU factorization exists with the realization $U \sim \begin{bmatrix} A_a & B_a \\ F & I \end{bmatrix}$, *where F is given by*

$$F = -\left(C_W R A_V' + D_W B_V'\right)(R^{\langle -1\rangle})^{-1}. \tag{12.24}$$

Conversely, if a strict LU factorization exists, then U will have a realization as given, and the solution R of the Lyapunov–Stein equation will be bounded and boundedly invertible.

Proof

Only if: suppose the strict LU factorization of $T = LU$ exists, and let W and V be defined as stated. From the definition of W, and assuming U normalized with $D_U := I$, it follows that U must be of the form (12.22) for some diagonal operator F (Proposition 12.6). Next, $T_o^{-1}L = VW^{-1}U^{-1}$, and V can be identified as the smallest inner operator that forces $W^{-1}U^{-1}$ back to lower (or causality if you wish) from the left. This makes V necessarily bi-inner, by Proposition 12.3. Given the expression for U in terms of the still unknown F and the fact that a realization for $W^{-1} = W'$ is given by

$$W^{-1} \sim \begin{bmatrix} A_W' & C_W' \\ B_W' & D_W' \end{bmatrix}, \tag{12.25}$$

we obtain

$$\begin{aligned} W^{-1}U^{-1} &= \left(D_W' + B_W'(I - Z'A_W')^{-1}Z'C_W'\right)\cdot\left(I - F[I - Z'(A_W' - C_W'F)]^{-1}Z'C_W^*\right) \\ &= D_W' + (B_W' - D_W'F)[I - Z'(A_W' - C_W'F)]^{-1}Z'C_W', \end{aligned} \tag{12.26}$$

where the necessary cancelations take place as expected. UW is causal minimal, because the reachability space of U is certainly contained in that of T_a and cannot be smaller (we skip the simple technical proof). Since $(UW)^{-1}$ is anti-causal, UW is a maximal phase operator, and Proposition 12.5 applies, guaranteeing the minimality of the given realization for $W^{-1}U^{-1}$, because

$$UW \sim_c \begin{bmatrix} A_W & B_W \\ C_W + FA_W & D_W + FB_W \end{bmatrix} = \begin{bmatrix} I & 0 \\ F & I \end{bmatrix}\begin{bmatrix} A_W & B_W \\ C_W & D_W \end{bmatrix} \tag{12.27}$$

whose inverse is

$$\begin{bmatrix} A_W' & C_W' \\ B_W' & D_W' \end{bmatrix}\begin{bmatrix} I & 0 \\ -F & I \end{bmatrix} = \begin{bmatrix} A_W' - C_W'F & C_W' \\ B_W' - D_W'F & D_W' \end{bmatrix}, \tag{12.28}$$

which is the realization given for $W^{-1}U^{-1}$ and hence is minimal. The inner V brings this minimal realization to lower from the left; hence, it must have a reachability pair that matches the observability pair of $W^{-1}U^{-1}$; see Chapter 7 on external factorizations. More precisely, there exists a boundedly invertible state transformation R such that (writing shorthand $R^{-\langle -1\rangle}$ for $(R^{\langle -1\rangle})^{-1}$)

$$\begin{bmatrix} RA'_V R^{-\langle -1\rangle} \\ B'_V R^{-\langle -1\rangle} \end{bmatrix} = \begin{bmatrix} A'_W & C'_W \\ B'_W & D'_W \end{bmatrix} \begin{bmatrix} I \\ -F \end{bmatrix}. \tag{12.29}$$

Inverting the unitary factor produces

$$\begin{cases} R^{\langle -1\rangle} &= A_W RA'_V + B_W B'_V, \\ F &= -\left(C_W RA'_V + D_W B'_V\right) R^{-\langle -1\rangle}, \end{cases} \tag{12.30}$$

as announced. This shows the “only if” part.

If: now we are given T from which W and V are derived as given before the statement of the theorem, and we assume that we have found a boundedly invertible R that solves the Lyapunov–Stein Eq. (12.23). We have to show that U as given in the theorem is the (normalized) right factor in the LU factorization. First, UW will be lower, from the definition of W and the reachability pair of U. Next consider U^{-1}. From Eq. (12.23) and the definition for F, we obtain that $A'_W - C'_W F$ is u.e.s. just like V. (The equation for R and the definition for F actually give the observability pair of U^{-1} as an upper operator.) Now it remains to be shown that $L := TU^{-1}$ is a lower operator with lower inverse. Let us, for that matter, *define* $L = T_o[V(UW)^{-1}]$; then the definition of V, from which R derives, results in the causality and the outerness of the factor $[V(UW)^{-1}]$. Hence, L itself is outer as the product of two outer factors, and the factorization $T = LU$ follows. □

$R = Y_2$ so obtained must be bounded (this is automatic, because R satisfies a convergent Lyapunov–Stein equation) and boundedly invertible (this has to be checked recursively while progressing). If this condition is satisfied with a uniform bound on the local norms, then the LU factorization exists.

The End Game

Step 4: Compute L, the Pivots and the Inverses of the Factors

At this point, we have already obtained the right factor U explicitly:

$$U = I + F(I - Z'A_a)^{-1}Z'B_a, \tag{12.31}$$

with $F = -\left(C_W RA'_V + D_W B'_V\right)\left(R^{\langle -1\rangle}\right)^{-1}$ and inverse

$$U^{-1} = I - F\left(I - Z'(A_a - B_a F)\right)^{-1} Z'B_a = I - FR^{\langle -1\rangle}(I - Z'A'_V)^{-1}Z'R^{-1}B_a. \tag{12.32}$$

The latter formula shows the boundedness of U^{-1} explicitly; provided V is inner, then A'_V is u.e.s. Hence, the bi-innerness of V assures the strict outerness of U.

To compute the left factor L from the work so far (it could also be computed directly by a dual procedure), we evaluate directly $L = T_u(W^{-1}U^{-1})$, in terms of realizations:

$$\begin{aligned} L &= \left(D_u + C_u(I - ZA_u)^{-1}ZB_u\right)\cdot\left(D'_W + B'_V(I - Z'A'_V)^{-1}R^{-1}C'_W\right) \\ &= (C_u Y R^{-1}C'_W + D_u D'_W) + C_u(I - ZA_u)^{-1}Z(A_u Y R^{-1}C'_W + B_u D'_W) \end{aligned} \tag{12.33}$$

(remark the occurrence of Y again), because $Y_2 = R$, and after introducing the decomposition of T_u, we obtain the realization

$$L \sim_c \left[\begin{array}{cc|c} A_c & B_c B_a' & (A_c Y_1 R^{-1} + B_c B_a')C_W' + B_c D_W D_W' \\ 0 & A_a' & A_a' C_W' + B_W D_W' \\ \hline C_c & DB_a' + C_a A_a' & (C_c Y_1 R^{-1} + DB_a' + C_a A_a')C_W' + (DD_W + C_a B_W)D_W' \end{array}\right]. \tag{12.34}$$

Since $A_a' C_W' + B_W D_W' = 0$, the states corresponding to A_a' are unreachable, and hence can be canceled out, and we find for L the minimal realization

$$L \sim_c \begin{bmatrix} A_c & A_c Y_1 R^{-1} C_W' + B_c \\ C_c & C_c Y_1 R^{-1} C_W' + D \end{bmatrix} \tag{12.35}$$

(using also the other equality $B_a' C_W' + D_W D_W' = I$). An explicit expression for the diagonal of pivots is hence $D_L := C_c Y_1 R^{-1} C_W' + D$, the gist of which is the ratio $Y_1 R^{-1} = Y_1 Y_2^{-1}$, where $Y = \begin{bmatrix} Y_1 \\ Y_2 \end{bmatrix}$ is the solution of the Lyapunov–Stein Eq. (12.21), which balances the contributions of the "poles" (given by W) and the "zeros" (given by V).

This is as far as we can get using only knowledge of the right factor, and the hypothesis that T is boundedly invertible. In many circumstances, one may not be sure about the latter, so a separate determination of the left factor and the diagonal constant in the factorization may be necessary. So, let us move to an attempted determination of $T = LdU$, with U as before (including the determination of $d = C_c Y_1 R^{-1} C_W' + D$), and then calculate the left factor L normalized with $D_L = I$ separately. The procedure is exactly the same as previously, now executed on T'. Using the dual case judiciously, we obtain in sequence, replacing W by w, V by v, T_u by $T_\ell := T'w$, T_o by the outer–inner factorization $T_\ell = T_p v$, in which the connecting quantity Y is replaced by y^*, and R by P (and assuming a realization in output normal form of the causal part of T),

$$T' = B_a' Z(I - A_a' Z)^{-1} C_a' + D' + B_c' Z'(I - A_c' Z')^{-'} C_c',$$

$$w = D_w + C_w (I - ZA_w)^{-1} Z B_w \sim_c \begin{bmatrix} A_c & B_w \\ C_c & D_w \end{bmatrix},$$

$$L' \sim_a \begin{bmatrix} A_c' & C_c' \\ G' & I \end{bmatrix},$$

$$P'^{\langle -1\rangle} = A_w P' A_v' + B_w B_v',$$

$$G' = -\left(C_w P' A_v' + D_w B_v'\right)\left(P'^{\langle -1\rangle}\right)^{-1},$$

$$\begin{bmatrix} A_\ell & B_\ell \\ C_\ell & D_\ell \end{bmatrix} = \left[\begin{array}{cc|c} A_a' & C_a' C_c & C_a' D_w, \\ 0 & A_c & B_w \\ \hline B_a' & D'C_c + B_c' A_c & D' D_w + B_c' B_w \end{array}\right],$$

$$d = C_w P^{-1} y_1 B_a + D, \tag{12.36}$$

with the outer–inner factorization

$$\begin{bmatrix} A_\ell y' & B_\ell \\ C_\ell y' & D_\ell \end{bmatrix} \cdot \left[\begin{array}{c|cc} C'_m & A'_v & C'_v \\ D'_m & B'_v & D'_v \end{array}\right] = \left[\begin{array}{c|cc} 0 & y'^{\langle -1 \rangle} & B_p \\ 0 & 0 & D_p \end{array}\right], \tag{12.37}$$

everything being, of course, dual from before. The same conditions as mentioned in the algorithm for U should apply here as well; in particular, C_m and D_m in the factorization should vanish, and $P = y_2$ should be square invertible at each index point with a uniform bound on the inverse. In addition, the factor d obtained from either the U or the L calculation should be invertible at each index point with a uniform bound as well. When all these conditions are satisfied, T will be boundedly invertible and possess a strict LU factorization (and conversely, as shown in Theorem 12.7).

12.3 A Tri-diagonal Example

As the mechanics of the diverse operations may seem a bit tricky, here is a nontrivial but simple example that already contains most of the intricacies and can easily be computed and verified by hand. Consider the half-infinite matrix in real arithmetic

$$T = \begin{bmatrix} \boxed{1} & b_0 & & \\ a_0 & 1 & b_1 & \\ & a_1 & 1 & \ddots \\ & & \ddots & \ddots \end{bmatrix}, \tag{12.38}$$

and let us develop the factorization algorithm for this case. We observe that $W = Z$ now (unless some $b_i = 0$, and the problem can be reduced, W is often very simple) and

$$T_u := TW = \begin{bmatrix} 1 & \boxed{b_0} & & \\ a_0 & 1 & b_1 & \\ & a_1 & 1 & \ddots \\ & & \ddots & \ddots \end{bmatrix}, \tag{12.39}$$

indicating a right shift of the main diagonal to make T_u lower. This shift on the right side of the matrix forces an input at the index point -1 and hence also a realization for that index point. Here is an immediate realization for T_u (assembling local realizations in a global block diagonal matrix):

$$T_u \sim_c \operatorname{diag}\left(\begin{bmatrix} | & 1 \\ \cdot & - \end{bmatrix}, \boxed{\left[\begin{array}{c|c} 1 & 0 \\ 0 & 0 \\ \hline 1 & b_0 \end{array}\right]}, \left[\begin{array}{cc|c} 0 & 1 & 0 \\ 0 & 0 & 1 \\ \hline a_0 & 1 & b_1 \end{array}\right], \ldots \right) \tag{12.40}$$

(first matrix is the realization for the index point -1), the general term for $k \geq 1$ being (in the input normal form)

$$\left[\begin{array}{cc|c} 0 & 1 & 0 \\ 0 & 0 & 1 \\ \hline a_{k-1} & 1 & b_k \end{array}\right]. \tag{12.41}$$

The outer–inner recursion $T_u = T_o V$ now produces
Step -1:

$$\begin{bmatrix} | & 1 \\ \cdot & - \end{bmatrix}\begin{bmatrix} - & \cdot \\ 1 & | \end{bmatrix} = \begin{bmatrix} 1 & | \\ - & \cdot \end{bmatrix}. \tag{12.42}$$

Hence, $Y_0 = \begin{bmatrix} - \\ 1 \end{bmatrix}$, $A_{V,-1} = -$, $T_{0,-1}$ empty.

Step 0:
With $n_0 := \sqrt{1 + b_0^2}$:

$$\left[\begin{array}{c|c} 1 & 0 \\ 0 & 1 \\ \hline 1 & b_0 \end{array}\right] \cdot \frac{1}{n_0}\left[\begin{array}{c|c} -b_0 & 1 \\ \hline 1 & b_0 \end{array}\right] = \frac{1}{n_0}\left[\begin{array}{c|c} -b_0 & 1 \\ 1 & b_0 \\ \hline 0 & n_0^2 \end{array}\right], \tag{12.43}$$

which produces $Y_1 = \frac{1}{n_0}\begin{bmatrix} -b_0 \\ 1 \end{bmatrix}$, $A_{V,0} = \frac{-b_0}{n_0}$, besides a realization at the index point 0 for T_o: $T_{o,0} = \left[\begin{array}{c|c} 1 & \frac{1}{n_0} \\ 0 & \frac{b_0}{n_0} \\ \hline 1 & n_0 \end{array}\right]$.

Step 1:
With $n_1 := \sqrt{(\frac{1-a_0b_0}{n_0})^2 + b_1^2}$:

$$\left[\begin{array}{c|c} \frac{1}{n_0} & 0 \\ 0 & 1 \\ \hline \frac{1-a_0b_0}{n_0} & b_1 \end{array}\right] \frac{1}{n_1}\left[\begin{array}{c|c} -b_1 & \frac{1-a_0b_0}{n_0} \\ \hline \frac{1-a_0b_0}{n_0} & b_1 \end{array}\right] = \frac{1}{n_1}\left[\begin{array}{c|c} -\frac{b_1}{n_0} & \frac{1-a_0b_0}{n_0^2} \\ \frac{1-a_0b_0}{n_0} & b_1 \\ \hline 0 & n_1^2 \end{array}\right], \tag{12.44}$$

which, again, produces $Y_2 = \frac{1}{n_1}\begin{bmatrix} -\frac{b_1}{n_0} \\ \frac{1-a_0b_0}{n_0} \end{bmatrix}$, $A_{V,2} = -\frac{b_1}{n_1 n_0}$ and a realization for T_o:

$T_{o,1} = \left[\begin{array}{cc|c} 0 & 1 & \frac{1-a_0b_0}{n_0^2 n_1} \\ 0 & 0 & \frac{b_1}{n_1} \\ \hline a_0 & 1 & n_1 \end{array}\right]$ at the index point 1.

The general step in the recursion is then, with $y_k := Y_{k,2} + a_{k-1}Y_{k,1}$ and $n_k := \sqrt{(y_k^2 + b_k^2)}$:

$$\left[\begin{array}{c|c} Y_{k,2} & 0 \\ 0 & 1 \\ \hline y_k & b_k \end{array}\right] \frac{1}{n_k}\left[\begin{array}{c|c} -b_k & y_k \\ \hline y_k & b_k \end{array}\right] = \frac{1}{n_k}\left[\begin{array}{c|c} -Y_{k,2}b_k & Y_{k,2}y_k \\ y_k & b_k \\ \hline 0 & n_k^2 \end{array}\right] \tag{12.45}$$

producing $Y_{k+1} = \frac{1}{n_k} \begin{bmatrix} -Y_{k,2} b_k \\ Y_{k,2} + a_{k-1} Y_{k,1} \end{bmatrix}$ and $A_{V,k} = \frac{-b_k}{n_k}$, besides the other quantities at the index point k. The recursion for Y turns out to be the simple

$$\begin{bmatrix} Y_{k+1,1} \\ Y_{k+1,2} \end{bmatrix} = \frac{1}{n_k} \begin{bmatrix} 0 & -b_k \\ a_{k-1} & 1 \end{bmatrix} \begin{bmatrix} Y_{k,1} \\ Y_{k,2} \end{bmatrix}, \tag{12.46}$$

in which one can recognize a linearization of the recursion for the pivot as it arises in the Gaussian elimination. All data are now available to determine U for this case: $F_k = -\frac{Y_{k+1,2}}{Y_{k+1,1}} = \frac{b_{k-1}}{1 - a_{k-1} F_{k-1}}$, with initial conditions $F_0 = -$ and $F_1 = b_0$, and the diagonal d is given by $d_k = 1 - a_k F_{k-1}$, equal to the pivot. Hence,

$$U \sim \operatorname{diag}\left(\boxed{\begin{bmatrix} | & 1 \\ | & 1 \end{bmatrix}}, \begin{bmatrix} 0 & 1 \\ b_0 & 1 \end{bmatrix}, \ldots, \begin{bmatrix} 0 & 1 \\ F_k & 1 \end{bmatrix}, \ldots \right) \tag{12.47}$$

as expected. The conditions for the existence of the factor now boil down to $R_k = Y_{k,2}$ and d_k to be bounded away from zero and R_k to be uniformly bounded as well.

12.4 Points for Discussion

1. *Beginning recursions:* The outer–inner recursion Eq. (12.20) of Step 2 works backwards. In the finitely indexed case, the recursion starts out with an empty Y. Given the original (block) matrix T_u and its realization, how does it proceed from there? In particular, what shall its dimension be after the first step? And the subsequent ones?

   One should also realize that the outer–inner recursion just considered is preceded by another (external) recursion, which works forward for the determination of W. The condition that the (finite matrix) T is invertible makes it necessarily square, but the subdivisions of the input and output spaces into time steps does not have to be uniform a priori. Can something be said about the relationship of these subdivisions and the existence of the LU factorization? (Notice that even when the central blocks are square, the LU factorization does not have to exist. For example, $T = \begin{bmatrix} 0 & 1 \\ 1 & 0 \end{bmatrix}$, seen as a system with scalar entries, does not have an LU factorization, but J is perfectly acceptable as a block 2 nonsingular diagonal entry.)
2. An interesting topic for further consideration is the relation between an infinitely indexed operator T and a finitely indexed suboperator, say $T_{-N:N,-N:N}$ for some large N, considering, for example, the full (block) Toeplitz case versus finite (block) Toeplitz central submatrices. Finite Toeplitz (block) matrices with realizations corresponding to their infinite LTI embedding have realizations that are time-variant at their endings, and one may study what the effects are of cutting-off endings, by considering variations in initial or final states, using system properties like (strong) reachability and/or observability to account for the restrictions on the time support.
3. LU factorization is not the best way to solve a system of equations. For that, QR-based methods are to be preferred, in particular in the "orthogonal Faddeev form"

that not only uses orthogonal transformations throughout but also avoids the back-substitution step [45]. The value of LU factorization has a different motivation: in many problems, it is needed as a central constituent of the solution. This is often the case for problems that lead to Fredholm- or Volterra-type equations, such as Wiener filtering.

4. The general LU-factorization problem gives rise to a quadratic Riccati equation (which one?). The method presented here reduces the solution to two coupled inner–outer factorizations. This issue has not been worked out yet in the literature and may provide an attractive way to deal with Riccati equations just as it provides an attractive way to deal with LU factorization.

12.5 Notes

1. There are many variants of LU-factorization methods and many improvements of them. A major issue is numerical stability. One source of instability is that, in most algorithms, one must divide by the pivot to reach subsequent stages, and when the pivot is zero, division is impossible, or else, when it is very small or very large, the division necessarily leads to uncertain results. These issues have been studied extensively in the literature (see, e.g., [64]). Another issue is the extensive use of inner product calculations, which, when not done properly leads to a serious loss of accuracy and hence numerical instability. (When one accumulates numbers that are vastly different in magnitude, one may loose the smallest ones due to rounding, while in a later stage of the computation, the larger ones get canceled out, e.g., $10^{10} + 1 - 10^{10} = 0$! when executed in sequence and in single precision on most computers.) Most of these instability issues disappear when one uses QR (and hence inner–outer) rather than LU in the Jacobi–Givens version, because the orthogonal transformations represented by simple rotations are intrinsically backward stable, at least when executed properly.

   To deal with numerical instability, one resorts classically to "pivoting," that is, rearranging rows and columns so as to create advantageous pivots, which will not create havoc in subsequent divisions. For example, one looks for the largest element in the column being handled and then exchanges rows to put it in the top position. This method is even standard in some numerical packages. It has the disadvantage that it does not factor the original matrix, which is what many applications of spectral factorization require. There is an additional reason why LU is not that advantageous for solving linear equations, namely that it leads, as a first approach, to two back-substitutions: when $LUx = b$, one solves first for an intermediate u in $Lu = b$ by forward back-substitution, and then for x in $Ux = u$ by back-substitution. The forward back-substitution requires another division with the pivots, and both back-substitutions require further inner product computations, both sources of numerical instability.

2. From the beginning of estimation theory, not to talk about Volterra and Wiener–Hopf theory, *spectral factorization* has played a central role, and it has occupied

mathematicians and signal-processing engineers ever since. The spectral factorization of a square, nonsingular and rational LTI transfer function $T(z)$ aims at factoring $T(z) = T_\ell(z)T_r(z)$, so that $T_\ell(z)$ is outer (i.e., it has all its poles and zeros outside the unit disk) and $T_r(z)$ conjugate outer (i.e., it has all its poles and zeros inside the unit disk). The classical way of doing this is to assign the causal poles and zeros to the $T_\ell(z)$ factor and relegate the anti-causal ones to $T_r(z)$ (see Fig. 12.1). This will only be possible consistently if the number of causal zeros equals the

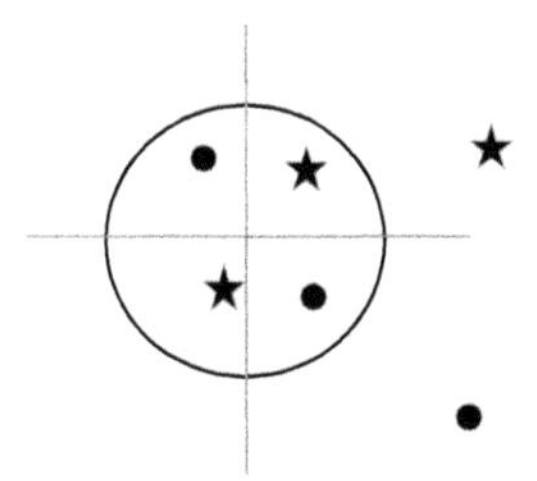

Figure 12.1 Equal numbers of poles and zeros inside the unit disk are needed for spectral factorization.

number of causal poles, and then the number of anti-causal poles and zeros are also equal (Since the total number of poles and zeros is necessarily the same, with multiplicities counted.) Strict spectral factorization therefore does not allow for poles nor zeros on the unit circle. When there are none, one says that there is *dichotomy*, a term that we believe has been introduced by Gohberg and Ben Artzi [14]. In the case where there are indeed poles and zeros on the unit circle, one may think of distributing them among the outer and conjugate-outer factors, but the theory becomes considerably more complex in that case. Dichotomy in the LTI case and matching of poles and zeros assure the boundedness of the L and U factors (and conversely). In our time-variant case, the matching is reflected in the fact that Eq. (12.20) yields a square R and hence a match of the state dimensions of W and V.

3. A particularly impressive treatment of the positive definite LTI case is given by Helson in his wonderful book on invariant subspaces [42]. He provides, as far as we know, the most appealing method and proofs for the case of positive definite matrix functions (the time-invariant or Wiener–Masani setting [54]). The more general case of nonpositive spectral factorization for rational matrix functions was extensively considered by the late Israel Gohberg and his collaborators [10, 38] and gave rise to numerous results. For the positive definite quasi-separable matrix case, a pretty thorough treatment is given in the book [29]. The remarkable property is that in the quasi-separable matrix setting, the classical Gaussian elimination in its LU-factorization form and spectral factorization are one and the same thing. A mathematical treatment of the theory is given in [25]; here we give a treatment geared toward numerical usage following [28]. An operator-theoretic treatment covering also the time-invariant case in an appealing way can be found in [39].

13 Matrix Schur Interpolation

Often a positive definite operator or matrix is only partially specified. This may be because only a limited number of measurements are present, or because one wants to approximate a positive definite operator with another positive definite operator that has a much lower algebraic complexity. The present chapter solves this problem in an elegant way for the important case where the known data are consistently ordered in a "staircase." The method is based on an algebraic analog of classical Schur interpolation in complex analysis (sometimes called the "z-domain"). The chapter is fully self-contained algebraically, not using system realization theory: it serves to introduce the basic techniques to be used in the subsequent chapters on interpolation, in particular the use of various types of so-called scattering matrices.

Menu

Hors d'oeuvre
Interpolating Positive Definite Matrices

Course 1
J-Orthogonalization

Course 2
Generalized Schur Interpolation

Course 3
Norm Approximation

Dessert
Array Processing

13.1 Introduction

Consider the following 3×3 matrix: $\begin{bmatrix} 1 & 4/5 & t \\ 4/5 & 1 & 4/5 \\ t & 4/5 & 1 \end{bmatrix}$, with t being a parameter. It is a positive definite matrix when $7/25 \leq t \leq 1$ (as can be seen from an LU factorization). Suppose we are allowed to give t any convenient value that keeps the matrix positive definite. It would seem that the easiest guess is $t = 0$, but that would result in a non positive-definite matrix. Anticipating our discussion, a better choice would be $t = 16/25$, resulting in

$$\begin{bmatrix} 1 & 4/5 & 16/25 \\ 4/5 & 1 & 4/5 \\ 16/25 & 4/5 & 1 \end{bmatrix} = \begin{bmatrix} 1 & 0 & 0 \\ 4/5 & 3/5 & 0 \\ 16/25 & 12/25 & 3/5 \end{bmatrix} \begin{bmatrix} 1 & 4/5 & 16/25 \\ 0 & 3/5 & 12/25 \\ 0 & 0 & 3/5 \end{bmatrix} \tag{13.1}$$

positive definite, with the inverse $(1/9)\begin{bmatrix} 25 & -20 & 0 \\ -20 & 41 & -20 \\ 0 & -20 & 25 \end{bmatrix}$, which is not only positive definite but also tri-diagonal (although in this small-dimensional case that is not very impressive). It can be seen that the value $t = 16/25$ maximizes the determinant and at the same time puts the entry corresponding to t equal to zero *in the inverse*, while keeping the result positive definite.

Positive definite matrices arise, for example, as covariances for stochastic processes. Let us assume X is a (real) stochastic $n \times n$ vector with zero-mean components; then its covariance matrix is given by

$$C := \mathbf{E} \begin{bmatrix} X_1 \\ X_2 \\ \vdots \\ X_n \end{bmatrix} \begin{bmatrix} X_1 & X_2 & \cdots & X_n \end{bmatrix} \tag{13.2}$$

(or else the X_i could also be "data vectors," and then C would be their correlation matrix). An interesting estimation problem is: *given the diagonal elements $C_{i,i}$ and the first off-diagonals $C_{i,i+1}$ for all relevant values of i, what is the best guess for the other covariances?* Often it is known that only close-by variables (i.e., with close indices) influence each other significantly, or that the accuracy of the further away estimates is not very good, so it makes sense to try to guess a good covariance matrix based on the limited amount of data. In the example just given, we saw that putting the unknown elements equal to zero may result in non positive-definite matrices, which cannot be covariances. This danger is very real: as soon as dimensions get large and the amount of data available become relatively small, ending up with nonsensical matrices is almost certain.

Maximum Entropy Interpolation

The solution to the just-mentioned completion problem, in terms of statistical analysis, is given by the notion of *maximum entropy interpolation.* The notion of entropy is a central notion in information theory, and for the present case, it is given by $\mathcal{E} := \ln \det C$. (For an introduction to the theory of entropy, we have to refer to a textbook on information theory, e.g., [17].) As entropy measures uncertainty, the estimate of C that uses only the known data would maximize $\mathcal{E}$, the information-theoretical measure for uncertainty. Suppose we have to estimate an unknown entry $t := C_{i,j}$ in C (understood with $i{\neq}j$), let us try to maximize the entropy and see what effect that would have. So we would require $\partial_t \mathcal{E} = 0$. How does $t = C_{i,j}$ enter in $\log\det C$? The classical Laplace expansion formula for the determinant along row i gives $\det C = \sum_{k=1}^{n} C_{i,k} M_{i,k}$, with $M_{i,k}$ being the minor of $C_{i,k}$, which is independent of all elements on row i (and column k), and in particular of $t = C_{i,j}$. It follows that $\partial_t \det C = M_{i,j}$ and subsequently, $\partial_t \mathcal{E} = \frac{\partial_t \det C}{\det C} = (C^{-1})_{j,i}$ by Cramer's rule for the elements of C^{-1}. Hence, a stationary value in $\log\det C$ is obtained when the entry $(C^{-1})_{j,i} = (C^{-1})_{i,j} = 0$ (in this simple case). In particular, in case the band consisting of the diagonal elements and the single first main off-diagonal is given, then the remaining off-diagonals *in the inverse* have to be put equal to zero to maximize the entropy. These considerations transform the entropy optimization problem in an interpolation problem, whereby the known entries in the original matrix are complemented with zero entries for the nonspecified positions in the inverse matrix. It will turn out that this interpolation problem has a unique solution, which is then also the solution of the optimization problem.

In the sequel, we shall not try to solve the general interpolation problem, in which all diagonal entries and certain randomly chosen further entries are specified. For the case where the specified entries form a (possibly nonuniform) staircase concentrated along the main diagonal (i.e., without gaps or rags), there is an attractive, closed form and efficient solution, whereby efficient means that it can be computed by a number of operations proportional to the number of known entries and not proportional to the overall matrix dimension squared or cubed, as is the case for many problems involving matrices, such as matrix inversion, or determining eigenvalues.

13.2 J-Orthogonalization

In this section, we introduce the connection between the maximum entropy interpolation problem just described, and a special type of orthogonalization, namely orthogonalization using J-unitary transformation matrices, with J being a sign matrix, rather than regular orthogonal transformations. The theory may appear to come out of the blue, but it turns out to be a powerful application to matrix calculus of the ideas from classical Schur interpolation theory. We give a straight description of the results without the motivational background of the more general underlying scattering theory, which we present in Chapter 14.

Let us work, for the time being, on a strictly positive definite matrix C with scalar entries of dimension $n \times n$. (We shall relax the "scalar entry" requirement later.) Connected to C there is a *Positive Real* or PR matrix Y, defined as twice the strictly lower part of C, keeping the same values on the diagonal. Hence, with our usual conventions, $C = \frac{1}{2}(Y + Y')$, with Y lower.

Many approximation problems can be solved through an orthogonalization procedure. In this chapter, we show that a special type of limited recursive orthogonalization is capable of solving the *matrix Schur interpolation problem* alias the *maximum entropy approximation problem*, based solely on matrix entries specified on the complement of a staircase in the original (strictly) positive definite matrix C.

Consider the 2×1 block matrix consisting of two lower matrices

$$E_1 := \begin{bmatrix} \frac{1}{2}(Y+I) \\ \frac{1}{2}(Y-I) \end{bmatrix}, \tag{13.3}$$

with Y being PR, as defined above, and let $J := \begin{bmatrix} I_n & \\ & -I_n \end{bmatrix}$ be a conformal sign matrix. Then

$$E_1' J E_1 = \frac{1}{2}(Y + Y') = C, \tag{13.4}$$

E_1 consists of (column) vectors that form what is known (by definition) as a (strict) J-positive space (notice that they are independent of each other and therefore necessarily form a basis). Let now $C = LL'$, with the Cholesky factor L lower triangular (as before); then the columns in $\overline{E}_1 := E_1 L^{-1}$ are J-orthogonal on each other and have what may be called a unit *J-norm*:

$$\overline{E}_1' J \overline{E}_1 = I. \tag{13.5}$$

The columns of E_1 and $\overline{E}_1$ produce n independent vectors with positive J-norms in a $2n$-dimensional space. One may subsequently wonder what a completion of the basis with further J-orthogonal vectors could be. This question is easily answered. Consider the *complementary*

$$E_2 := \begin{bmatrix} \frac{1}{2}(Y'-I) \\ \frac{1}{2}(Y'+I) \end{bmatrix}. \tag{13.6}$$

Then one can easily verify that $E_2' J E_1 = 0$ and $E_2' J E_2 = -C$. In words, the columns of E_2 are J-orthogonal on the columns of E_1 and are independent of each other (i.e., form a basis). Let, moreover, $C = M'M$ be the "conjugate" Cholesky factorization and consider

$$\overline{E}_2 = E_2 M^{-\prime}. \tag{13.7}$$

Then the columns in $\overline{E}_2$ form a basis of an n-dimensional subspace of $\mathbf{R}^{2n}$ (or $\mathbf{C}^{2n}$) whose base vectors are J-orthogonal, have J-norm -1 and are J-orthonormal on the columns of E_1 and E_2. The joint block matrix

$$\overline{E} := \begin{bmatrix} \overline{E}_1 & \overline{E}_2 \end{bmatrix} \tag{13.8}$$

is J-isometric of full dimension:

$$\overline{E}' J \overline{E} = J, \tag{13.9}$$

so that $\overline{E}$ is invertible as well, and

$$\overline{E}^{-1} = J \overline{E}' J. \tag{13.10}$$

We have specifically

$$\begin{bmatrix} \frac{1}{2}(Y+I) \\ \frac{1}{2}(Y-I) \end{bmatrix} = \overline{E} \begin{bmatrix} L \\ 0 \end{bmatrix}, \tag{13.11}$$

which interprets as a new kind of QR factorization of the left-hand side (containing original data) into a J-unitary factor $\overline{E}$ and a block lower "remainder." This is nothing but a *J-orthogonalization of the data contained in E*, much like a QR factorization, which entails the (regular) orthogonalization of the original data. It turns out that a J-orthogonalization is a bit (but not much) harder to compute than a regular orthogonalization. We give the details in the following section.

J-unitary matrices play an important role in interpolation theory, because, as will appear in Sections 13.3 and 13.4, one can perform *partial* QR factorizations with Q a J-unitary matrix. It turns out to be practical to define a standard form for J-unitary matrices. To be consistent with the literature, we shall use as *standard Θ matrix*, the inverse of the J-unitary matrix we derived in the previous paragraphs, which is a J-unitary matrix in its own right.

Definition 13.1 A "standard Θ matrix" is a J-unitary matrix of the form

$$\Theta := \begin{bmatrix} M^{-\prime}\frac{1}{2}(I+Y^{\prime}) & M^{-\prime}\frac{1}{2}(I-Y^{\prime}) \\ L^{-1}\frac{1}{2}(I-Y) & L^{-1}\frac{1}{2}(I+Y) \end{bmatrix}, \tag{13.12}$$

in which Y is lower, $C = \frac{1}{2}(Y + Y^{\prime}) = LL^{\prime} = M^{\prime}M$, with L and M being the left and right lower Cholesky factors of C respectively.

In the present chapter (and some subsequent ones), we put the additional assumption that the main diagonal of C is normalized to I; there are cases where normalization is not possible, but convenient when it is.

This standard Θ matrix has the following central property:

$$\Theta \begin{bmatrix} \frac{1}{2}(Y+I) \\ \frac{1}{2}(Y-I) \end{bmatrix} = \begin{bmatrix} M \\ 0 \end{bmatrix}. \tag{13.13}$$

In other words, the J-unitary transformation Θ *eliminates* $\frac{1}{2}(Y - I)$. In the following section, we investigate how this procedure can be done partially (similar to a Givens elimination but now using elementary J-unitary transformations called *hyperbolic rotations* instead of elementary unitary rotations). But before that, we show how any J-unitary Θ defines the related quantities Y, L, M and C.

Consider therefore a 2×2 block J-unitary matrix Θ with its first block row consisting of upper matrices and the second block row consisting of lower ones. Then we show that the lower matrix $L^{-1} = \Theta_{21} + \Theta_{22}$ is nonsingular and that the lower matrix $Y = L(\Theta_{21} + \Theta_{22})$ is PR:

Proposition 13.2 *Let* $\Theta = \begin{bmatrix} \Theta_{11} & \Theta_{12} \\ \Theta_{21} & \Theta_{22} \end{bmatrix}$ *be a J-unitary matrix such that Θ_{11} and Θ_{12} are upper and Θ_{21} and Θ_{22} are lower. Then $L^{-1} := \Theta_{21} + \Theta_{22}$ is nonsingular lower, $Y := L(\Theta_{22} - \Theta_{21})$ is lower and strictly positive real (i.e., $\frac{1}{2}(Y + Y^{\prime}) \gg 0$), and Θ has the form (13.12), with L and M satisfying $\frac{1}{2}(Y + Y^{\prime}) = L^{\prime}L = MM^{\prime}$ and in which both L and M are lower matrices.*

Proof

By J-unitarity, we have $\Theta_{22}\Theta_{22}' = I + \Theta_{21}\Theta_{21}'$. Hence, $\Theta_{22}\Theta_{22}' \geq I$ and Θ_{22} is nonsingular. Dividing by Θ_{22} left and Θ_{22}' right, we find, with $S := -\Theta_{22}^{-1}\Theta_{21}$ (the "−" sign in the definition is by convention, to be motivated later),

$$I - SS' = \Theta_{22}^{-1}\Theta_{22}^{-\prime} \gg 0. \tag{13.14}$$

This in turn makes S strictly contractive and $I - S$ nonsingular. Next, we have $\Theta_{22}(I - S) = \Theta_{21} + \Theta_{22} = L^{-1}$ is nonsingular (and, of course, lower). Putting $Y := L(\Theta_{22} - \Theta_{21}) = (I + S)(I - S)^{-1}$ makes Y positive real (proof: just compute $Y + Y'$), and the other identifications follow readily, also by J-unitarity. □

One may then wonder in which cases a standard J-unitary Θ-matrix can be used to eliminate the lower block in a 2×1 stack of conformal lower matrices, say $\begin{bmatrix} \Gamma \\ \Delta \end{bmatrix}$. Suppose we have indeed achieved such an elimination, that is, we have found a standard Θ resulting, for some lower Γ_t, in

$$\Theta \begin{bmatrix} \Gamma \\ \Delta \end{bmatrix} = \begin{bmatrix} \Gamma_t \\ 0 \end{bmatrix}. \tag{13.15}$$

Then what is a characteristic property of $\begin{bmatrix} \Gamma \\ \Delta \end{bmatrix}$? Premultiplying the equation with its transpose times J and sing the J-unitarity $\Theta' J \Theta = J$ of Θ, we find

$$\Gamma'\Gamma - \Delta'\Delta = \Gamma_t'\Gamma_t. \tag{13.16}$$

If we assume, in addition, that Γ is invertible, then we may define $S := \Delta\Gamma^{-1}$, and we find that S *must be contractive, that is,* $I - S'S \gg 0$. We shall call such a $\begin{bmatrix} \Gamma \\ \Delta \end{bmatrix}$ admissible:

Definition 13.3 We say that a pair of $n \times n$ matrices $\begin{bmatrix} \Gamma \\ \Delta \end{bmatrix}$, with Γ nonsingular and both Γ and Δ lower, is *admissible* if and only if $S = \Delta\Gamma^{-1}$ is strictly contractive.

13.3 The Schur Partial Factorization Algorithm

We are now ready to consider the recursive elimination of entries in an admissible pair of causal blocks using hyperbolic transformations, much in the style of using Givens rotations in a unitary environment. The basic recursive J-unitary elimination steps work as follows (assuming Δ_0 strictly lower). We start out with a first admissible pair

$$\begin{bmatrix} \Gamma_0 \\ \Delta_0 \end{bmatrix} = \begin{bmatrix} \frac{1}{2}(Y + I) \\ \frac{1}{2}(Y - I) \end{bmatrix} \tag{13.17}$$

and annihilate an entry in Δ_0 on the first lower off-diagonal, to produce a new admissible pair $\begin{bmatrix} \Gamma_1 \\ \Delta_1 \end{bmatrix}$, in which both Γ_1 and Δ_1 are lower, but one entry in Δ_1 has been made zero, using a to-be-defined *elementary* standard Θ matrix, that is, a Θ matrix whose top block row consists of upper and bottom block row of lower matrices. It will turn out that this procedure can be repeated without fill-ins so long as the remaining, not eliminated, entries keep a staircase form. We explain this procedure in detail further in this section, after a few preliminaries.

Given a (real or complex) coefficient ρ with $|\rho| < 1$, a 2×2 elementary J-unitary or hyperbolic transformation has the form

$$\theta(\rho) = \frac{1}{\sqrt{1-|\rho|^2}} \begin{bmatrix} 1 & \rho' \\ \rho & 1 \end{bmatrix} \tag{13.18}$$

(easy check). As in the theory of Givens elimination, the trick will be to embed the elementary hyperbolic rotation in a larger 2×2 block matrix, keeping the general properties needed, namely the structure of the Θ matrices, the admissibility of $\begin{bmatrix} \Gamma \\ \Delta \end{bmatrix}$'s and the (gradually decreasing) staircase form of the entries of the subsequent Δ's.

Definition 13.4 A *staircase support* for strictly lower matrices is a set of indices $\sigma = \{(i,j)\}$, with $1 \leq i,j \leq n$, $j < i$ such that $\forall i : (i,j) \in \sigma \Rightarrow (0 < k < j \Rightarrow (i,k) \in \sigma) \wedge (i < k \leq n \Rightarrow (k,j) \in \sigma)$, that is, indices left and below a given element in σ are also in σ.

The complement of the staircase set in the set of lower indices $((i,j) : i \geq j)$ is a "staircase band" and will be denoted as τ (see Fig. 13.1).

An elementary elimination step then starts with an admissible $\begin{bmatrix} \Gamma_i \\ \Delta_i \end{bmatrix}$, in which Δ_i is in a staircase form and already partially zero, and proceeds eliminating a further entry in it, to produce a new, admissible pair $\begin{bmatrix} \Gamma_{i+1} \\ \Delta_{i+1} \end{bmatrix}$, with Δ_{i+1} again in a staircase form, using a standard but elementary Θ matrix (which we denote using a lower case θ to indicate that it belongs to an elementary step):

$$\begin{bmatrix} \Gamma_{i+1} \\ \Delta_{i+1} \end{bmatrix} = \theta_{i+1} \begin{bmatrix} \Gamma_i \\ \Delta_i \end{bmatrix}. \tag{13.19}$$

We will show soon that (1) θ_{i+1} has indeed the form of a standard Θ matrix and (2) the procedure can be repeated recursively, until all entries in Δ are eliminated, reducing

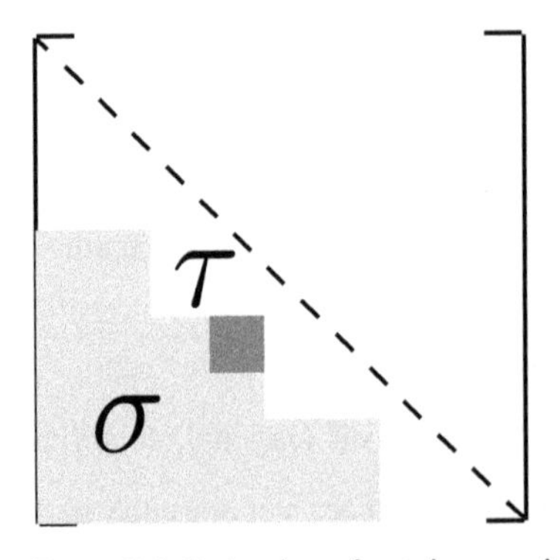

Figure 13.1 Reduction of a staircase with a corner.

the staircase in the process, that is, eliminating at each step a corner entry located on the boundary of the staircase.

As a simple example, let $\begin{bmatrix} \Gamma \\ \Delta \end{bmatrix} = \left[\begin{array}{cc} 1 & \\ & 1 \\ \hline 0 & \\ 1/2 & 0 \end{array}\right]$ and let us determine Θ and the other related quantities. We only have to eliminate the single entry $\Delta_{2,1}$, and Θ will have the form $\Theta = \left[\begin{array}{cc|cc} c & & & s \\ & 1 & & \\ \hline & & 1 & \\ s & & & c \end{array}\right]$ with $|c|^2 - |s|^2 = 1$. Hence,

$$\Theta \begin{bmatrix} \Gamma \\ \Delta \end{bmatrix} = \left[\begin{array}{cc} c + s/2 & \\ 0 & 1 \\ \hline 0 & \\ s + c/2 & 0 \end{array}\right], \tag{13.20}$$

and we find $s = -c/2$, $c = 2/\sqrt{3}$ (since $c^2 - s^2 = 1$), $s = -1/\sqrt{3}$, and hence $\Theta = \left[\begin{array}{cc|cc} 2/\sqrt{3} & & & -1/\sqrt{3} \\ & 1 & & \\ \hline & & 1 & \\ -1/\sqrt{3} & & & 2/\sqrt{3} \end{array}\right]$, $L = \begin{bmatrix} 1 & \\ 1/2 & \sqrt{3}/2 \end{bmatrix}$, $Y = \begin{bmatrix} 1 & \\ 1 & 1 \end{bmatrix}$, $C = \begin{bmatrix} 1 & 1/2 \\ 1/2 & 1 \end{bmatrix}$, and, of course, $S = \begin{bmatrix} 0 & \\ 1/2 & 0 \end{bmatrix}$ – the matrix we started with.

One further preliminary and helpful observation (also to understand why the staircase form is important) is: suppose the recursive elimination procedure as described is successful; then, after a few steps, we shall have

$$\begin{bmatrix} \Gamma_i \\ \Delta_i \end{bmatrix} = \Theta_a \begin{bmatrix} \Gamma_0 \\ \Delta_0 \end{bmatrix}, \tag{13.21}$$

for $\Theta_a = \theta_i \cdots \theta_2 \theta_1$ having indeed the form of a standard Θ-matrix (to be shown soon) and Δ_i having a staircase support. If A is an arbitrary lower matrix, then $A\Delta_i$ will have at most the same staircase support as Δ_i. (The proof is elementary: the ith row in $A\Delta_i$ only takes a linear combination of rows in Δ_i that lie *on and above* the entry being computed, because A is a lower matrix.)

An elementary Schur step will eliminate an element (i, j) on a corner of the staircase σ_1 in Δ_1 without "fill-ins" using an hyperbolic rotation and produce a new Δ_2, so that the staircase set is diminished by one element, producing a new smaller $\sigma_2 := \sigma_1 \backslash (i, j)$ – see Fig. 13.1 – but remains a staircase.

So, let us use real calculus (for simplicity) and assume recursively that we dispose of an admissible pair $\begin{bmatrix} \Gamma_1 \\ \Delta_1 \end{bmatrix}$, with Δ_1 having support σ_1, with the goal to transform it to a new admissible pair $\begin{bmatrix} \Gamma_2 \\ \Delta_2 \end{bmatrix}$ with support $\sigma_2 = \sigma_1 \backslash (i, j)$ for Δ_2 using an elementary hyperbolic rotation that eliminates the corner element $\Delta_{1;i,j}$. Let us define $\rho_{i,j} := -\Delta_{1;i,j}/\Gamma_{1;j,j}$ and then $\theta^{(i,j)} = I_{2n}$, *except* for the entries $[\theta^{(i,j)}_{1,1}]_{j,j} = [\theta^{(i,j)}_{2,2}]_{i,i} :=$

$1/\sqrt{1-\rho_{i,j}^2}$ and $[\theta_{2,1}^{(i,j)}]_{i,j} = [\theta_{1,2}^{(i,j)}]_{j,i} := \rho_{i,j}/\sqrt{1-\rho_{i,j}^2}$. So, the recursive claim is that

$$\begin{bmatrix} \Gamma_2 \\ \Delta_2 \end{bmatrix} = \theta^{(i,j)} \begin{bmatrix} \Gamma_1 \\ \Delta_1 \end{bmatrix} \tag{13.22}$$

produces a new admissible pair Γ_2, Δ_2 so that the support of Δ_2 is now reduced at the position (i, j) (and is, of course, a staircase again). We check:

1. Only the jth row of Γ_1 gets changed by $\theta^{(i,j)}$, and it is modified by the ith row of Δ_1, which remains lower, thanks to the staircase (both rows are multiplied with appropriate factors).
2. The support of Δ_2 is reduced by the choice of $\rho_{i,j}$.
3. $\Delta_2\Gamma_2^{-1}$ is a causal, contractive scattering matrix, because $\theta^{(i,j)}$ is J-unitary, and hence,

$$\Gamma_2'\Gamma_2 - \Delta_2'\Delta_2 = \begin{bmatrix} \Gamma_1' & \Delta_1' \end{bmatrix} (\theta^{(i,j)})' J \theta^{(i,j)} \begin{bmatrix} \Gamma_1 \\ \Delta_1 \end{bmatrix} = \Gamma_1'\Gamma_1 - \Delta_1'\Delta_1, \tag{13.23}$$

making $\Delta_2\Gamma_2^{-1}$ (strictly) contractive, while the lower form is already evident.

The full Schur algorithm applied on a C and a Y whose diagonal entries are 1, starts with the admissible pair $\begin{bmatrix} \Gamma_0 \\ \Delta_0 \end{bmatrix} := \begin{bmatrix} \frac{1}{2}(Y+I) \\ \frac{1}{2}(Y-I) \end{bmatrix}$ and gradually reduces the strictly lower support of Δ_0 to the desired staircase, starting, for example, with the corner position $(2, 1)$, then $(3, 2)$ (or any other $(j + 1, j)$), etc.... It is easy to see that one can reach any staircase in this fashion. After some k steps, one obtains

$$\begin{bmatrix} \Gamma_k \\ \Delta_k \end{bmatrix} = \theta^{(i_k,j_k)} \cdots \theta^{(i_1,j_1)} \begin{bmatrix} \Gamma_0 \\ \Delta_0 \end{bmatrix}, \tag{13.24}$$

in which $\begin{bmatrix} \Gamma_k \\ \Delta_k \end{bmatrix}$ is still admissible, and Δ_k has the desired staircase σ (with complement τ) as its support.

The global $\Theta_\tau := \theta^{(i_k,j_k)} \cdots \theta^{(i_1,j_1)}$ now appears to have the necessary properties to solve the interpolation problem:

1. It is again J-unitary as a product of J-unitary matrices (easy check!).
2. The support of $\Theta_{\tau;2,1}$ and $\Theta_{\tau;2,2}$ is τ (and dual for $\Theta_{\tau;1,1}$ and $\Theta_{\tau;1,2}$), and hence also the support $L_\tau^{-1} = \Theta_{\tau;2,1} + \Theta_{\tau;2,2}$ (and M_τ^{-1}!). This we check recursively: suppose we have arrived at a support σ for Δ_k with complement τ and corresponding global Θ_τ, and suppose we eliminate the next corner element (i, j) $(i > j)$, to obtain the new global $\Theta_n := \theta^{(i,j)}\Theta_\tau$; then we have to show that the support of this new global Θ matrix is based on $\tau \cup (i, j)$, assuming that the support for Θ_τ is based on τ (by which we mean that the support of the first block row entries is τ' and that of the second block entries is τ). Let us specify $\theta^{(i,j)}$ a bit more: d_i is a unit diagonal, in which the ith element has been changed to $c_{i,j}$, and $e_{i,j}$ is an $n \times n$ matrix with 1 in the (i, j)th position, and otherwise zero. Then $\theta^{(i,j)} = \begin{bmatrix} d_j & s_{i,j}e_{j,i} \\ s_{i,j}e_{i,j} & d_i \end{bmatrix}$. Observe

further that for any matrix M, $e_{i,j}M$ is zero except for the ith row that is now been set equal to the original jth row. Thus, we have

$$\begin{aligned}\Theta_n &:= \theta^{(i,j)}\Theta_\tau \\ &= \begin{bmatrix} d_j\Theta_{\tau;1,1} + s_{i,j}e_{j,i}\Theta_{\tau;2,1} & d_j\Theta_{\tau;1,2} + s_{i,j}e_{j,i}\Theta_{\tau;2,2} \\ s_{i,j}e_{i,j}\Theta_{\tau;1,1} + d_i\Theta_{\tau;2,1} & s_{i,j}e_{i,j}\Theta_{\tau;2,1} + d_i\Theta_{\tau;2,2} \end{bmatrix}. \end{aligned} \tag{13.25}$$

The entries that are multiplied with the diagonals d_i and d_j already have the correct support. Concerning the other four, consider, for example, $e_{j,i}\Theta_{\tau;2,1}$: it moves the ith row of $\Theta_{\tau;2,1}$ to the jth row in the new $\Theta_{n;1,1}$, so that the position (i, j) arrives on the diagonal, and all positions right of it become upper. Notice also that in $\Theta_{\tau;2,1}$ the element in position (i, j) is zero (it does not belong to the actual τ), and hence it will not influence the diagonal of $\Theta_{n;1,1}$. Similar things happen for the other three block entries *mutatis mutandis*. Actually, and as already remarked, more is true: the main diagonal of $\Theta_{n;1,1}$ is simply the product of d_j with the existing main diagonal; of $\Theta_{n;2,2}$ the product of d_i with the main diagonal of $\Theta_{\tau;2,2}$, and the other main diagonals remain zero. (We shall exploit this result in the norm approximation of the Section 14.4.)

As derived in Proposition 13.2, the result is that Y_τ derived from Θ_τ is causal, and $C_\tau = \frac{1}{2}(Y_\tau + Y_\tau')$ is strictly positive definite. Next we have to show that it also solves the maximum entropy interpolation problem, namely that Y_τ is equal to C on τ and C_τ^{-1} is zero on σ. By the same token, C_τ solves the maximum entropy interpolation problem for data given on the staircase induced diagonal band τ.

13.4 The (Generalized) Schur Interpolation Property

Let us assume that, starting from a covariance matrix C and related lower PR matrix Y, assumed known (or specified) only on the complement τ of a staircase σ (a staircase induced band around the diagonal), we have derived, by successive elimination with hyperbolic transformations, a standard Θ-matrix Θ_τ and an admissible remainder $\begin{bmatrix} \Gamma_\sigma \\ \Delta_\sigma \end{bmatrix}$, such that the support of Δ_σ is the staircase σ. By Proposition 13.2, there are lower matrices Y_τ, L_τ and M_τ, and a positive definite matrix $C_\tau = \frac{1}{2}(Y_\tau + Y_\tau') = L_\tau L_\tau' = M_\tau' M_\tau$ such that

$$\Theta_\tau = \begin{bmatrix} M_\tau^{-\prime}\frac{1}{2}(I + Y_\tau') & M_\tau^{-\prime}\frac{1}{2}(I - Y_\tau') \\ L_\tau^{-1}\frac{1}{2}(I - Y_\tau) & L_\tau^{-1}\frac{1}{2}(I + Y_\tau) \end{bmatrix}. \tag{13.26}$$

Theorem 13.5 gives the interpolation property.

Theorem 13.5 *C_τ interpolates C on $\tau \cup \tau'$, and C_τ^{-1} has support on $\tau \cup \tau'$, and is hence a maximum entropy interpolation for C based on the data supported by τ.*

Proof

On the one hand, by the elimination procedure, we have

$$\begin{bmatrix} \Gamma_\sigma \\ \Delta_\sigma \end{bmatrix} = \Theta_\tau \begin{bmatrix} \frac{1}{2}(Y+I) \\ \frac{1}{2}(Y-I) \end{bmatrix}; \tag{13.27}$$

on the other hand, we also have

$$\begin{bmatrix} M_\tau \\ 0 \end{bmatrix} = \Theta_\tau \begin{bmatrix} \frac{1}{2}(Y_\tau+I) \\ \frac{1}{2}(Y_\tau-I) \end{bmatrix}. \tag{13.28}$$

Taking differences, we find

$$\begin{bmatrix} \Gamma_\sigma - M_\tau \\ \Delta_\sigma \end{bmatrix} = \Theta_\tau \begin{bmatrix} \frac{1}{2}(Y-Y_\tau) \\ \frac{1}{2}(Y-Y_\tau) \end{bmatrix} = \begin{bmatrix} M_\tau^{-\prime} \\ L_\tau^{-1} \end{bmatrix} (Y - Y_\tau). \tag{13.29}$$

The second block row gives

$$Y - Y_\tau = L_\tau \Delta_\sigma. \tag{13.30}$$

The support of $L_\tau \Delta_\sigma$ is σ, because L_τ is a lower matrix and left multiplication of Δ_σ by a lower triangular matrix does not change the support. Hence, Y_τ interpolates Y on σ. Moreover, $C_\tau = \frac{1}{2}(Y_\tau + Y_\tau') = L_\tau L_\tau'$ by Proposition 13.2, therefore C_τ^{-1} is zero on σ so that C_τ is a maximum entropy interpolation for C based on the data with support on τ. □

Some Further Properties*

The determination of Θ_τ allows us to generate *all* PR interpolations, using independent parameters. To show this, we start out with a simple observation. Instead of working with Y as the quantity to be interpolated, given data on the complement of a staircase denoted by τ, we can equivalently work on the corresponding *Cayley transform* $S = (Y+I)^{-1}(Y-I)$, which is strictly contractive, so that $(I-S)^{-1}$ exists and $Y = (I-S)^{-1}(I+S)$. Now,

Lemma 13.6 *The data of Y on the support τ determine uniquely the data of S on the same support τ, and vice versa.*

Proof

Just as in Proposition 13.2, the premultiplication or postmultiplication of $Y-I$ or $S+I$, with a lower triangular matrix maps data on τ to data on τ one-to-one and onto, thanks to the staircase structure of τ. □

It follows that an interpolation of Y on τ corresponds to an interpolation of S on τ and vice versa. We can write, starting the same elimination as before but now with the data $\gamma_0 := I$ and $\delta_0 := S$:

$$\Theta_\tau \begin{bmatrix} I \\ S \end{bmatrix} = \begin{bmatrix} \gamma_\sigma \\ \delta_\sigma \end{bmatrix} := \begin{bmatrix} I \\ S_L \end{bmatrix} \gamma_\sigma, \tag{13.31}$$

whereby $S_L = \delta_\sigma \gamma_\sigma^{-1}$ is again a strictly contractive matrix, equal to $\Delta_\sigma \Gamma_\sigma^{-1}$, and has support on σ. Hence, after some manipulation and dropping the index τ, $S_L \Theta_{1,1} - \Theta_{2,1} = (\Theta_{2,2} - S_L \Theta_{1,2})S$. Thanks to the J-unitarity of Θ_τ, $\Theta_{2,2} - S_L \Theta_{1,2}$ is invertible:

Lemma 13.7 $\Theta_{1,2}\Theta_{2,2}^{-1}$ *is strictly contractive and so is* $S_L\Theta_{1,2}\Theta_{2,2}^{-1}$ *when* S_L *is contractive.*

Proof

Since $\Theta_{1,2}'\Theta_{1,2} - \Theta_{2,2}'\Theta_{2,2} = -I$ and $\Theta_{2,2}$ is invertible, we have

$$I - \Theta_{2,2}^{-\prime}\Theta_{1,2}'\Theta_{1,2}\Theta_{2,2}^{-1} = \Theta_{2,2}^{-\prime}\Theta_{2,2}^{-1}. \tag{13.32}$$

Hence, $\Theta_{1,2}\Theta_{2,2}^{-1}$ is strictly contractive, and so is $S_L\Theta_{1,2}\Theta_{2,2}^{-1}$ as the product of two (strictly) contractive matrices. □

It follows that $\Theta_{2,2} - S_L\Theta_{1,2} = (I - S_L\Theta_{1,2}\Theta_{2,2}^{-1})\Theta_{2,2}$ is invertible, and we find $S = (\Theta_{2,2} - S_L\Theta_{1,2})^{-1}(S_L\Theta_{1,1} - \Theta_{2,1})$. It turns out that this bi-linear expression characterizes *all* possible interpolations of the given data on τ that produce a strictly contractive S or strictly PR Y. We formulate this as a proposition.

Proposition 13.8 *All strictly contractive interpolations of the given data on the staircase complement* τ *are given by*

$$S = (\Theta_{2,2} - S_\sigma\Theta_{1,2})^{-1}(S_\sigma\Theta_{1,1} - \Theta_{2,1}), \tag{13.33}$$

in which S_σ *is any strictly contractive matrix with support on* σ.

Proof

The previously derived recursion shows that if S is such that it is strictly contractive and interpolates the data given on τ, then the recursion produces Θ_τ and results in a contractive $S_\sigma = \gamma_\sigma^{-1}\delta_\sigma$ with support on σ.

For the converse, one has to show that if S_σ is any contractive matrix with support on σ, then S as given by (13.33) is causal contractive. A simple proof now rests on the observation that both terms in the ratio are lower and invertible: $\Theta_{1,1}$ and $\Theta_{1,2}$ have support τ' so that the products $S_c\Theta_{1,1}$ and $S_c\Theta_{1,2}$ are lower when S_c has support on σ. Since $\Theta_{2,2} - S_\sigma\Theta_{1,2}$ is invertible when S_σ is contractive, the inverse is lower when S_c has support on σ. Hence, it follows that S is both contractive and lower. □

Remark 1: Using the J-unitarity of Θ_τ, in particular, using $\Theta_{2,1}\Theta_{1,1}' = \Theta_{2,2}\Theta_{1,2}'$, the invertibility of $\Theta_{2,2}$ and $\Theta_{1,1}$, and $\Theta_{1,1}\Theta_{1,1}' - \Theta_{1,2}\Theta_{1,2}' = I$, one finds

$$S = -\Theta_{2,2}^{-1}\Theta_{2,1} + (\Theta_{2,2} - S_\sigma\Theta_{1,2})^{-1}S_\sigma\Theta_{1,1}^{-\prime}. \tag{13.34}$$

$\Theta_{1,1}^{-\prime}$ is lower with support on τ. However, $\Theta_{1,2}\Theta_{2,2}^{-1}$ is necessarily of mixed causality and has support on $\tau \cup \tau'$.

Remark 2: We shall see in further chapters on constrained interpolation that the theorem generalizes dramatically. In the proof given here, and the (more complex) proofs given for the more general cases, heavy use is made of the properties of the Θ and S matrices involved. These matrices are characteristic of what is known as *scattering theory*, which we discuss in Chapter 14 and further chapters as well.

Parametrization

The next (important) observation is that, given a staircase, the corresponding $\rho_{i,j}$ *parametrize* the solution, that is, produce a set of independent numbers that characterize the solution (as well as the original data). They are called *reflection coefficients*, for reasons that will be discussed in Chapter 14. If the data on a support τ are given, then the corresponding $\{\rho_{i,j}\}|_\tau$ are fixed (of course, with $|\rho_{i,j}| < 1$). The reflection coefficients for data on σ can be chosen at will, with the only restriction that such a $\rho_{i,j}$ must be smaller than 1 in magnitude. In other words, it is *free* in the set of coefficients with magnitude smaller than 1. Hence, it parametrizes the solution. One can show that this parametrization is also one-to-one: two different parametrizations will produce different solutions. It takes some reasoning to show this: any parametrization is determined respecting staircase structures, but a final staircase can be constructed in many ways. It turns out that any of these will produce the same set of reflection coefficients, provided the staircase structure is indeed respected at each step, or, in other words, the determination of reflection coefficients is not affected by the order, in which the staircase gets reduced, as long as it remains a staircase after each step.

13.5 Practical Realizations

Normalization

It is standard numerical practice to work on a normalized covariance $\underline{C} = D^{-1/2}CD^{-1/2}$, in which D is the main diagonal of C, if possible.[1] Θ is then derived from $\underline{C}$, and in that case, the main diagonal, of $\underline{Y}$ is I so that $\underline{Y} - I$ has zero main diagonal, and any partial matrix Θ_a derived from it through Schur steps will have the $(2, 1)$ block entry strictly lower. Since $\underline{C} = \underline{L}\underline{L}' = \underline{M}'\underline{M}$ (and likewise for an interpolation based on data in a staircase), we have $L = D^{1/2}\underline{L}$ and $M = \underline{M}D^{1/2}$ (and likewise for the interpolating approximation), and we may further remark that if $\underline{C}_a$ interpolates $\underline{C}$ with respect to a staircase, so will $C_a := D^{1/2}\underline{C}_a D^{1/2}$ interpolate C on the same related data. Although it is possible (and not too difficult) to develop an unnormalized algorithm using also unnormalized rotations, we shall only work out fully normalized examples. This just simplifies the theory at the cost of supplementary (normalization) operations. So we assume C to be normalized in the following sections.

Array Processing of the Θ Matrix

Let us start out with representing an elementary hyperbolic rotation graphically (Fig. 13.2). It has two modes of operations: the "vectoring mode" where it computes its characteristic reflection coefficient ρ using the incoming data, and the "rotating mode" when it applies its hyperbolic rotation to incoming data (understood is that the vectoring mode also *applies* the rotation to the incoming data it has already used to compute

[1] Entries in the covariance matrix may have widely different dimensions, as they may relate different types of quantity with each other, such as temperature with density or distance with luminosity. Normalizing the autocorrelations puts them on equal footing.

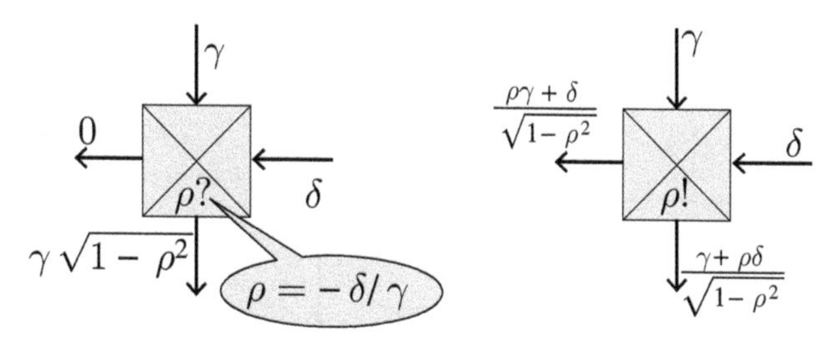

Figure 13.2 Hyperbolic rotors in vectoring and rotating modes (see text).

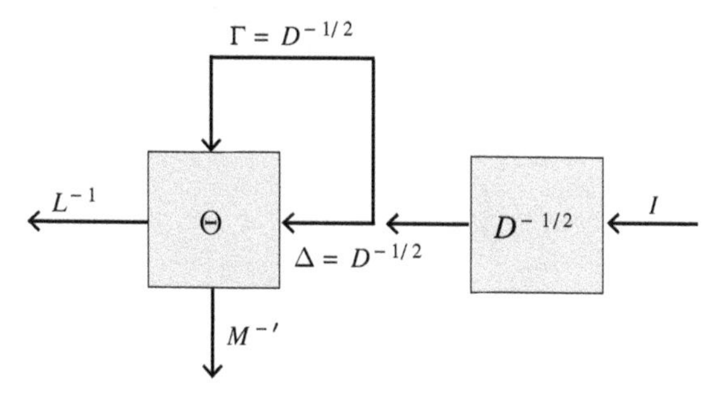

Figure 13.3 Generating L^{-1} from the Θ matrix.

ρ – see Fig. 13.2). For array processing purposes, we let the Γ-data flow from right to left and the Δ-data from top to bottom (see further the completed schemas). In the realization diagrams, we assume that the rotor actually knows whether it is in vectoring or rotating mode. (The former is usually at the beginning of the matrix operations, and in the latter, it already knows its ρ.)

The purpose of the processing is either to produce a low-complexity "model" for a stochastic process whose covariance is C (see for this the section on the Levinson–Schur method) or to produce an approximate inverse (sometimes called a "pre-conditioner" – see the notes at the end of the chapter). In the first case, it is the matrix L_a; and in the second case, it is the matrix L_a^{-1} derived from a Θ_a that has been computed recursively. This can be easily done simply by manipulating the computation schema that has determined Θ_a. Indeed, we have

$$\Theta_a \begin{bmatrix} I \\ I \end{bmatrix} = \begin{bmatrix} M_a^{-'} \\ L_a^{-1} \end{bmatrix}. \tag{13.35}$$

This immediately leads to the schemas shown in Fig. 13.3, in which also the normalization is included. The flow graph realization of the Θ matrix itself is just an assembly of the elementary hyperbolic rotations previously discussed in an array schema. This is perhaps best illustrated with a 4×4 example, in which the following (staircase-based) hyperbolic rotations are executed: (2, 1), (3, 2), (4, 3) for the full first lower off-diagonal and then, for example, just the element (3, 1) of the second lower off-diagonal. The resulting data flow graph is shown in Fig. 13.4.

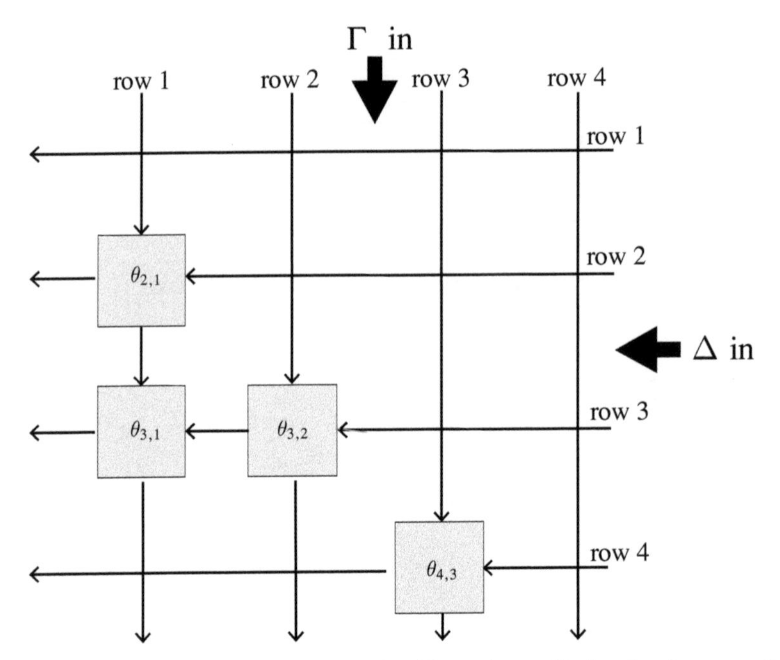

Figure 13.4 A 4×4 elementary example with a staircase. Notice that the arrow directions have been chosen so as to keep the original matrix form of the Θ matrix in the flow graph. Rows are inputted from low to high order; matrix elements that are not used have not been presented in the input stacks. The first operation of each θ-matrix is a "vectoring" operation, the next a rotation – but only $\theta_{2,1}$ has to perform that operation because of the simple structure of the staircase here. The schema is ready for use when all reflection coefficients have been computed.

Generating the L Matrix by Arrow Inversion

Inverting one of the arrow directions in an elementary hyperbolic rotation turns it into a circular rotation. This is shown in Fig. 13.5. By systematically reversing the horizontal arrows in Figs. 13.3 and 13.4, an array processor using circular rotations is obtained which generates L, see Fig. 13.6.

The Toeplitz Case

In the case the original (finite) covariance or positive definite matrix is Toeplitz, then the schema of reflection coefficients will be Toeplitz as well, when full diagonals are taken out one at a time, of course. This can easily be seen from the computation schema for Θ.

Computational Complexity

The computation of the Θ matrix corresponding to an approximate C_a of (average) bandwidth b involves $\frac{1}{2}nb$ hyperbolic vectoring computations (where b is the number of side diagonals involved and $b \ll n$, which is usually the case). This results in a computational complexity ranging from $\frac{1}{2}nb$ to $2nb$ depending on how the hyperbolic rotations are executed (e.g., this could be done using a CORDIC processor, or in an unnormalized fashion, in both cases resulting in the lowest complexity number, or

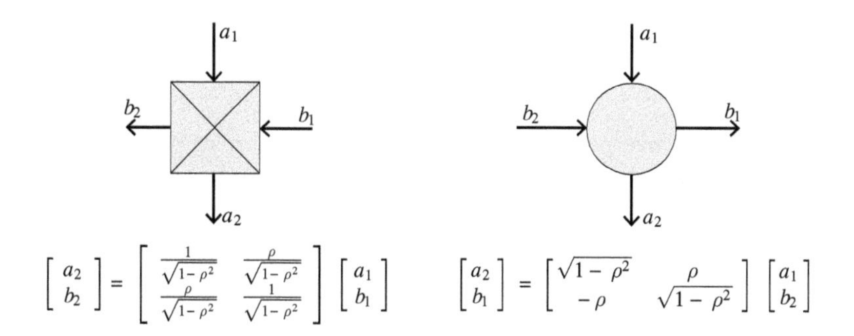

Figure 13.5 Inverting arrows in one direction turns a hyperbolic rotation into a circular one.

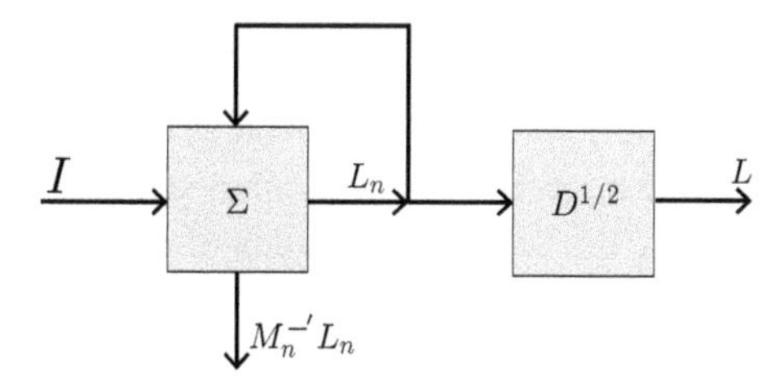

Figure 13.6 Generating L using a unitary operator Σ derived from Θ.

just by straight normalized arithmetic, resulting in the highest number of operations). Using the resulting L^{-1} or L to compute a matrix–vector product then has the same complexity. In the Toeplitz case, the original Θ matrix can be computed with complexity b, but the computations of matrix–vector products will again have a complexity of order nb. These levels of complexity are very near the algebraic minimum (which is determined here by the number of algebraically independent real data items).

13.6 Approximation in Norm

Matrix Schur interpolation as discussed in this chapter is an example of constrained interpolation, that is, interpolation with given data and with a constraint on the result, either a norm constraint or a positivity constraint. We shall discuss those in more generality using state-space models in Chapter 15. However, such (constrained) interpolation problems lead to much stronger approximation results than just maximum entropy, which we now discuss and which already appear even in the (deceivingly elementary) case considered here. Because of their elegance and connection with other theories, we briefly explore them next.

To get a good grasp on error norms, we define a new type of extended or *matrix inner product*. Suppose M and N are two square matrices (say of dimensions $n \times n$), let Π_0 be the projection on the main diagonal, and let C be a strictly positive definite normalized matrix; then we define $\langle M, N \rangle_C := \Pi_0(N'CM)$. That this may be considered a *matrix inner product* is due to the fact that the quadratic form $(M, N)_C := \frac{1}{n}\mathrm{Trace}(\langle M, N \rangle_C)$ is a true inner product (of Frobenius type). Orthogonality in the matrix inner product is a stronger property than in the usual inner product. $\langle M, N\rangle_I = 0$ states that the ranges of M and N are orthogonal to each other for the natural inner product.

Next, Schur interpolation produces some special matrix subspaces based on staircase-type supports σ and τ. A lower matrix with (given) support σ (such as Δ) is said to belong to the matrix subspace $\mathcal{S}$, and likewise for $\mathcal{T}$ with respect to the corresponding τ. We also denote their complements in the upper matrices as $\mathcal{S}'$ and $\mathcal{T}'$. Some special, immediate properties of these spaces are:

1. $\mathcal{S}\mathcal{T} \subset \mathcal{S}$ as well as $\mathcal{T}\mathcal{S} \subset \mathcal{S}$ (by $\mathcal{S}\mathcal{T}$ is meant the set obtained by taking a product of a matrix in $\mathcal{S}$ with a matrix in $\mathcal{T}$);
2. for the natural Hilbert–Schmidt or Frobenius matrix inner product (i.e., with $C = I$), $\mathcal{S} \perp \mathcal{T}$ and $\mathcal{S} \perp \mathcal{T}'$ (check the diagonal products and use the complementarity of σ and τ or look at natural bases for these spaces).

Let us further denote by $\Pi_{\mathcal{S}}$ and $\Pi_{\mathcal{T}}$ the projections of $n \times n$ matrices on the respective spaces, and let $C_\tau = L_\tau L_\tau' = M_\tau' M_\tau$ be the Schur interpolation of $C = LL' = M'M$ based on τ. Then also the following properties of the matrix inner products hold:

1. For all F_1 and F_2 in $\mathcal{T}_\tau$, one has $\langle F_1, F_2\rangle_C = \langle F_1, F_2\rangle_{C_\tau}$ (Proof: let $C - C_\tau := \Delta_\sigma + \Delta_\sigma'$ with $\Delta_\sigma \in \mathcal{S}_\sigma$. Then
$$\Pi_0(F_2'(C - C_\tau)F_1) = \Pi_0(F_2'\Delta_\sigma F_1 + F_2'\Delta_\sigma' F_1) = 0, \tag{13.36}$$
because both $\Delta_\sigma F_1$ and $\Delta_\sigma F_2$ are in $\mathcal{S}_\sigma$ and, e.g., $\Pi_0(F_2'\Delta_\sigma F_1)$ is the matrix inner product of an element in $\mathcal{T}$ with and element in $\mathcal{S}$);
2. $\|M_\tau\|_C^{-1} = 1$, or for the matrix innerproduct: $\langle M_\tau^{-1}, M_\tau^{-1}\rangle_C = I$ (Proof: $\langle M_\tau^{-1}, M_\tau^{-1}\rangle_C = \langle M_\tau^{-1}, M_\tau^{-1}\rangle_{C_\tau} = I$);
3. let F be lower and $G \in \mathcal{T}_\tau$. Then
$$\left\langle F, M^{-1}G\right\rangle_C = \left\langle \Pi_{\mathcal{T}} F, M^{-1}G\right\rangle_C. \tag{13.37}$$
(Proof: the difference $R = F - \Pi_{\mathcal{T}} F$ has $\langle R, M^{-1}G\rangle_C = \Pi_0(G'(MR))$, whose factors have disjoint supports.)

This then leads to the following propositions (using $M_0 := \Pi_0 M$ and $M_{\tau 0} := \Pi_0 M_\tau$):

Proposition 13.9 *$M_\tau^{-1} M_{\tau 0}^{-1}$ is the projection of $M^{-1} M_0^{-1}$ on $\mathcal{T}_\tau$ in the C-metric. The matrix norm of the orthogonal complement is given by $M_0^{-2} - M_{\tau 0}^{-2}$.*

Proof

We show that the difference $M^{-1} M_0^{-1} - M_\tau^{-1} M_{\tau 0}^{-1}$ is orthogonal on any $G \in \mathcal{T}_\tau$:
$$\begin{aligned} &\left\langle M^{-1} M_0^{-1} - M_\tau^{-1} M_{\tau 0}^{-1}, G\right\rangle_C \\ &\quad = \Pi_0\left(G'M'MM^{-1}M_0^{-1} - G'CM_\tau^{-1}M_{\tau 0}^{-1}\right) \\ &\quad = G_0'M_0'M_0^{-1} - G_0'M_{\tau 0}'M_{\tau 0}^{-1} = 0, \end{aligned} \tag{13.38}$$
in which we have used the property
$$\left\langle M_\tau^{-1} M_{\tau 0}^{-1}, G\right\rangle_C = \left\langle M_\tau^{-1} M_{\tau 0}^{-1}, G\right\rangle_{C_\tau}, \tag{13.39}$$

denoting the fact that $\Pi_0 M$ and $\Pi_0 M_\tau$ are real and the diagonals commute. To compute the error, one uses projection theory, from which it follows that the error is simply given by

$$\begin{aligned}
&\left\langle M^{-1}M_0^{-1} - M_\tau^{-1}M_{\tau 0}^{-1}, M^{-1}M_0^{-1}\right\rangle_C \\
&\quad = \Pi_0\left(M_0^{-\prime}M^{-\prime}M^{\prime}M\left(M^{-1}M_0^{-1} - M_\tau^{-1}M_{\tau 0}^{-1}\right)\right) \\
&\quad = M_0^{-\prime}M_0^{-1} - \Pi_0\left(M_0^{-\prime}MM_\tau^{-1}M_{\tau 0}^{-1}\right) \\
&\quad = M_0^{-2} - M_{\tau 0}^{-2},
\end{aligned} \tag{13.40}$$

where the diagonal projector just propagates when the product involves one-sided matrices of the same type (upper or lower). □

The final result produces an approximation in a strong norm, which is remarkable, since the Schur recursion only solves an interpolation problem.

Proposition 13.10 $\widehat{M_\tau} := M_0^{-1}M_{\tau 0}M_\tau$ *provides a strong approximation to* M *in the sense that for the Frobenius norm*

$$\|I - M\widehat{M_\tau}^{-1}\|_F^2 = \text{Trace}\left(I - M_0^2 M_{\tau 0}^{-2}\right). \tag{13.41}$$

Proof

Using the general relation $\langle F,F\rangle_C = \langle MF, MF\rangle_I$ and the previous proposition, we have

$$\left\langle M_0^{-1} - MM_\tau^{-1}M_{\tau 0}^{-1}, M_0^{-1} - MM_\tau^{-1}M_{\tau 0}^{-1}\right\rangle_I = M_0^{-2} - M_{\tau 0}^{-2}. \tag{13.42}$$

Pre- and postmultiplying with M_0 gives

$$\left\langle I - MM_\tau^{-1}M_{\tau 0}^{-1}M_0, I - MM_\tau^{-1}M_{\tau 0}^{-1}M_0\right\rangle_I = I - M_{\tau 0}^{-2}M_0^2. \tag{13.43}$$

Taking traces both sides finally gives the result. □

We see that a diagonally normalized version of the Schur interpolation based on the support τ gives a (relative) Frobenius approximation to the Cholesky factor M. Only the normalization depends on, in principle, unknown data. The closer the $M_{\tau 0}^{-1}M_0$ to I the better the approximation. This normalization term, which also appears in the error formula, can be expressed directly in terms of the reflection coefficients one would obtain if one would push the interpolation to the full end, whereby all elements in the original Δ are eliminated. This is a consequence of the fact, already observed, that the main diagonal of M^{-1} consists exclusively of products of the elementary d_i – the diagonals of the elementary factors. Consider Θ_τ and its M_τ^{-1}. Suppose that at row i, the support τ starts at $j_i < i$; then all elements between (i, j_i) and (i,i) belong to τ, while the other elements on row i do not belong to τ because of the staircase structure. It follows that $\left[M_\tau^{-1}\right]_{i,i} = \prod_{j=j_i}^{i} \frac{1}{\sqrt{1-\rho_{i,j}^2}} \left(= M_{\tau 0}^{-1}\right)$. A similar expression holds for M_0^{-1}, now with τ being the full lower support and the set of reflection coefficients being expanded to all lower entries. Hence,

$$\left[M_{\tau 0}^{-1}M_0\right]_{i,i} = \prod_{j=1}^{j_i-1}\sqrt{1-\rho_{i,j}^2}, \tag{13.44}$$

and the approximation shall be good when all these additional $\rho_{i,j} \sim 0$ (all located on σ). We finally obtain for the Frobenius error,

$$\left\| I - M\widehat{M_\tau}^{-1} \right\|_F^2 = n - \sum_{i=1}^{n} \prod_{j=1}^{j_i - 1} \left(1 - \rho_{i,j}^2\right), \tag{13.45}$$

where entries may disappear and the sum would be n when all $\rho_{i,j} = 0$.

13.7 Discussion

1. **Preconditioning**
   In many iterative methods to solve systems of linear equations, a good preconditioner is essential. In the case of positive definite matrices, it is equally essential that the preconditioner be positive definite itself. An iterative method to solve a system of equations $Ax = b$ requires (1) a low-complexity multiplication Ax and (2) a pre-conditioner P which is such that the multiplication of a vector with P^{-1} is cheap as well, and such that P^{-1} is a good approximation of A^{-1}. It is not a good idea to derive P by putting a number of diagonals in A equal to zero, since the result may not be positive definite (and often will not be in large matrices), hence may be close to singular as well, if not with an inverse far away from A^{-1}. The Schur interpolation technique provides for an ideal preconditioner, which fulfills all the requirements: minimal complexity and close approximation!
2. **Entropy**
   The standard definition for "entropy" of a collection $\{X_k\}_{k=1\cdots m}$ of continuous, zero-means and jointly Gaussian variables with probability distribution $p(x) = \frac{1}{\sqrt{(2\pi)^m \det C}} e^{-\frac{1}{2}x'C^{-1}x}$ is the expectation

$$\mathbf{E}\left[-\ln\left(p(x)\right)\right]. \tag{13.46}$$

   This then evaluates to $\frac{1}{2}\ln\left[(2\pi e)^m \det C\right]$ – see [17, p. 230]. This will be maximum together with $\ln\det C$, the form we used, and $\det C$ as well. The function $\ln\det C(\cdot)$ in terms of an entry $c_{i,j}$ of C is convex but very flat, so that its maximum is ill defined (it is actually of the form $\ln\left(ax^2 + bx + c\right)$, with $a < 0$). The result generalizes: when one optimizes on a set of entries, the function "det" to be maximized will be a multivariable quadratic polynomial, which is necessarily convex, and hence a convex optimization problem results, for which efficient algorithms exist, with the issue being the computation of the determinant in function of multiple unknown entries.
3. **Beyond a Single Principal Band Structure**
   The reasoning we applied to show that maximum entropy corresponds to a zero entry in the inverse applies no matter which structure the support for interpolation has, provided the main diagonal is included. This leads to the important result that no matter what this pattern is, there will be a maximal entropy solution, *provided the pattern originates from a matrix that is known to be positive definite*, and the resulting maximum entropy interpolating matrix will have the same pattern of zeros

in its inverse. This seems actually a very strong result, the catch being that the maximum entropy interpolating matrix is hard to compute, except for staircase patterns, or patterns derived from staircases after reshuffling, where we showed that a direct computation with minimal algebraic complexity is indeed possible.

4. **The Multiband Case**
   In the multiband case, an extension of the Schur method may lead to useful approximations with low computational complexity. For an extensive discussion of this case, we refer to the literature [53].
5. **Causal Embeddings**
   In the subsequent chapters on interpolation, we shall use chain scattering matrices Θ that correspond to inner scattering operators Σ (these have to be causal by definition.) The simple Θ matrices used in this chapter do not have that property but can easily be brought into what is called the "lossless" form through a simple re-indexing. This can be done by premultiplication with an inner U. Such a U follows from an external factorization of $M^{-1} := \Delta' U$ in Eq. 13.12, or, in other words, from an input normal form for M^{-1}. Rather than deriving a general expression for U with much bookkeeping, let us just look at a representative example (the procedure generalizes to arbitrary staircases). Consider the M^{-1} matrix

$$M^{-1} = \begin{bmatrix} m_{1,1} & & & & \\ m_{2,1} & m_{2,2} & & & \\ m_{3,1} & m_{3,2} & m_{3,3} & & \\ & & & m_{4,4} & \\ & & & m_{5,4} & m_{5,5} \end{bmatrix}. \tag{13.47}$$

A realization in the input normal form is easily found as

$$\operatorname{diag}\left(\left[\begin{array}{c|c|c} & & 1 \\ \hline & & m_{1,1} \end{array}\right], \left[\begin{array}{c|c} 1 & 0 \\ 0 & 1 \\ \hline m_{2,1} & m_{2,2} \end{array}\right], \left[\begin{array}{cc|c} - & - & - \\ \hline m_{3,1} & m_{3,2} & m_{3,3} \end{array}\right], \left[\begin{array}{c|c|c} & & 1 \\ \hline & & m_{4,4} \end{array}\right], \left[\begin{array}{c|c} - & - \\ \hline m_{5,4} & m_{5,5} \end{array}\right] \right), \tag{13.48}$$

yielding the realization for the *staircase shift matrix* U:

$$\operatorname{diag}\left(\left[\begin{array}{c|c|c} & & 1 \\ \hline & \cdot & - \end{array}\right], \left[\begin{array}{c|c} 1 & 0 \\ 0 & 1 \\ \hline - & - \end{array}\right], \left[\begin{array}{cc|c} - & - & - \\ \hline 1 & 0 & 0 \\ 0 & 1 & 0 \\ 0 & 0 & 1 \end{array}\right], \left[\begin{array}{c|c|c} & & 1 \\ \hline & \cdot & - \end{array}\right], \left[\begin{array}{c|c} - & - \\ \hline 1 & 0 \\ 0 & 1 \end{array}\right] \right). \tag{13.49}$$

Using the shifted version $U_a M_a^{-\prime}$ instead of $M_a^{-\prime}$ in the expressions for Θ_a and the corresponding Σ_a yields a corresponding Σ_a matrix that is causal and a Θ_a that is J-unitary and may be called *lossless*, because it represents a causal and unitary (inner) scattering system (it so happens that Θ_a is causal as well). Remarkably, this consists only of index bookkeeping, numerically the matrices remain the same! The lossless forms are the ones that we shall use systematically in the chapter on constrained interpolation.

13.8 Notes

1. The nonstationary Schur algorithm was originally presented, to the best of our knowledge, by E. Deprettere in [22]. The norm approximation theory was added to this in [27]. The theory can be extended in an approximate fashion to handle multidimensional data [53] (like a 2D Poisson problem) and has been extensively used for very large-scale-integrated (VLSI) circuit modeling, as part of an efficient layout-to-circuit extractor [67].
2. An interesting alternative viewpoint on maximum entropy extensions (like what the Schur algorithm does) is shown in [35], in which partial inversions are used to determine the extension, much in the style of the original Levinson inversion method.

14 The Scattering Picture

In the previous chapters, we have time and again seen the important role played by orthogonal transformations. They have appeared in various types of operations, in particular factorizations and embeddings. A crucial property of these transformations as we encountered them is that they themselves are realized by a causal (or sometimes anti-causal) system with a low state degree, derived from the state-space properties of the original system. In Chapter 13, we discovered that J-unitary systems are instrumental in building low-rank approximations of a positive real system that are positive real themselves and meet interpolation conditions of the Schur type, namely, interpolating the first few terms of a series expansion in Z of the transfer function. Unitary and related J-unitary systems enjoy an impressive physical interpretation as *scattering operators* that classically represent the transmission and reflection of *waves* in what can be seen as a lossless, layered medium. This chapter is devoted to presenting the full scattering picture and deriving further properties of the unitary and J-unitary systems involved. These properties are then used extensively in the two subsequent chapters on interpolation.

Menu

Hors d'oeuvre
The Physical Scattering Picture

Course 1
Fractional Transformation

Course 2
Chain Scattering Matrices

Course 3
Indefinite Metrics

Dessert
Completion of a J-Isometric Basis

14.1 Introduction

An outline of the physical scattering picture is as follows. Envisage a partially transparent but nonabsorbing transmission medium (e.g., a stack of glassy slabs) that partitions

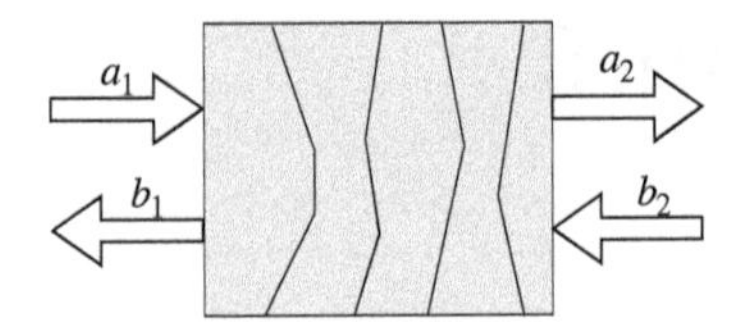

Figure 14.1 A rough sketch of the scattering picture.

space in two sides, with left and right, called side 1 and side 2. At each side, there are incident waves and *ports* through which the waves enter the medium, which then partly propagate through the medium, and are partly reflected as well, and leave the medium back into space through the ports. Because the medium is assumed nonabsorbing, no energy gets dissipated in the process, so that part of the incident energy (whether coming from side 1 or side 2) gets reflected back to the incident side and another part is transmitted. The medium is assumed further to be linear, so that the effects of various incident waves are simply additive, and, as it befits a physical situation, the propagation is assumed causal as well. See Fig. 14.1 for a rough sketch.

In the case we consider, we assume all waves (incident and outgoing) to be signals of the type used in the book up to now, that is, belonging to some time-variant $\ell_2^{\mathbf{k}}$ space, with the quadratic norm squared interpreted as "energy." With reference to Fig. 14.1, the incident waves at the left side are combined in a global signal $a_1 \in \ell_2^{\mathbf{m}_1}$, and at the right side in $b_2 \in \ell_2^{\mathbf{n}_2}$, while the outgoing waves (reflected and transmitted) are represented respectively by $a_2 \in \ell_2^{\mathbf{m}_2}$ and $b_1 \in \ell_2^{\mathbf{n}_1}$. Energy is defined simply as the ℓ_2 norm, and the conservation of energy forces

$$\|a_1\|_2^2 - \|b_1\|_2^2 = \|a_2\|_2^2 - \|b_2\|_2^2 \tag{14.1}$$

or, to put it differently, global incident energy equals global outgoing energy. This actually means that all energy that gets stored in the medium by incident waves eventually leaks out again through one of the ports.

A basic "physical" assumption about the setup is that all incident waves in $\ell_2^{\mathbf{m}_1} \oplus \ell_2^{\mathbf{n}_2}$ are allowed and that they produce unique outgoing waves, respecting, moreover, causality. This allows one to define the *scattering map*:

$$\Sigma\colon \ell_2^{\mathbf{m}_1} \oplus \ell_2^{\mathbf{n}_2} \to \ell_2^{\mathbf{m}_2} \oplus \ell_2^{\mathbf{n}_1} : \begin{bmatrix} a_2 \\ b_1 \end{bmatrix} = \Sigma\left(\begin{bmatrix} a_1 \\ b_2 \end{bmatrix}\right). \tag{14.2}$$

The further assumption is linearity:

$$\begin{bmatrix} a_2 \\ b_1 \end{bmatrix} = \begin{bmatrix} \Sigma_{1,1} & \Sigma_{1,2} \\ \Sigma_{2,1} & \Sigma_{2,2} \end{bmatrix} \begin{bmatrix} a_1 \\ b_2 \end{bmatrix}, \tag{14.3}$$

in which the $\Sigma_{i,j}$ are causal and contractive transfer functions belonging to a linear space of bounded, causal (i.e., lower) transfer functions, which we call generically $\mathcal{L}^{\mathbf{n}\times\mathbf{m}}$. *Losslessness* means "conservation of energy" and implies that Σ is isometric:

$$\Sigma'\Sigma = I. \tag{14.4}$$

For our treatments and usage of the scattering setup, we make the extra assumption that Σ is actually unitary (we discuss the meaning of this extra assumption in the "Notes" section of this chapter) and restrict ourselves even to systems for which $\Sigma_{2,2}$ is outer (also this issue is discussed further in the "Notes"); hence $\Sigma_{2,2}^{-1}$ exists as defined on a dense domain of $\ell_2^{\mathbf{n}_2}$, and for ease of treatment, we start by assuming $\Sigma_{2,2}^{-1}$ simply causal and bounded. (In Chapter 16, we relax the condition that $\Sigma_{2,2}$ is outer, allowing it to be bounded invertible of mixed causality, but in this chapter and the subsequent chapter on constrained interpolation, we keep $\Sigma_{2,2}^{-1}$ bounded and causal.)

Assuming the existence of $\Sigma_{2,2}^{-1}$, the next observation is the existence of the *chain scattering map*

$$\Theta : \ell_2^{\mathbf{m}_1} \oplus \ell_2^{\mathbf{n}_1} \to \ell_2^{\mathbf{m}_2} \oplus \ell_2^{\mathbf{n}_2} : \begin{bmatrix} \Theta_{1,1} & \Theta_{1,2} \\ \Theta_{2,1} & \Theta_{2,2} \end{bmatrix} = \begin{bmatrix} \Sigma_{1,1} - \Sigma_{1,2}\Sigma_{2,2}^{-1}\Sigma_{2,1} & \Sigma_{1,2}\Sigma_{2,2}^{-1} \\ -\Sigma_{2,2}^{-1}\Sigma_{2,1} & \Sigma_{2,2}^{-1} \end{bmatrix} \quad (14.5)$$

(as can be computed easily from Eq. 14.3).

Under the going assumptions, we see that Θ is causal and bounded when $\Sigma_{2,2}^{-1}$ is causal and bounded. (An extension to Σ outer with $\Sigma_{2,2}^{-1}$ unbounded is possible but requires quite a bit more machinery and only occurs in special infinitely indexed cases, like LTI selective filtering.)

Θ being well defined, the unitarity of Σ is now easily expressible in terms of the Θ matrix:

Proposition 14.1 *Under the going assumptions, Σ is unitary if and only if Θ is J-unitary:*

$$\Theta' \begin{bmatrix} I^{\mathbf{n}_2} & \\ & -I^{\mathbf{m}_2} \end{bmatrix} \Theta = \begin{bmatrix} I^{\mathbf{m}_1} & \\ & -I^{\mathbf{n}_1} \end{bmatrix}, \Theta \begin{bmatrix} I^{\mathbf{m}_1} & \\ & -I^{\mathbf{n}_1} \end{bmatrix} \Theta' = \begin{bmatrix} I^{\mathbf{n}_2} & \\ & -I^{\mathbf{m}_2} \end{bmatrix}. \quad (14.6)$$

Also this proposition is easy to prove. The "J" in J-unitary refers to the traditional notation for a sign matrix. In this case, there is a whole collection of sign matrices with appropriate dimensions: at every index point, there are two relevant sign matrices, at ports 1 and 2, with the "+" sign corresponding to the incident energy and the "–" sign corresponding to the outgoing energy. In the subsequent treatments, we shall often omit these dimensional indications, as their values are already determined by the transfer operators they act upon, and we can write $J_1 := \begin{bmatrix} I^{\mathbf{m}_1} & \\ & -I^{\mathbf{n}_1} \end{bmatrix}$ and similarly J_2. From this follows the invertibility of Θ as well:

$$\Theta^{-1} = J_1 \Theta' J_2, \quad (14.7)$$

a relation that makes Θ in many respects like Σ.

The Σ to Θ transformation is reversible in the same way as the original transformation:

$$\begin{bmatrix} \Sigma_{1,1} & \Sigma_{1,2} \\ \Sigma_{2,1} & \Sigma_{2,2} \end{bmatrix} = \begin{bmatrix} \Theta_{1,1} - \Theta_{1,2}\Theta_{2,2}^{-1}\Theta_{2,1} & \Theta_{1,2}\Theta_{2,2}^{-1} \\ -\Theta_{2,2}^{-1}\Theta_{2,1} & \Theta_{2,2}^{-1} \end{bmatrix}, \quad (14.8)$$

but it may now be remarked that, given Θ J-unitary and causal, the existence of a corresponding outer $\Sigma_{2,2} = \Theta_{2,2}^{-1}$ is not difficult. This is due to the J-unitarity,

$$\Theta_{2,2}'\Theta_{2,2} = I + \Theta_{1,2}'\Theta_{1,2}; \ \Theta_{2,2}\Theta_{2,2}' = I + \Theta_{2,1}\Theta_{2,1}', \quad (14.9)$$

that guarantees the existence of an invertible and contractive $\Sigma_{2,2} = \Theta_{2,2}^{-1}$. However, given a causal J-unitary Θ, the already assumed causal $\Theta_{2,2}$ must be outer as well in order to produce a causal Σ. (As already mentioned, a more general case will be considered in Chapter 16.) But even in the case that $\Sigma_{2,2}$ is not causal, Σ is still unitary, and its component parts are contractive.

An important, somewhat symmetrical, representation for Θ in terms of Σ uses the J-unitarity:

$$\Theta = \begin{bmatrix} \Sigma_{1,1}^{-\prime} & \Sigma_{1,2}\Sigma_{2,2}^{-1} \\ \Sigma_{1,2}^{\prime}\Sigma_{1,1}^{-\prime} & \Sigma_{2,2}^{-1} \end{bmatrix} \tag{14.10}$$

This representation shows that when $\Sigma_{2,2}$ is chosen outer and Θ causal, $\Sigma_{1,1}$ becomes automatically *maximum phase*: it has an anti-causal inverse. This property plays an important role in interpolation theory.

14.2 Fractional Transformation

What happens when a two-sided scattering matrix Σ or chain scattering matrix Θ is loaded by a contractive transfer operator S_L; see Fig. 14.2 (often called a *passive load*)? Of interest is the relation between the *input scattering operator* S and the load S_L, and how the intermediate Σ/Θ affects S. An easy derivation shows

$$\begin{cases} S &= \Sigma_{2,1} + \Sigma_{2,2}S_L(I - \Sigma_{1,2}S_L)^{-1}\Sigma_{1,1} \\ &= (\Theta_{2,2} - S_L\Theta_{1,2})^{-1}(S_L\Theta_{1,1} - \Theta_{2,1}), \\ A_2 &= (I - \Sigma_{1,2}S_L)^{-1}\Sigma_{1,1}, \end{cases} \tag{14.11}$$

provided the inverses exist, which we now show for the conditions we settled on. Since $\Sigma_{2,2}$ is assumed invertible, there is a (perhaps large) positive number M such that $\Sigma_{2,2}^{-1}\Sigma_{2,2}^{-\prime} \le M \cdot I$, and hence an $\epsilon > 0$ such that $I - \Sigma_{1,2}^{\prime}\Sigma_{1,2} \ge \epsilon \cdot I$, making $\Sigma_{1,2}$ strictly contractive (surely $\|\Sigma_{1,2}\| \le \sqrt{1-\epsilon}$ by definition of the operator norm; hence $\Sigma_{1,2}S_L$ is strictly contractive when S_L is contractive (for $\|\Sigma_{1,2}S_L\| \le \|\Sigma_{1,2}\|\|S_L\|$), and hence $(I-\Sigma_{1,2}S_L)$ is invertible. The same holds for $(\Theta_{2,2}-S_L\Theta_{1,2}) = (I-S_L\Sigma_{1,2})\Theta_{2,2}^{-1}$.

However, more is the case: $(I - \Sigma_{1,2}S_L)$ is not only invertible; it is causally invertible. *This will appear to be the crucial mathematical property!* The property follows from the following lemma, applied to $\Sigma_{1,2}S_L$.

Lemma 14.2 *Let s be causal and strictly contractive; then $I - s$ is outer with bounded (and causal) inverse.*

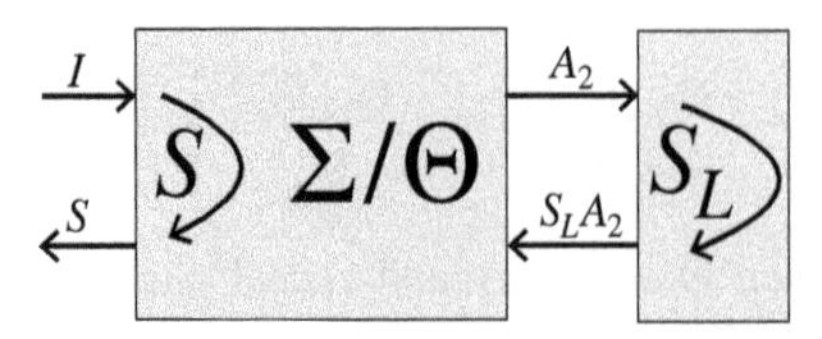

Figure 14.2 A two-sided lossless scattering or chain scattering matrix loaded in a passive load S_L.

Sketch of a standard proof: The Neumann series $I + s + s^2 + \cdots$ converges in norm to a causal operator t for which $t(I - s) = (I - s)t = I$ holds. Hence, $t = (I - s)^{-1}$. $1 - s$ hence has a (unique) causal bounded inverse and is therefore outer by definition. □

The Connection with Interpolation

Formula (14.11) gives the relation between S and S_L as produced by Σ or Θ, which we write shorthand as

$$S = \sigma_\Sigma(S_L)(= \sigma_\Theta(S_L)), \tag{14.12}$$

and its interpretation goes as follows. $\Sigma_{2,2}$ is the "reverse gain" for a signal that travels back from the side of the load to the left port. We require it to be outer, so, in this model, there is an instantaneous connection between output and input. Also $(I - \Sigma_{1,2}S_L)$ is seen to be outer by Lemma 14.2. On the other hand, $\Sigma_{1,1}$ appears to be *maximal phase*: its inverse is anti-causal (see Chapter 8 for the definition of "maximum phase operators").

An outer–inner factorization of $\Sigma_{1,1}$ gives $\Sigma_{1,1} = \Sigma^{\mathrm{o}}_{1,1}U_f$, with U_f an inner operator characterizing the "maximal phase" of the forward gain $\Sigma_{1,1}$ (and the "transmission zeros" in the LTI case!). Collecting the causal factors of the first equation in (14.11) in a single $\Sigma_c = \Sigma_{2,2}S_L(I - \Sigma_{1,2}S_L)^{-1}\Sigma^{\mathrm{o}}_{1,1}$, we find

$$S - \Sigma_{2,1} = \Sigma_c U_f. \tag{14.13}$$

The classical "Schur interpolation," generalized to LTV systems, turns out to be the special case where the inner function $U_f = Z^n$ for a shift Z^n, in which case the formula shows that $\Sigma_{2,1}$ interpolates the series expansion of S up to order $n - 1$. More generally, one may say that $\Sigma_{2,1}$ *right interpolates* S *modulo* U_f. This is the main motivation for exploring the connection between scattering and interpolation, and in Chapter 15 we develop the connection further, using state-space properties of Θ, which we now derive.

14.3 State-Space Properties of the Chain Scattering Matrix

Preliminary definition. In this and following chapters, we use a new shorthand convention: a boldface to denote the global matrix realizing a given operator. For example, $\mathbf{S} := \begin{bmatrix} A & B \\ C & D \end{bmatrix}$ is a realization for the operator S (previously also written as $S \sim_c \mathbf{S}$). Realizations are not unique, so that $\mathbf{S}$ as given is not unique and may be redefined if needed.

To properly define the characteristic system properties and operators in a global way, we have to use global input and output spaces, that is, a global input and output space for each index point i viewed as dividing past, present and future for consideration at that point. These global spaces were defined in Section 7.4. Referring to that section, we hence consider Hilbert–Schmidt or Frobenius spaces of stacked input spaces of the type $\mathcal{X}_2^{\mathbf{m}} = \oplus_{i=-\infty:\infty}[\ell_2^{\mathbf{k}}]^i$, whereby each index i gets an independent input space with reference i, for which the index i is thought as representing its actual presence, that is, individually. The global strict past input space of an operator $T \in \mathcal{L}^{\mathbf{n}\times\mathbf{m}}$ is then the subspace $\mathcal{U}_2^{\mathbf{m}} = \oplus_{i=-\infty:\infty}[\Pi^{\leq i}\ell_2^{\mathbf{m}}]$, where $\Pi^{\leq i} = \Pi_{\mathrm{p}}$ is the projection on the past with respect to the index i, and the global future space $\mathcal{L}_2^{\mathbf{n}} = \oplus_{i=:}[\Pi^{\geq i}\ell_2^{\mathbf{n}}]$, with $\Pi^{\geq i} = \Pi_{\mathrm{f}}$ the projection on the future with respect to the index i. For further information, we refer to Section 7.4 with the added remark that there the global spaces were called, not incorrectly, "Frobenius spaces," but here, more commonly, Hilbert–Schmidt spaces.

Let now for any series $\mathbf{k} = [k_i]_{i=-\infty:\infty}$, $\mathcal{X}_2^{\mathbf{k}} = [\mathcal{U}_2^{\mathbf{k}}]^- \oplus \mathcal{L}_2^{\mathbf{k}}$ denote the orthogonal decomposition of the stacked Hilbert–Schmidt space (one entry for every index point) into the global subspace of past signals $\mathcal{U}_2^{\mathbf{k}-} := \mathcal{U}_2^{\mathbf{k}}Z'$ represented by strictly upper signals of dimension $\mathbf{k}$, (notice that $\mathcal{X}_2^{\mathbf{k}}Z' = \mathcal{X}_2^{\mathbf{k}}$; i.e., Z' just shifts identical column spaces $\ell_2^{\mathbf{k}}$ one notch to the right – a right multiplication with Z' moves $\mathcal{U}_2^{\mathbf{k}}$ into the strict past $\mathcal{U}_2^{\mathbf{k}-}$, leaving all the row dimensions undisturbed), and the global subspace of future signal $\mathcal{L}_2^{\mathbf{k}}$ represented by lower signals, while $\mathcal{D}_2^{\mathbf{k}}$ is a Hilbert space of diagonal (instantaneous) elements, equal to $\oplus_{i=-\infty}^{\infty}\mathbf{R}^{k_i}$. A bounded causal operator $T \in \mathcal{L}^{\mathbf{n}\times\mathbf{m}}$ then maps $\mathcal{X}_2^{\mathbf{m}} \to \mathcal{X}_2^{\mathbf{n}}$ as well as the restriction $\mathcal{L}_2^{\mathbf{m}} \to \mathcal{L}^{\mathbf{n}}$ and is represented by a block lower matrix. Connected to T, there are two "natural" state spaces: $\mathcal{H}_T$ in the strict past and $\mathcal{H}_{o,T}$ in the future. $\mathcal{H}_{o,T} := \Pi_{\mathrm{f}}(T[\mathcal{U}_2^{\mathbf{m}}]^-)$ is the "space of natural future responses" from strict past signals, while $\mathcal{H}_T := \Pi_{\mathrm{sp}}(T'\mathcal{L}_2^{\mathbf{n}})$ is the "space of Nerode equivalence classes" and can be defined as the space of natural strict past responses for the adjoint system T'. Connected with these spaces are the so-called null spaces, as specific orthogonal complements: $\mathcal{K}_T := [\mathcal{U}_2^{\mathbf{m}}]^- \ominus \mathcal{H}_T$ (in the strict past) and $\mathcal{K}_{o,T} := \mathcal{L}_2^{\mathbf{n}} \ominus \mathcal{H}_{o,T}$. Notice that $\mathcal{H}_T$ and $\mathcal{H}_{o,T}$ do not have to be closed, so we have only $[\mathcal{U}_2^{\mathbf{m}}]^- = \mathcal{K}_T \oplus \overline{\mathcal{H}_T}$ and $\mathcal{L}_2^{\mathbf{n}} = \overline{\mathcal{H}}_{o,T} \oplus \mathcal{K}_{o,T}$ as decompositions of the strict past input space and the future output space respectively. $\mathcal{K}_T$ is the kernel of the Hankel operator (which now is a three-dimensional tensor) defined as $H\colon [\mathcal{U}_2^{\mathbf{m}}]^- \to \mathcal{L}_2^{\mathbf{n}} : H_T = \Pi_{\mathcal{L}_2^{\mathbf{n}}}(T|_{[\mathcal{U}_2^{\mathbf{m}}]^-})$ (with $\Pi_{\mathcal{L}_2^{\mathbf{n}}}$ orthogonal projection on $\mathcal{L}_2^{\mathbf{n}}$), while $\mathcal{K}_{o,T}$ is the kernel of H'. (The subscript "T" may be dropped when clear from the context!)

A minimal realization is obtained by choosing a sliced basis either in $\mathcal{H}$ or, equivalently, in $\mathcal{H}_o$ (see Chapter 4 and Section 7.4). When an orthonormal basis in $\mathcal{H}$ is chosen, then a realization in the input normal form results, and vice versa, in the output normal form, when an orthonormal basis in $\mathcal{H}_o$ is selected. (One can indeed choose a *sliced basis* that is the direct sum of individual bases, one in each $[\ell_2]^i$, because both $\mathcal{H}$ and $\mathcal{H}_o$ are right D-invariant, and hence the direct sum of their projections on individual slices, one per index point.)

Let us assume a minimal realization in the input normal form $\mathbf{T} = \begin{bmatrix} A & B \\ C & D \end{bmatrix}$ with state dimension sequence η and $[A\; B]$ co-isometric, and assuming $\ell_A < 1$ (the spectral radius of ZA), one has $\mathcal{H}_T = B'Z'(I - A'Z')^{-1}\mathcal{D}_2^{\eta}$ and $\mathcal{H}_{o,T} = C(I - ZA)^{-1}\mathcal{D}_2^{\eta}$. If the realization is unitary, then the basis for $\mathcal{H}_{o,T}$ is isometric; both the reachability Gramian $\Pi_0[(I - ZA)^{-1}ZBB'Z'(I - A'Z')^{-1}] = I$ and the observability Gramian $\Pi_0[(I - A'Z')^{-1}C'C(I - ZA)^{-1}] = I$ (Π_0 being the orthogonal projection on the main diagonal), and the global operator T is unitary as well as the realization $\mathbf{T}$, thanks to the (sufficient) assumption $\ell_A < 1$.

An important property of an inner U with unitary realization $\mathbf{U}$ is that U maps $\mathcal{H}_U$ isometrically on $\mathcal{H}_{o,U}$ with the inverse isometric map U': $\mathcal{H}_{o,U} = U\mathcal{H}_U$ and $\mathcal{H}_U = U'\mathcal{H}_{o,U}$. This can be verified by direct computation on the state-space realization: assuming $\mathbf{U}$ unitary and $\ell_A < 1$, one finds

$$(D + C(I - ZA)^{-1}ZB)B'Z'(I - A'Z')^{-1} = C(I - ZA)^{-1} \tag{14.14}$$

and dually.

We now show that *this property carries over to causal J-unitary systems with a slight modification.*

Let us first restrict ourselves to what are called *causal J-inner operators*. Those are, by definition, *causal, J-unitary operators Θ whose corresponding Σ is inner*. We have

Proposition 14.3 *Let $\Sigma = \begin{bmatrix} \Sigma_{1,1} & \Sigma_{1,2} \\ \Sigma_{2,1} & \Sigma_{2,2} \end{bmatrix}$ be inner, with $\Sigma_{2,2}$ having a bounded causal inverse (it is outer); then the corresponding Θ is causal and J-unitary. Conversely, let $\Theta = \begin{bmatrix} \Theta_{1,1} & \Theta_{1,2} \\ \Theta_{2,1} & \Theta_{2,2} \end{bmatrix}$ be bounded causal, J-unitary, and such that $\Theta_{2,2}$ has a causal inverse; then the corresponding Σ is inner.*

Let

$$\Sigma = \left[\begin{array}{c|cc} A & B_1 & B_2 \\ \hline C_1 & D_{1,1} & D_{1,2} \\ C_2 & D_{2,1} & D_{2,2} \end{array}\right]. \tag{14.15}$$

Then a corresponding realization $\mathbf{\Theta}$ for Θ is obtained simply by arrow reversal (see Fig. 14.3), after remarking the next property:

Lemma 14.4 *Under the running assumptions (existence and causality of $\Sigma_{2,2}^{-1}$), $D_{2,2}$ is square invertible and its inverse $D_{2,2}^{-1}$ is square and bounded.*

Proof

Preliminary remark: We use the technical property that if the bounded and minimal $T = D + C(I - ZA)^{-1}ZB$ with $\ell_A < 1$ has the bounded, causal inverse $T^{-1} = \delta + \gamma(I - Z\alpha)^{-1}Z\beta$ with $\alpha = A - BD^{-1}C$, then it is known from [29, Proposition 13.2], that also $\ell_\alpha < 1$ and $(I - Z\alpha)^{-1}$ is hence bounded.

The proof then reduces to

$$I = TT^{-1} = (D + Z(\text{causal}))(\delta + Z(\text{causal}) = D\delta + Z(\text{causal}). \tag{14.16}$$

Hence, $D\delta = I$. Likewise, $\delta D = I$, since also $T^{-1}T = I$, and, finally, $\delta = D^{-1}$. □

A realization for Θ follows from the Σ/Θ transformation:

$$\mathbf{\Theta} = \left[\begin{array}{c|cc} A - B_2D_{2,2}^{-1}C_2 & B_1 - B_2D_{2,2}^{-1}C_1 & B_2D_{2,2}^{-1} \\ \hline C_1 - D_{1,2}D_{2,2}^{-1}C_2 & D_{1,1} - D_{1,2}D_{2,2}^{-1}D_{2,1} & D_{1,2}D_{2,2}^{-1} \\ -D_{2,2}^{-1}C_2 & -D_{2,2}^{-1}D_{2,1} & D_{2,2}^{-1} \end{array}\right]. \tag{14.17}$$

An easy consequence of this construction is

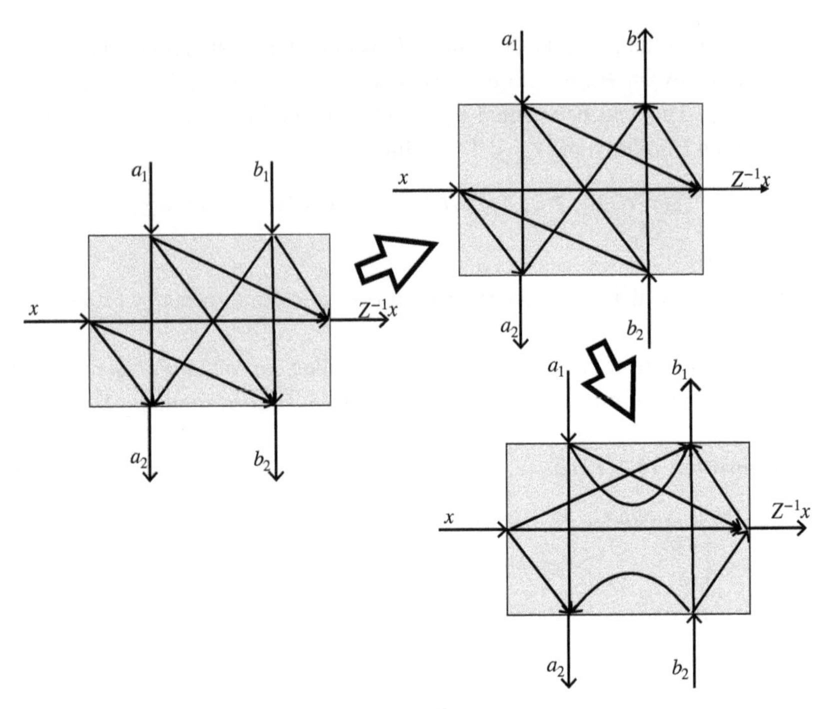

Figure 14.3 Reversal of arrows carrying outgoing energy.

Proposition 14.5 *The realization of* $\mathbf{\Theta}$ *in (14.17) is* J*-unitary:* $\mathbf{\Theta}'J_2\mathbf{\Theta} = J_1$, $\mathbf{\Theta}J_1\mathbf{\Theta}' = J_2$ *for*

$$J_1 := \left[\begin{array}{c|cc} I_\eta & & \\ \hline & I_{\mathbf{m}_1} & \\ & & -I_{\mathbf{m}_2} \end{array}\right], \quad J_2 := \left[\begin{array}{c|cc} I_{\eta^-} & & \\ \hline & I_{\mathbf{n}_1} & \\ & & -I_{\mathbf{n}_2} \end{array}\right], \tag{14.18}$$

in which $[\eta^-]_k = \eta_{k+1}$ *– a shift back, and necessarily,* $\mathbf{m}_2 = \mathbf{n}_2$ *and* $\eta + \mathbf{m}_1 = \eta^- + \mathbf{n}_1$.

The property follows directly from the unitarity of the realization Σ. □

As a further result, we can now characterize the natural state spaces pertaining to a J-inner Θ with J-unitary realization $\mathbf{\Theta} = \left[\begin{array}{c|cc} \alpha & \beta_1 & \beta_2 \\ \hline \gamma_1 & \delta_{1,1} & \delta_{1,2} \\ \gamma_2 & \delta_{2,1} & \delta_{2,2} \end{array}\right]$. We put this as a proposition for later reference. From the definitions, $\mathcal{H}_\Theta = \Pi_{\text{sp}}\Theta' \begin{bmatrix} \mathcal{L}_2^{\mathbf{n}_1} \\ \mathcal{L}_2^{\mathbf{n}_2} \end{bmatrix}$ and $\mathcal{H}_{o,\Theta} = \Pi_f\Theta \begin{bmatrix} [\mathcal{U}_2^{\mathbf{m}_1}]^- \\ [\mathcal{U}_2^{\mathbf{m}_2}]^- \end{bmatrix}$ (with Π_{sp} being the projection on the strict past and Π_f on the future), and the null-spaces $\mathcal{K}_\Theta = \begin{bmatrix} [\mathcal{U}_2^{\mathbf{m}_1}]^- \\ [\mathcal{U}_2^{\mathbf{m}_2}]^- \end{bmatrix} \ominus \mathcal{H}_\Theta$ and $\mathcal{K}_{o,\Theta} = \begin{bmatrix} \mathcal{L}_2^{\mathbf{n}_1} \\ \mathcal{L}_2^{\mathbf{n}_2} \end{bmatrix} \ominus \mathcal{H}_{o,\Theta}$.

Notation: for brevity's sake, we write, for example, $\mathcal{L}_2^{\mathbf{m}_1} \oplus \mathcal{L}_2^{\mathbf{m}_2} = \begin{bmatrix} \mathcal{L}_2^{\mathbf{m}_1} \\ \mathcal{L}_2^{\mathbf{m}_2} \end{bmatrix}$ and drop the indices when clear from the context.

Proposition 14.6 *When Θ is causal, J-unitary and bounded, then*

$$\mathcal{K}_\Theta = J_1\Theta'[\mathcal{U}_2^{\mathbf{n}_1\oplus\mathbf{n}_2}]^-,\ \mathcal{K}_{o,\Theta} = J_2\Theta\mathcal{L}_2^{\mathbf{m}_1\oplus\mathbf{m}_2}, \tag{14.19}$$

$$\mathcal{H}_\Theta = \Theta' J_2\mathcal{H}_{o,\Theta},\ \mathcal{H}_{o,\Theta} = \Theta J_1\mathcal{H}_\Theta. \tag{14.20}$$

Remark: Θ is only assumed to be causal here; it may not be J-inner (in which case, Σ is still unitary but not causal).

Proof

First, we show that the orthogonal complement $\mathcal{K}_\Theta$ of $\mathcal{H}_\Theta$ in $[\mathcal{U}_2^{\mathbf{m}_1\oplus\mathbf{m}_2}]^-$ is given by $J_1\Theta'[\mathcal{U}_2^{\mathbf{m}_1\oplus\mathbf{m}_2}]^-$. $J_1\Theta'[\mathcal{U}_2^{\mathbf{m}_1\oplus\mathbf{m}_2}]^-$ is a closed space because both Θ and Θ^{-1} are bounded. Moreover, for any $u\in\mathcal{H}_\Theta$ and any $v\in J_1\Theta'\mathcal{U}_2^-$, we have, by definition, $y_1\in\mathcal{L}_2$ and $u_1\in\mathcal{L}_2$, with $u=\Theta'y_1+u_1$, and $y_2\in[\mathcal{U}_2]^-$ with $v=J_1\Theta'y_2$. It follows that the (Hilbert–Schmidt) inner product

$$\begin{aligned}(u,v) &= (\Theta'y_1+u_1, J_1\Theta'y_2)\\ &= (\Theta'y_1, J_1\Theta'y_2)\\ &= (y_1, \Theta J_1\Theta' J_2 y_2)\\ &= (y_1, \Theta\Theta^{-1}y_2)\\ &= (y_1,y_2)=0.\end{aligned} \tag{14.21}$$

Hence $J_1\Theta'\mathcal{U}_2^-\subset\mathcal{K}_\Theta$. Conversely, let $v\in\mathcal{K}_\Theta$; then $(y:=\Theta v)\in\mathcal{U}_2^-$ (by definition), and $v=\Theta^{-1}y=J_1\Theta'(J_2y)\in J_1\Theta'\mathcal{U}_2^-$, showing that $\mathcal{K}_\Theta\subset J_1\Theta'\mathcal{U}_2^-$ as well. A dual proof holds for $\mathcal{K}_{o,\Theta}$. This shows the first part of the proposition.

As to the second part, consider $u\in\mathcal{H}_\Theta$, and $y=\Theta J_1u$. Then $y\in\mathcal{L}_2$, since for all $y_1\in\mathcal{U}_2^-$ and by the previous derivation $(u,J_1\Theta'y_1)=0$, and hence $(\Theta J_1u,y_2)=(y,y_2)=0$. Reversing the operation, $\Theta'J_2y=\Theta'J_2\Theta J_1u=u$, shows the isomorphism between $\mathcal{H}_\Theta$ and $\mathcal{H}_{o,\Theta}$, given by $y=\Theta J_1u\Leftrightarrow u=\Theta'J_2y$. □

Towards Interpolation

Written out in terms of a J-orthogonal state-space realization, the relation between the reachability and observability bases is given by

Proposition 14.7

$$\Theta J_1\begin{bmatrix}\beta_1'\\ \beta_2'\end{bmatrix}Z'(I-\alpha'Z')^{-1}=\begin{bmatrix}\gamma_1\\ \gamma_2\end{bmatrix}(I-Z\alpha)^{-1}. \tag{14.22}$$

Proof

By direct calculation,

$$\begin{aligned}&\Theta J_1\beta'Z'(I-\alpha'Z')^{-1}\\ &=(\delta+\gamma(I-Z\alpha)^{-1}Z\beta)J_1\beta'Z'(I-\alpha'Z')^{-1}\\ &=\delta J_1\beta'Z'(I-\alpha'Z')^{-1}+\gamma(I-Z\alpha)^{-1}Z\beta J_1\beta'Z'(I-\alpha'Z')^{-1}\\ &=\delta J_1\beta'Z'(I-\alpha'Z')^{-1}+\gamma(I-Z\alpha)^{-1}Z\alpha+\gamma+\gamma\alpha'Z'(I-\alpha'Z')^{-1}\\ &=(\delta J_1\beta'+\gamma\alpha')Z'(I-\alpha'Z')^{-1}+\gamma(I-Z\alpha)^{-1}\\ &=\gamma(I-Z\alpha)^{-1},\end{aligned} \tag{14.23}$$

where use has been made of J-orthogonality: $\beta J_1 \beta' = 1 - \alpha\alpha'$ and $\delta J_1 \beta' + \gamma\alpha' = 0$. □

The second block row of Eq. (14.22) gives

$$(\Theta_{2,1}\beta_1' - \Theta_{2,2}\beta_2')(Z - \alpha')^{-1} = \gamma_2(I - Z\alpha)^{-1}. \tag{14.24}$$

The input scattering operator derived from the Σ/Θ transformation is $\Sigma_{2,1} = -\Theta_{2,2}^{-1}\Theta_{2,1}$, and we find, after premultiplying with the causal and bounded $\Theta_{2,2}^{-1}$,

$$\Sigma_{2,2}\beta_1' = -\beta_2' - \left[\Theta_{2,2}^{-1}\gamma_2(I - Z\alpha)^{-1}\right](Z - \alpha'). \tag{14.25}$$

This formula may be interpreted as describing an interpolation, whereby the "value of $\Sigma_{2,2}\beta_1'$ at α'" is requested to be equal to $-\beta_2'$, and whereby $\Sigma_{2,2}$ is constrained to be causal and contractive. As a further observation, the choice of basis for $\mathcal{H}_\Theta$ does not really matter, since Θ maps $\mathcal{H}_\Theta$ to causality in any case (although not necessarily J-isometrically). With a different basis $\mathcal{H}_\Theta = b'(Z - A')^{-1}\mathcal{D}_2^{\eta}$, and $\Sigma_{2,2}b_1' = -b_2' + \Delta_r(Z - A)$ for some causal remainder Δ_r, and $S = \Sigma_{2,2}$ solves the interpolation problem "$Sb_1' = -b_2'$ modulo $(Z - A')$," with S causal and contractive (sometimes called *passive* or *lossy*). These statements have to be given meaning for the more general LTV case, which we do in Chapter 15.

The Connection with the Classical Case*

Let us look at how interpolation works in the LTI case. For example, in the scalar (LTI) Nevanlinna–Pick case, the constrained interpolation asks for a contractive function $S(z)$:

1. analytic in the open unit disk $\mathbf{D}$ of the complex plane (a so-called H_∞ function);
2. that meets n interpolation conditions $S(a_i) = s_i$, $i = 1 : n$ for distinct $a_i \in \mathbf{D}$. In that case we would have:
3. a (non–J-unitary) basis for $\mathcal{H}_\Theta$ given by $b'(z - A')^{-1}$, with
4. $b_1' = [1\ 1\ \cdots\ 1]$, $b_2' = -[s_1\ s_2\ \cdots\ s_n]$ and
5. $A' = \begin{bmatrix} a_1 & & & \\ & a_2 & & \\ & & \ddots & \\ & & & a_n \end{bmatrix}$

so that the classical Nevanlinna–Pick interpolation problem looks as follows in our formalism:

$$S(z)\ \begin{bmatrix} 1 & 1 & \cdots & 1 \end{bmatrix} = \begin{bmatrix} s_1 & s_2 & \cdots & s_n \end{bmatrix} + \begin{bmatrix} S_{r,1}(z) & S_{r,2}(z) & \cdots S_{r,n}(z) \end{bmatrix} \begin{bmatrix} z - a_1 & & & \\ & z - a_2 & & \\ & & \ddots & \\ & & & z - a_n \end{bmatrix}, \tag{14.26}$$

with the remainders $S_{r,k}$ analytic and contractive in the unit complex disk ($S_{r,k} \in H_\infty$). An $S = \Sigma_{2,2}$ will solve the problem if and only if there is a regular state transformation R that converts the reachability data $[A \;\; b]$ to a J-co-isometric basis $[\alpha \;\; \beta] = [R^{-1}AR \;\; R^{-1}b]$, or, equivalently, if and only if there exists a strictly positive definite $P = RR'$ satisfying $P = APA' + b_1b_1' - b_2b_2'$. In the Nevanlinna–Pick case, the P that solves this equation is seen to be the *Pick matrix* $P = \left[\frac{1-\bar{s}_i s_j}{1-\bar{a}_i a_j}\right]_{i=1:n,j=1:n}$ (by series development), and the problem has a solution if and only if P is positive definite.

In the case of elementary Schur interpolation $S(z) = s_0 + zs_1 + \cdots + z^n s_n + \mathcal{O}(z^{n+1})$, analytic in $\mathbf{D}$ and contractive, the data become:

1. $b_1' = [1\;0\;\cdots\;0]$, $b_2' = [s_0\;s_1\;\cdots\;s_n]$,
2. $A' = \begin{bmatrix} 0 & 1 & & \\ & \ddots & \ddots & \\ & & 0 & 1 \\ & & & 0 \end{bmatrix}$,

$$\begin{aligned} S(z)\begin{bmatrix}1 & 0 & \cdots & 0\end{bmatrix} &= \begin{bmatrix}s_0 & s_1 & \cdots & s_n\end{bmatrix} \\ &+ \begin{bmatrix}S_{r,1}(z) & S_{r,2}(z) & \cdots S_{r,n}(z)\end{bmatrix}\begin{bmatrix} z & -1 & & \\ & \ddots & \ddots & \\ & & z & -1 \\ & & & z \end{bmatrix}, \end{aligned} \tag{14.27}$$

with the Pick matrix now $P = [\delta(i-j) - \bar{s}_i s_j]_{i=0:n,j=0:n}$.

A new situation arrises when the Pick matrix is nonsingular indefinite. This situation gives rise to a new type of interpolation, called Schur–Takagi interpolation. It turns out that this new type provides for an effective method of *constrained model reduction*, that is, of finding low-degree (causal) approximations to a (causal) system that satisfy a strong norm constraint on the approximation error.

In the following chapters, we show how interpolation may be defined for general LTV systems, and how formula (14.22) covers all important cases of constrained interpolation problems for LTV systems. Chapter 15 treats the classical case where the Pick matrix is positive definite, and Chapter 16 discusses the Schur–Takagi case and the related model reduction theory. In view of their importance for these developments, basic properties of indefinite metrics are considered in Section 14.4.

14.4 Indefinite Metrics

In the previous sections, we saw that J-unitary matrices play a central role in scattering theory and in particular their canonical input and output spaces. In Chapter 15 on constrained, Nevanlinna–Pick-type interpolation, these spaces will be found to be J-positive, meaning that they have positive definite Gramians for a well-defined indefinite metric on them, while in the chapter on constrained model reduction or

Schur–Takagi-type interpolation, the corresponding Gramian turns out to be nonsingular and indefinite. So it pays to study elementary properties of indefinite metrics, which is what we propose to do in this section.

First, we look at finitely indexed, nonsingular indefinite metrics, characterized by signature matrices of the type $J = \begin{bmatrix} I_p & \\ & -I_q \end{bmatrix}$ and of dimension $p+q$. (We generalize later to infinite direct sums of such spaces, one for each time index point, but the main definitions and properties remain the same.) So let $\mathcal{X}$ be a vector space of dimension $p+q$, endowed with the J-metric (in addition to the natural metric, of course). To distinguish the two metrics, we use the qualification "J" to the relevant operations. So, for $u, v \in \mathcal{X}$, $(u,v) = v'u$ is the natural inner product and $(u,v)_J := v'Ju$ the indefinite J-inner product. Similarly, we say that v is *J-orthogonal* on u if $(u,v)_J = v'Ju = 0$. Here are some further definitions:

1. A vector $u \in \mathcal{X}$ is said to be *isotropic* (of course, with respect to J), if $(u,u)_J = u'Ju = 0$; it is J-orthogonal on itself.
2. Let $\mathcal{H} \subset \mathcal{X}$ be a subspace of $\mathcal{X}$; then $\mathcal{H}$ is said to be *J-positive* if $\forall x \in \mathcal{H} : (x,x)_J \geq 0$, *J-negative* if $\forall x \in \mathcal{H} : (x,x)_J \leq 0$, *strictly positive* (or *uniformly positive*) if $\exists(\epsilon > 0)\forall(x \in \mathcal{H}) : (x,x)_J \geq \epsilon^2\|x\|^2$ (we write $(x,x)_J \gg 0$), and *strictly or uniformly negative*: $\exists(\epsilon > 0)\forall(x \in \mathcal{H}) : -(x,x)_J \geq \epsilon^2\|x\|^2$ $(-(x,x)_J \gg 0)$ (notice $\exists\forall \neq \forall\exists$: ϵ is independent of x!).
3. $\mathcal{H}$ is *J-neutral* if $\forall x \in \mathcal{H} : (x,x)_J = 0$.

Some first examples:

1. In $\mathbf{R}^2$, with $J = \begin{bmatrix} 1 & \\ & -1 \end{bmatrix}$, we have col[3 1] J-positive, col[1 3] J-negative, and col[1 1] isotropic. $u = \text{col}[\sqrt{2}\ 1]$ *and* $v = \text{col}[1\ \sqrt{2}]$ form a J-orthonormal basis for the whole space, and so do $u = \text{col}[1\ 0]$ *and* $v = \text{col}[0\ 1]$. Also, $u = \text{col}[1\ 1]$ *and* $v = \text{col}[1\ -1]$, consisting of isotropic vectors, form a basis for the whole space, but they do not form a neutral subspace, since u is not J-orthogonal on v.
2. A nontrivial neutral subspace in $\mathbf{R}^{2+2}$, with $J = \begin{bmatrix} I_2 & \\ & -I_2 \end{bmatrix}$ is generated by the basis $u = \text{col}[1\ 0 \mid 1\ 0]$, $v = \text{col}[0\ 1 \mid 0\ 1]$. In a neutral subspace, all vectors are J-orthogonal on each other and on themselves.
$\bigvee(u = \text{col}[3\ 0 \mid 1\ 0],\ v = \text{col}[0\ 1 \mid 0\ 1])$ is positive but not strict, and
$\bigvee(u = \text{col}[3\ 0 \mid 1\ 0],\ v = \text{col}[1\ 0 \mid 1\ 0])$ is of rank 2 and indefinite.

We say that a space $\mathcal{H} \subset \mathcal{X}$ is a *Krein space* if it decomposes into the *J-direct sum* of a strictly positive subspace $\mathcal{H}_1$ and a strictly negative subspace $\mathcal{H}_2$ that is J-orthogonal on $\mathcal{H}_1$, and we write $\mathcal{H} = \mathcal{H}_1 \oplus_J \mathcal{H}_2$ for the original $J = \begin{bmatrix} I_p & \\ & -I_q \end{bmatrix}$. If, in addition, one of these is finite dimensional, then the Krein space is called a *Pontryagin space*. Let us, at this point, assume both decomposing spaces to be finite dimensional, corresponding to a nondegenerate, but indefinite metric, and let m_1 and m_2 be the dimensions of $\mathcal{H}_1$ and $\mathcal{H}_2$, respectively. (Later we shall generalize to spaces where the finite indefinite metric is local but combines to global Krein spaces with dimensions $\mathbf{m}_1$ and $\mathbf{m}_2$, characterizing discrete-time LTV systems.) A few immediate properties are not hard to derive:

1. A finite, nondegenerate space $\mathcal{H} = \mathcal{H}_1 \oplus_J \mathcal{H}_2$ has a J-orthogonal basis $[\xi_1\ \xi_2]$, with ξ_1 consisting of m_1 vectors and ξ_2 of m_2 vectors.
   Proof: Suppose x_1 is a basis for $\mathcal{H}_1$; then because $\mathcal{H}_1$ is strictly positive, $M_1 := x_1' J x_1 \gg 0$. Let $M_1 = T_1' T_1$, with T_1 nonsingular (Cholesky factorization), and $\xi_1 := x_1 T_1^{-1}$; then ξ_1 is an orthonormal basis for $\mathcal{H}_1$. A similar construction holds for ξ_2. Moreover, $\xi_1 \perp_J \xi_2$ by definition, so ξ is a J-orthonormal basis for $\mathcal{H}$. □
2. A finite-dimensional nondegenerate space $\mathcal{H} = \mathcal{H}_1 \oplus_J \mathcal{H}_2$ does not have a neutral subspace $\mathcal{H}_0 \subset \mathcal{H}$ (by definition J-orthogonal on the whole space).
   Proof: Since $\mathcal{H}$ is assumed a Krein space, there is, by the previous property, a J-orthonormal basis $\xi := [\xi_1\ \xi_2]$ with the signature matrix $J_2 = \mathrm{diag}[I_{m_1}\ -I_{m_2}]$ such that $\xi' J \xi = J_2$. Suppose $x \in \mathcal{H}_0$; then $x = \xi a$ for some vector a, and $\xi' J x = 0$ since x is assumed J-orthogonal on all vectors in $\mathcal{H}$. Hence, $\xi' J \xi a = J_2 a = 0$ and $a = 0$. □
3. Suppose $\mathcal{H}$ is a Krein space defined as the J-direct sum $\mathcal{H}_1 \oplus_J \mathcal{H}_2$, with $\mathcal{H}_1$ strictly positive of dimension m_1 and $\mathcal{H}_2$ strictly negative of dimension m_2, and suppose $\mathcal{H} = \overline{\mathcal{H}_1} \oplus_J \overline{\mathcal{H}_2}$ is another J-direct sum defined as the J-direct sum with dimensions $\overline{m}_1$ and $\overline{m}_2$ respectively; then the corresponding dimensions match: $m_1 = \overline{m}_1$ and $m_2 = \overline{m}_2$; the data m_1 and m_2 are hence characteristic of $\mathcal{H}$, which we may call an $(m_1 + m_2)$ Krein space.
   Proof: Suppose that $\xi = [\xi_1\ \xi_2]$ with signature J_2 and $\overline{\xi} = [\overline{\xi}_1\ \overline{\xi}_2]$ with signature $\overline{J}_2$ are two different orthonormal bases, then there is a non-singular T such that $\overline{\xi} = \xi T$, and $\overline{J}_2 = \overline{\xi}' J \overline{\xi} = T' \xi' J \xi T = T' J_2 T$. Similarly $J_2 = T \overline{J}_2 T'$. Hence T is J-unitary and $\overline{J}_2 = J_2$, since they are congruent (and ordered). □
4. Suppose $\mathcal{H} \subset \mathcal{X}$ is an $m = m_1 + m_2$ Krein space. Let $J_2 = \begin{bmatrix} m_1 & \\ & -m_2 \end{bmatrix}$, x be a basis for $\mathcal{H}$, and $x = \begin{bmatrix} \xi_1 & \xi_2 \end{bmatrix} T$. Then T is, of course, nonsingular. Let $M = x' J x$ be the J-Gramian of the basis x. *Then the signature of M is J_2* since $M = x' J x = T' J_2 T$ and M is hence congruent to J_2

14.5 Completion of a J-Isometric Basis

Preliminary

In anticipation of further use, we discuss the completion of a set of rows to a full J-isometric basis (the column case is similar and dual). Let $\begin{bmatrix} \alpha \\ \beta \end{bmatrix}$ be a partial J-isometric basis, meaning that for three nonsingular sign matrices J_1, J_2 and j_1, the indefinite isometry

$$\alpha' J_1 \alpha + \beta' J_2 \beta = j_1 \tag{14.28}$$

holds. For ease of treatment, we may allow the sign matrices J_1, J_2 and j_1 to be unordered, although we keep them diagonal: a re-ordening to the normal order requires an identical permutation of both rows and columns. We show how to complete this J-co-isometric basis into a full J-unitary basis. (For practical reasons, we keep the use of unordered sign matrices to a bare minimum!)

Because j_1 is assumed nonsingular, the columns of $\begin{bmatrix} \alpha \\ \beta \end{bmatrix}$ are linearly independent, and let the columns of $\begin{bmatrix} c \\ d \end{bmatrix}$ form a basis for the orthogonal complement of $\bigvee \begin{bmatrix} J_1 & \\ & J_2 \end{bmatrix} \begin{bmatrix} \alpha \\ \beta \end{bmatrix}$, which is of dimension $\#J_1 + \#J_2 - \#j_1$. The matrix $\begin{bmatrix} \alpha & c \\ \beta & d \end{bmatrix}$ is square nonsingular (as it is the product of two invertible matrices, one of which is J). Hence,

$$\begin{bmatrix} \alpha' & \beta' \\ c' & d' \end{bmatrix} \begin{bmatrix} J_1 & \\ & J_2 \end{bmatrix} \begin{bmatrix} \alpha & c \\ \beta & d \end{bmatrix} = \begin{bmatrix} j_1 & \\ & N \end{bmatrix}, \tag{14.29}$$

with $N := \begin{bmatrix} c' & d' \end{bmatrix} \begin{bmatrix} J_1 & \\ & J_2 \end{bmatrix} \begin{bmatrix} c \\ d \end{bmatrix}$ nonsingular, because all matrices at the left-hand side of Eq. (14.29) are square nonsingular. Let now j_2 be the inertia of N, so that there exists a square invertible matrix r such that $N = r' j_2 r$. It follows that

$$\begin{bmatrix} \gamma \\ \delta \end{bmatrix} := \begin{bmatrix} c \\ d \end{bmatrix} r^{-1} \tag{14.30}$$

completes the given J-co-isometric basis into a full J-unitary basis

$$\mathbf{\Theta} := \begin{bmatrix} \alpha & \gamma \\ \beta & \delta \end{bmatrix}, \tag{14.31}$$

and necessarily $\#J_{1,+} + \#J_{2,+} = \#j_{1,+} + \#j_{2,+}$ as well as $\#J_{1,-} + \#J_{2,-} = \#j_{1,-} + \#j_{2,-}$ because of the congruence induced by the relation $\mathbf{\Theta}' J \mathbf{\Theta} = j$.

A Square-Root Algorithm to Compute the Completion

We try to emulate the QR factorization, $T = \begin{bmatrix} Q_1 & Q_2 \end{bmatrix} \begin{bmatrix} R \\ 0 \end{bmatrix}$, which produces an orthogonal basis for the columns of T and R right invertible (a minimal row basis), now in the form ΘR with Θ J-unitary, and R having only a minimal set of independent rows. (We shall see that we get R in the form $\begin{bmatrix} R_1 \\ 0 \\ \hline R_2 \\ 0 \end{bmatrix}$ according to an ordered sign partitioning of the relevant sign matrix.) We start out with a set of m columns $\begin{bmatrix} A \\ B \end{bmatrix}$ of dimension $(p + q) \times m$, which is such that

$$\begin{bmatrix} A' & B' \end{bmatrix} J \begin{bmatrix} A \\ B \end{bmatrix} = A'A - B'B \gg 0, \tag{14.32}$$

with $J = \begin{bmatrix} 1_p & \\ & -1_q \end{bmatrix}$. The goal is to produce a J-unitary matrix Θ, and a minimal set of independent rows R (completed with zero rows as needed), such that (notice $Q^{-1} = J\Theta'J$)

$$J\Theta'J \begin{bmatrix} A \\ B \end{bmatrix} = R. \tag{14.33}$$

The situation is more complex than in the case of QR factorization, but we shall see that under the condition of nonsingularity, a conclusive algorithm can be derived using elementary transformations directly on $\begin{bmatrix} A \\ B \end{bmatrix}$ much like in the QR case, although there is one instance that may force additional computational complexity, namely when isotropic columns occur in the proceedings. The procedure consists of both J-unitary operations on the rows of $\begin{bmatrix} A \\ B \end{bmatrix}$ and base preservation operations on its columns. To develop some feel for it, here is a first example.

An Example

Consider

$$\begin{bmatrix} A \\ B \end{bmatrix} = \begin{bmatrix} 1 & 0 \\ 0 & 1 \\ \hline 1 & 1 \\ 0 & 1 \end{bmatrix}, \tag{14.34}$$

which is easily seen to be J-nonsingular, with $J := \begin{bmatrix} I_2 & \\ & -I_2 \end{bmatrix}$.

The first column is isotropic, necessitating an interchange of columns, producing a first column operation with matrix $r_1^{-1} = \begin{bmatrix} 0 & 1 \\ 1 & 0 \end{bmatrix}$. The first vector now is $\mathrm{col}\begin{bmatrix} 0 & 1 \mid 1 & 1 \end{bmatrix}$, and we proceed with $\begin{bmatrix} u_1' & \\ & u_2' \end{bmatrix}$: unitary transformations on the rows of the top and bottom blocks individually. (These transformations are, of course, J-unitary.) This yields the transformed set of columns

$$\begin{bmatrix} A_1 \\ B_1 \end{bmatrix} := \begin{bmatrix} 1 & 0 \\ 0 & 1 \\ \hline \sqrt{2} & \frac{1}{\sqrt{2}} \\ 0 & -\frac{1}{\sqrt{2}} \end{bmatrix}. \tag{14.35}$$

A first 2×2 hyperbolic transformation can now take place on the first column, putting the first column in the direction of a unit vector, and to be applied on rows 1 and 3, using the reduced sign matrix $j := \begin{bmatrix} 1 & \\ & -1 \end{bmatrix}$,

$$j\theta'_{13}j\begin{bmatrix}1\\ \sqrt{2}\end{bmatrix} := \begin{bmatrix}\sqrt{2} & -1\\ -1 & \sqrt{2}\end{bmatrix}\begin{bmatrix}1\\ \sqrt{2}\end{bmatrix} = \begin{bmatrix}0\\ 1\end{bmatrix}. \tag{14.36}$$

Embedding this 2×2 elementary hyperbolic transformation in the full matrix as $J\Theta'_{13}J = \left[\begin{array}{cc|cc}\sqrt{2} & & -1 & \\ & 1 & & \\ \hline -1 & & \sqrt{2} & \\ & & & 1\end{array}\right]$, the block is now transformed to $\left[\begin{array}{cc}0 & -\frac{1}{\sqrt{2}}\\ 0 & 1\\ \hline 1 & 1\\ 0 & -\frac{1}{\sqrt{2}}\end{array}\right]$.

Applying a transformation $r'_2 := \begin{bmatrix}1 & -1\\ 0 & 1\end{bmatrix}$ makes the two columns orthogonal to each other: $\left[\begin{array}{cc}0 & -\frac{1}{\sqrt{2}}\\ 0 & 1\\ \hline 1 & 0\\ 0 & -\frac{1}{\sqrt{2}}\end{array}\right]$. An orthogonal transformation u'_3 applied on rows 1 and 2 of this matrix now produces the next basis $\left[\begin{array}{cc}0 & -3/2\\ 0 & 0\\ \hline 1 & 0\\ 0 & -\frac{1}{\sqrt{2}}\end{array}\right]$. Finally, a further 2×2 hyperbolic transformation $j\theta'_{14}j$ applied on rows 1 and 4 produces a vector in the direction of the first coordinate vector:

$$j\theta'_{14}j\begin{bmatrix}-3/2\\ -\frac{1}{\sqrt{2}}\end{bmatrix} := \frac{2}{\sqrt{7}}\begin{bmatrix}3/2 & -\frac{1}{\sqrt{2}}\\ -\frac{1}{\sqrt{2}} & 3/2\end{bmatrix}\begin{bmatrix}-3/2\\ -\frac{1}{\sqrt{2}}\end{bmatrix} = \begin{bmatrix}-\frac{\sqrt{7}}{2}\\ 0\end{bmatrix}, \tag{14.37}$$

followed by a normalization of the second column, and a reordering with a right factor $r_4^{-1} = \begin{bmatrix} & 1\\ -\frac{2}{\sqrt{7}} & \end{bmatrix}$ produces the final basis $\left[\begin{array}{cc}1 & 0\\ 0 & 0\\ \hline 0 & 1\\ 0 & 0\end{array}\right]$ times a nonsingular column transformation matrix R. It now remains to embed all the intermediate steps in $\begin{bmatrix}I_2 & \\ & -I_2\end{bmatrix}$-unitary matrices and compute their product in sequence to finally obtain

$$\begin{bmatrix}A\\ B\end{bmatrix} = \Theta\left[\begin{array}{c}R_{1,:}\\ 0\\ \hline R_{2,:}\\ 0\end{array}\right], \tag{14.38}$$

whereby the J-orthonormal columns of Θ are divided between a positive subspace generated by the first column of $\begin{bmatrix}\Theta_{1,1}\\ \Theta_{2,1}\end{bmatrix}$ and a negative subspace generated by the first column of $\begin{bmatrix}\Theta_{1,2}\\ \Theta_{2,2}\end{bmatrix}$, while the other columns generate a space that is J-orthogonal to the column space of $\begin{bmatrix}A\\ B\end{bmatrix}$. Concluding the example, we have found a minimal representation for the mixed sign matrix $A'A - B'B$ as $R'_{1,:}R_{1,:} - R'_{2,:}R_{2,:}$ with a matrix R of dimension at most m. This may seem like not much gain on a simple example, but the result is truly equivalent to QR factorization, now in an indefinite matrix (and under the condition of nonsingularity).

The ΘR algorithm consists of the following operations (which form a complete set, always leading to a result):

1. *Crowding:* applying unitary transformations on the rows of both A and B separately to "crowd" the next vectors in line for processing so that they contain only one nonzero entry. This operation drastically reduces the number of 2×2 hyperbolic transformation needed to the bare minimum, thereby largely increasing numerical stability and control.
2. *Elementary hyperbolic transformations:* reducing a 2 dimensional J-nonsingular vector to a coordinate vector, either $\begin{bmatrix} 1 \\ 0 \end{bmatrix}$ or $\begin{bmatrix} 0 \\ 1 \end{bmatrix}$.
3. *Re-orderings:* column transformations that restore order in the subsequent bases.
4. *Elimination of isotropic columns:* if an isotropic vector remains after crowding, one has to search for a subsequent column that is not orthogonal to it. If such a column cannot be found, then the original system has to be declared singular, with a neutral space spanned by the offending isotropic vector. If a non-J-orthogonal subsequent vector is indeed found (as in the example above), then an exchange of the two vectors solves the problem for this step and the algorithm can resume.

The potential occurrence of isotropic columns can, in large cases, substantially increase the numerical complexity of the algorithm and reduce its numerical stability, similar to what happens in Gaussian elimination when pivoting is necessary.

14.6 Indefinite Metrics in Scattering Theory

Scattering theory deals with "left-side" waves $a_1 \in \ell_2^{\mathbf{m}_1}$ and $b_1 \in \ell_2^{\mathbf{n}_1}$, and "right-side" waves $a_2 \in \ell_2^{\mathbf{m}_2}$ and $b_2 \in \ell_2^{\mathbf{n}_2}$. These wave spaces have a natural J-metric, which we denote with J_1 and J_2. They form ordered series of input and output spaces. For each index point k, there is a local indefinite input space $\mathbf{R}^{m_{1,k}} \oplus \mathbf{R}^{n_{1,k}}$ and output space $\mathbf{R}^{m_{2,k}} \oplus \mathbf{R}^{n_{2,k}}$. However, Θ also maps the full input series to the full output series in a J-unitary way. When the system is finitely indexed, then, by reordering of the indices, there is an equivalent J-unitary map of the form

$$\Theta \sim \mathbf{R}^{\oplus \mathbf{m}_1} \oplus \mathbf{R}^{\oplus \mathbf{n}_1} \to \mathbf{R}^{\oplus \mathbf{m}_2} \oplus \mathbf{R}^{\oplus \mathbf{n}_2} \tag{14.39}$$

and dimensions have to match globally, since Θ is globally a square matrix:

$$\sum \mathbf{m}_1 + \sum \mathbf{n}_1 = \sum \mathbf{m}_2 + \sum \mathbf{n}_2. \tag{14.40}$$

In interpolation theory, one assumes, moreover, that $\Theta_{2,2}$ (and $\Sigma_{2,2}$) are square and invertible, so that globally also $\sum \mathbf{n}_1 = \sum \mathbf{m}_2$, and hence also $\sum \mathbf{m}_1 = \sum \mathbf{n}_2$. This does not have to be the case locally, and in each special case, the local dimensional relations have to be analyzed, leading to important information about the system. In the case of infinitely indexed systems, the situation is considerably more subtle, but the finitely indexed case generalizes to infinitely indexed when all operators under consideration are bounded (in particular Θ). We shall pay extra attention to this case when appropriate in order to keep LTI on board as a special case of LTV, since LTI is necessarily infinitely indexed.

14.7 Notes

1. The importance and usefulness of a scattering approach in circuit and system theory go back to the mid-1900s, in particular the article of Youla, Castriota and Carlin [20]. Much has been motivated by the steady increase in frequency of communication systems and the fact that wave propagation along transmission lines or in the ether started to play a major role. Electronic circuits at the receiving side have to take in as much signal energy as possible and, conversely, radiate out as much useful signal energy as possible at the emission side, and it involves maximizing energy transmission (either incoming or outgoing, or adapting one to the other. This is called *impedance matching*, which turns out to be essentially an interpolation problem [43]). As a consequence, not voltages or currents are the important quantities to be optimized as signal carriers, but the transmitted energy, which is a combination of both, mediated by an impedance, as is the case with wave propagation along a transmission line or to the "ether" via an antenna. This situation is similar to what happens in acoustics: most music instruments have devices to adapt the sound production to the characteristic impedance of the surrounding air (at normal atmospheric pressure), while our ear lobes are our very useful biological impedance matching acoustic input devices. It turned out that scattering parameters are not only natural parameters for signal transmission but are also much easier to measure at higher frequencies. Measuring frequency-dependent transmission and reflection parameters are the standard way to characterize the performance of high frequency devices.
2. Some J-unitary and causal chain scattering matrices Θ considered previously were J-inner, that is, they had a causally invertible (outer) $\Theta_{2,2} = \Sigma_{2,2}^{-1}$ corresponding to an outer, strictly contractive $\Sigma_{2,2}$ and therefore corresponded themselves to an inner Σ. This is the case needed for standard constrained interpolation cases of Nevanlinna–Pick- or Schur-type interpolation, to be treated in Chapter 15. However, one has to relax some of the conditions in order to cover other major applications. One case is "selective filtering" (mostly LTI – the LTV selective case has, as far as we know, not been investigated so far), in which $\Sigma_{2,2}$ is still outer, but $\Sigma_{2,2}^{-1}$ (and hence Θ) is unbounded. In another direction, which we consider in Chapter 16, the outerness condition (but not the invertibility condition) on $\Sigma_{2,2}$ is relaxed, so that $\Sigma_{2,2}$ is allowed to have an anti-causal part, while $\Theta_{2,2} = \Sigma_{2,2}^{-1}$ remains bounded and causal. As we shall see, this allows one to accommodate interpolation data

that do not meet the positive definite Pick condition, but where the Pick matrix is nonsingular indefinite. This case was first considered in scalar complex function calculus independently by Schur and Takagi, and then later generalized to complex matrix functions by Adamyan, Arov and Krein – known as the "AAK theory." The theory neatly generalizes to the LTV situation and is presented in Chapter 16.

15 Constrained Interpolation

In this chapter, we treat the problem of determining a causal system that matches a well-chosen coherent set of prescribed data and is restricted in norm (contractive). In classical analysis, this question is known as an interpolation problem, characterized as of Nevanlinna–Pick, Schur or Hermite–Fejer type. Classically, the data to be interpolated consist of prescribed values of the transfer operator (and potentially some of its derivatives) at specific points of the unit disk in the complex plane. In our time-variant theory, there is no "complex plane," in which we can perform valuations of the transfer operator, so a generalization of the notion of valuation has to be defined, in such a way that it reduces to the classical definition when the system happens to be time invariant. Amazingly, there is a natural way of defining such a general valuation. It has been called the W-transform due to its origin in reproducing kernel calculus. After having defined the generalized valuation method, the chapter proceeds to solve the equivalents of the classical interpolation problems in the new time-variant setting. One remarkable result is the reduction of the classical interpolation problems to a single unifying framework.

Menu

Hors d'oeuvre
The W-transform

Appetizer
A Smorgasbord of Interpolation Problems

Main Dish
The Great Unification

Dessert
The Pick condition

15.1 Valuations: The *W*-Transform

Let T be a [bounded] causal (or equivalently lower) operator, and let W be a (block) diagonal operator conformal with the shift Z. Let us look for a diagonal operator $T^{\wedge}(W)$ such that

$$T = T^{\wedge}(W) + T_r(Z - W), \tag{15.1}$$

in which T_r is again a [bounded] causal operator. We shall show that $T^\wedge(W)$ is uniquely defined when W is u.e.s., that is, when the spectral radius ℓ_W of $Z'W$ is less than 1 (a condition that is always satisfied in the finitely indexed case), and we may state

Definition 15.1 $T^\wedge(W)$ is by definition the valuation of T at W, and called "T at W." $T^\wedge(W)$, seen as a function of the diagonal W, is also called the W-transform of T.

The diagonal operator $T^\wedge(W)$ is best characterized in terms of an expansion of T in shifted main diagonals: $T = \sum_{i=0}^{\infty} T_{[i]} Z^i$, in which $T_{[i]} := \Pi_0(T(Z')^i$, with Π_0 the projection on the main diagonal and $T_{[i]} = 0$ for $i < 0$. A causal input $u \in \mathcal{L}_2^{\mathbf{m}}$ will have a similar diagonal expansion $u = \sum_{i=0:\infty}(u_{[i]} Z^i)$ (with $u_{[i]} = 0$ for $i < 0$!), and applying T to it produces the output $y = \sum_{i=0:\infty}(y_{[i]} Z^i)$ with

$$
\begin{aligned}
y_{[i]} = \Pi_0(y(Z')^i) &= \Pi_0 \left\{ \left[\sum_{k=0:} (T_{[k]} Z^k) \right] \left[\sum_{\ell=0:} (u_{[\ell]} Z^\ell) \right] (Z')^i \right\} \\
&= \Pi_0 \left\{ \sum_{k=0:,\ell=0:} T_{[k]} u_{[\ell]}^{\langle k \rangle} Z^{k+\ell-i} \right\} \\
&= \sum_{k=0:} T_{[k]} u_{[i-k]}^{\langle k \rangle},
\end{aligned} \tag{15.2}
$$

generalizing the classical transfer input–output *convolution*.

We find for the W-transform:

Proposition 15.2 *Given $T = \sum_{i=0}^{\infty} T_{[i]} Z^i$ a [bounded] causal operator decomposed in its diagonals $T_{[i]}$ and W a conformal diagonal operator,*[1] *$T^\wedge(W)$ is unique and given by*

$$
\begin{aligned}
T^\wedge(W) &= T_{[0]} + T_{[1]} W + T_{[2]} W^{\langle 1 \rangle} W + T_{[3]} W^{\langle 2 \rangle} W^{\langle 1 \rangle} W + \cdots \\
&= \sum_{k=0}^{\infty} \left(T_{[k]} \prod_{i=0}^{(k-1)\leftarrow} W^{\langle i \rangle} \right),
\end{aligned} \tag{15.3}
$$

in which $\prod_0^{-1\leftarrow}(\cdot) := I$, and $\prod_0^{0\leftarrow} W = W$, and $\prod^{\leftarrow}$ indicates a product ordered from right to left, following the indicated indexing.

Proof

The equation to be shown is equivalent to

$$
\left(T - T^\wedge(W)\right)(I - Z'W)^{-1} = T_r Z, \tag{15.4}
$$

in which $(I - Z'W)^{-1}$ is well defined and bounded by hypothesis. Identifying powers on both sides, one finds the expressions for $T^\wedge(W)$ and T_r without difficulty. □

[1] If $T : \mathcal{L}_2^{\mathbf{m}} \to \mathcal{L}_2^{\mathbf{n}}$, then $W : \mathcal{D}_2^{\mathbf{m}} \to \mathcal{D}_2^{Z\mathbf{m}}$, like Z, while $T_{[1]}$ maps $\mathcal{D}_2^{Z\mathbf{m}} \to \mathcal{D}_2^{\mathbf{n}}$.

Properties of the *W*-transform

The W-transform is weaker than the standard LTI valuation (traditionally called the z-transform), but it has the necessary properties to act as a generalization capable of interpolation theory. Here are the main properties for further use.

Proposition 15.3 *Let T, T_1 and T_2 be bounded causal operators and W a diagonal operator with $\ell_W < 1$, all conformal as needed; then the following holds:*

1. *The chain rule:* $(T_2T_1)^\wedge W = (T_2T_1^\wedge(W))^\wedge(W)$;
2. $T^\wedge(W) = \Pi_0(T(I - Z'W)^{-1})$;
3. $T^\wedge(W) = 0 \Leftrightarrow T(Z - W)^{-1} \in c$.

Proof

By definition and direct evaluation: One uses linearity, and for statement 1, the property $ZT = T^{\langle 1\rangle}Z$, so that $(ZT)^\wedge(W) = (T^\wedge(W))^{\langle 1\rangle}W$. Convergence is no problem so long as W is u.e.s. and the T's are bounded. □

The first property in particular makes the use of the W-transform a bit tricky. If $T^\wedge(W) = t$, where t is a conformal diagonal, then we can say "T interpolates t at W," just as in the case of an LTI transfer function $T(z)$, one says that $T(z)$ interpolates t at a point w, when $T(w) = t$, where $T(z)$ is a matrix function and t is a conformal matrix. Translating the z-calculus to Toeplitz matrix calculus, one sees that the W-transform generalizes the classical complex evaluation calculus to the matrix or LTV case. (This also motivates the term W-transform.)

15.2 A Potpourri of Constrained Interpolation Problems

Here is a first list of classical constrained interpolation problems, translated (or, better, generalized) to the matrix/LTV situation:

1. *Schur interpolation:* Given the first $n + 1$ components of a series $S_{[0]} + S_{[1]}Z + \cdots + S_{[n]}Z^n$, find a causal and contractive operator S that continues the series, if it exists, and otherwise, existence conditions in terms of the given data.
2. *Nevanlinna–Pick interpolation:* Find a causal contractive operator S that matches n given valuations $S^\wedge(W_i) = s_i$ $(i = 1 : n)$, if it exists, and otherwise, existence conditions.
3. *Tangential Nevanlinna–Pick:* Given (of course, conformal as needed) diagonal operators ξ, η and W, find a causal contractive S, such that $(S\xi)^\wedge(W) = \eta$, if it exists, and otherwise, existence conditions.

It turns out that all these cases are special cases of one generic problem, which all can be seen to be of type 3, with proper adjustment of the operators ξ, η and W. So, we first investigate the solution of Formulation 3 and will then show how the other types are covered by it (and some more as well).

A reformulation of the condition $(S\xi)^\wedge(W) = \eta$ is simply

$$(S\xi - \eta)(Z - W)^{-1} \in \text{causal}, \tag{15.5}$$

in which it is understood that $(Z - W)^{-1} = (I - Z'W)^{-1}Z' \in$ anti-causal with a convergent Neumann series in Z', thanks to the u.e.s. assumption $\ell_W < 1$, and in which S is requested to be causal and contractive. As usual, we deal first with the case where S is required to be strictly contractive.

There is a natural equivalence of interpolation problems, formulated in Proposition 15.4:

Proposition 15.4 *Let K be a suitably conformal bounded block diagonal operator with bounded inverse; then the interpolation data $\{K^{-\langle 1\rangle}WK, \xi K, \eta K\}$ are equivalent to $\{W, \xi, \eta\}$.*

Proof

Using the hypothesis on K, $(S\xi)^\wedge W = \eta$ If and only if $S\xi = \eta + r(Z - W)$, with $r \in c$. Hence, $S\xi K = \eta K + r(Z - W)K = \eta K + rK^{\langle 1\rangle}(Z - K^{-\langle 1\rangle}WK)$, yielding $(S\xi K)^\wedge(K^{-\langle 1\rangle}WK) = \eta K$ and vice versa. □

Solution Strategy

Expressions like $\xi(Z - W)^{-1} = \xi(I - Z'W)^{-1}Z'$ or $\eta(Z - W)^{-1} = \cdots$ look very much like the definition of the reachability spaces of a causal operator. (Remember: for the system $\{A, B, C, D\}$, one form of the reachability operator. is $B'(Z - A')^{-1}$.) In Chapter 14 we already indicated how the J-positivity of the reachability space of an inner chain scattering operator induces an interpolation problem. Here we reverse the argument: we start out from the interpolation problem and construct the related chain scattering operator, so that $\begin{bmatrix} \xi \\ -\eta \end{bmatrix}(Z - W)^{-1}$ becomes the reachability space of an inner chain scattering operator related to the solution. Observe, at first, that the interpolation implies the following property (abbreviating "$\in$ causal" to "$\in c$" and similarly for anti-causal, annotated as "$\in a$"):

Proposition 15.5 *If a strictly contractive solution $S \in c$ exists for the tangential Nevanlinna–Pick interpolation problem and the pair $\{W, \xi\}$ is strictly reachable, then*

$$P := \Pi_0(Z' - W')^{-1}(\xi'\xi - \eta'\eta)(Z - W)^{-1} \gg 0. \tag{15.6}$$

Proof

Assuming a solution S to the stated interpolation problem, we have (with Π_{sp} the projection on the strict past)

$$\Pi_{\text{sp}}((S\xi - \eta)(Z - W)^{-1}) = 0. \tag{15.7}$$

Also (with Π_f the projection on present and future),

$$\Pi_f((S\xi - \eta)(Z - W)^{-1}) = \Pi_f(S\xi(Z - W)^{-1}), \tag{15.8}$$

because $\eta(Z-W)^{-1} \in a$. Adding the two gives

$$\eta(Z-W)^{-1} = \Pi_{\text{sp}}(S\xi(Z-W)^{-1}). \tag{15.9}$$

This equation is of the form $X = Y_1$, with Y_1 such that $S\xi(Z-W)^{-1} = Y_1 + Y_2$ and $Y_1 x \perp Y_2 x$ for any conformal $x \in \mathcal{D}_2$. Hence, $\|Yx\|_F^2 \geq \|Xx\|_F^2$, equivalent[2] to $\Pi_0(Y'Y) \geq \Pi_0(X'X)$ or

$$\Pi_0((Z'-W')^{-1}\xi'S'S\xi(Z-W)^{-1}) \geq \Pi_0(\eta'(Z'-W')^{-1}\eta'\eta(Z-W)^{-1}). \tag{15.10}$$

Since $I - S'S \gg 0$ by assumption, we find finally $P \gg 0$, using the strict non singularity of $\xi(Z-W)^{-1}$. □

Remark: P satisfies and, assuming W u.e.s., is determined by the forward Lyapunov–Stein equation

$$P^{\langle -1\rangle} = \xi\xi' - \eta\eta' + W'PW, \tag{15.11}$$

since, with $X := \xi'\xi - \eta'\eta$,

$$(Z'-W')^{-1}X(Z-W)^{-1} = (Z'-W')^{-1}W'P + P + PW(Z-W)^{-1}, \tag{15.12}$$

as one finds out by premultiplying with $(Z'-W')$ and postmultiplying with $(Z-W)$.

Definition 15.6 The matrix P is called the Pick matrix for the interpolation problem.

The strict positive definiteness of the Pick matrix is hence a necessary condition for the solution of the strict tangential Nevanlinna–Pick problem. The additional condition of strict reachability on the data pair $\{W', \xi'\}$ may appear artificial at this point, but it is a condition of well-posedness of the interpolation data, to be considered at the end of this section.

In Sections 15.3 and 15.4, we show that the positivity of the Pick matrix is also a sufficient condition. However, before continuing on this issue, let us see how the Schur and the regular Nevanlinna–Pick interpolation problems are special cases of the tangential problem:

"Standard" Nevanlinna–Pick: Put

$$W := \begin{bmatrix} W_1 & & & \\ & W_2 & & \\ & & \ddots & \\ & & & W_n \end{bmatrix}, \quad \begin{aligned} \xi &:= \begin{bmatrix} I & I & \cdots & I \end{bmatrix} \\ \eta &:= \begin{bmatrix} s_1 & s_2 & \cdots & s_n \end{bmatrix}. \end{aligned} \tag{15.13}$$

Schur: Put

$$W := \begin{bmatrix} 0 & I & & 0 \\ & \ddots & \ddots & \\ & & 0 & I \\ 0 & & & 0 \end{bmatrix}, \quad \begin{aligned} \xi &:= \begin{bmatrix} I & 0 & \cdots & 0 \end{bmatrix} \\ \eta &:= \begin{bmatrix} S_{[0]} & S_{[1]} & \cdots & S_{[n]} \end{bmatrix}. \end{aligned} \tag{15.14}$$

[2] Because $\|a\|_F^2 = \text{trace}(a'a)$.

A combination of the two cases leads to what is known as a *Hermite–Fejer* interpolation problem. First, suppose we "would like to do a Schur interpolation of order m_k" in a diagonal point V_k different from zero; then we would have to put powers of $Z - V_k, k = 1 : m_k$ instead of powers of Z. Remarking that (more accurately) $Z^n = Z^{\langle n \rangle} Z^{\langle n-1 \rangle} \dots Z$, it turns out that in the V_k case, the correct set of powers and coefficients to be interpolated at V_k then has the form $S[\xi_k]_{[0]} \sim [\eta_k]_{[0]} + [\eta_k]_{[1]}(Z - V)^{\langle 1 \rangle} + \cdots + [\eta_k]_{[m_k]}(Z - V)^{\langle m_k \rangle}$, and the corresponding interpolation data become

$$W_k := \begin{bmatrix} V_k & I & & 0 \\ & \ddots & \ddots & \\ & & V_k^{\langle m_k - 1 \rangle} & I \\ 0 & & & V_k^{\langle m_k \rangle} \end{bmatrix}, \quad \begin{matrix} \xi_k := \begin{bmatrix} [\xi_k]_{[0]} & 0 & \cdots & 0 \end{bmatrix} \\ \eta_k := \begin{bmatrix} [\eta_k]_{[0]} & [\eta_k]_{[1]} & \cdots & [\eta_k]_{[m_k]} \end{bmatrix}. \end{matrix} \tag{15.15}$$

These may be stacked again for different k's and we obtain the full Hermite–Fejer interpolation problem with

$$W := \begin{bmatrix} W_1 & & & \\ & W_2 & & \\ & & \ddots & \\ & & & W_n \end{bmatrix}, \quad \begin{matrix} \xi := \begin{bmatrix} \xi_1 & \xi_2 & \cdots & \xi_n \end{bmatrix}, \\ \eta := \begin{bmatrix} \eta_1 & \eta_2 & \cdots & \eta_n \end{bmatrix}, \end{matrix} \tag{15.16}$$

where the W_i, ξ_i and η_i all have the form given in (15.15).

All these constrained interpolation problems are instances of just one, namely: find a causal (strictly) contractive S such that $(S\xi)^\wedge(W) = \eta$, with W u.e.s. and the other data conformal. An additional condition requires the data $\{W', \xi'\}$ to be strictly reachable, that is, to have the reachability Gramian $\Pi_0((Z - W)^{-\prime}\xi'\xi(Z - W)^{-1})$ strictly nonsingular. This extra condition is sufficient (although not always necessary) to make the interpolation well posed. One can see why, by looking at simple examples: the Nevanlinna–Pick interpolation problem requiring $S^\wedge(V) = s_1$ as well as $S^\wedge(V) = s_2$, is evidently ill posed when $s_1 \neq s_2$, and the conditions are redundant when $s_1 = s_2$. In both cases, the $\{W, \xi\} = \left\{ \begin{bmatrix} V & \\ & V \end{bmatrix}, \begin{bmatrix} 1 & 1 \end{bmatrix} \right\}$ and the corresponding reachability Gramian is $\Pi_0((Z' - W')^{-1} \begin{bmatrix} 1 & 1 \\ 1 & 1 \end{bmatrix} (Z - W)^{-1}) = \begin{bmatrix} v & v \\ v & v \end{bmatrix}$ with[3] $v = \Pi_0((Z' - V')^{-1}(Z - V)^{-1}) = I + V^{\langle 1 \rangle\prime} V^{\langle 1 \rangle} + V^{\langle 1 \rangle\prime} V^{\langle 2 \rangle\prime} V^{\langle 2 \rangle} V^{\langle 1 \rangle} + \cdots$, hence singular. *Ill posed* is what happens generically when the reachability Gramian of $\{W', \xi'\}$ is singular.

[3] Π_0 spreads over the block matrix, like Z does! This is perhaps an unfortunate side-effect of global block decomposition.

15.3 The Pick Condition

The generic constrained interpolation problem is therefore:

Given bounded diagonal operators W, ξ and η, with W u.e.s., ($\ell_W < 1$), find a causal contractive operator S such that $(S\xi)^\wedge(W) = \eta$, or equivalently, such that

$$\Pi_{sp}(S\xi - \eta)(Z - W)^{-1}) = \Pi_{sp}\left(\begin{bmatrix} S & I \end{bmatrix}\begin{bmatrix} \xi \\ -\eta \end{bmatrix}(Z - W)^{-1}\right) = 0. \tag{15.17}$$

The problem is only meaningful if the diagonal operators mentioned are conformal, that is, Z and W have equal dimensions, ξ and η have matching column dimensions, and $\{W', \xi'\}$ form a reachability pair (or, equivalently, the observability pair of an anti-causal system).

To formulate the interpolation theorem, we need some notions and notations from Chapter 14. Motivation comes from the properties of chain scattering matrices. The interpolation data written in the form

$$\begin{bmatrix} \xi \\ -\eta \end{bmatrix}(Z - W)^{-1} \tag{15.18}$$

look very much like the reachability operator of a causal matrix (the minus sign is used for "historical" reasons). With the additional property that the Pick matrix is J-positive and, using the generic signature matrix $J = \begin{bmatrix} I & \\ & -I \end{bmatrix}$, it defines the characteristic *reachability matrix of a causal chain-scattering matrix* Θ defined and studied in Chapter 14. We summarize the relevant construction of the Θ with this reachability operator.

A chain scattering operator Θ relates a pair of inputs $\begin{bmatrix} a_1 \\ b_1 \end{bmatrix}$ (called an incident wave and a reflected wave, respectively) to a pair of outputs $\begin{bmatrix} a_2 \\ b_2 \end{bmatrix} = \Theta \begin{bmatrix} a_1 \\ b_1 \end{bmatrix}$, and a load S_L for which $b_2 = S_L a_2$, with a_2 incident and b_2 reflected; then this setup defines a relation between b_1 and a_1. This relation turns out to be univocal, hence a function, and we may write $b_1 = S a_2$, in which S is determined by Θ and S_L, which we then write as $S = \sigma_\Theta(S_L)$. In the application here, the matrix Θ will be (in the terminology of Chapter 14) J-inner and causal, and S_L causal contractive. As shown in Chapter 14, the relation between S_L and S is bilinear:

$$S = \sigma_\Theta(S_L) = (-S_L\Theta_{1,2} + \Theta_{2,2})^{-1}(-\Theta_{2,1} + S_L\Theta_{1,1}). \tag{15.19}$$

Alternatively, and using the inner scattering matrix Σ: $\begin{bmatrix} a_2 \\ b_1 \end{bmatrix} \to \begin{bmatrix} a_1 \\ b_2 \end{bmatrix}$, whereby

$$\Sigma = \begin{bmatrix} \Theta_{1,1} - \Theta_{1,2}\Theta_{2,2}^{-1}\Theta_{2,1} & \Theta_{1,2}\Theta_{2,2}^{-1} \\ -\Theta_{2,1}\Theta_{2,2}^{-1} & \Theta_{2,2}^{-1} \end{bmatrix} = \begin{bmatrix} \Theta_{1,1}^{-\prime} & \Theta_{1,1}^{-\prime}\Theta_{2,1}^{\prime} \\ -\Theta_{2,1}\Theta_{2,2}^{-1} & \Theta_{2,2}^{-1} \end{bmatrix} \tag{15.20}$$

(remark that in this case $\Sigma_{2,2} = \Theta_{2,2}^{-1}$ is outer and $\Sigma_{1,1} = \Theta_{1,1}^{-\prime}$ is maximal phase, since $\Theta_{1,1}$ is causal and its inverse anti-causal), one has

$$S = \sigma_\Theta(S_L) = \Sigma_{2,1} + \Sigma_{2,2}S_L(I - \Sigma_{1,2}S_L)^{-1}\Sigma_{1,1}, \tag{15.21}$$

and (anticipating the results) we shall have $(S\xi)^\wedge W = \eta$ when $(\Sigma_{2,1}\xi)^\wedge W = \eta$ and $(\Sigma_{1,1}\xi)^\wedge W = 0$ – or, in the classical circuit theory terminology, W defines the collection of "transmission zeros" of Σ.

The reachability space of Θ is by definition $\mathcal{H}_\Theta = \Pi_{\text{sp}}\Theta'\mathcal{L}_2$; hence, in this case, $\mathcal{H}_\Theta = J\begin{bmatrix} \xi \\ \eta \end{bmatrix}(Z - W)^{-1}\mathcal{D}_2 = J\Theta'\mathcal{H}_{o,\Theta}$ (with appropriate dimensions, which we have not made explicit), and in which $\mathcal{H}_{o,\Theta} \subset \mathcal{L}_2$ is the space of natural responses of Θ. Dually, $\mathcal{H}_{o,\Theta} = \Pi_f\Theta\mathcal{U}_2^-$ (by definition), and in Proposition 14.6, we showed that $\mathcal{H}_{o,\Theta} = \Theta J\mathcal{H}_\Theta$ and dually $\Theta' J\mathcal{H}_{o,\Theta} = \mathcal{H}_\Theta$.

The interpolation property can simply be expressed in terms of $\mathcal{H}_\Theta$:

$$\begin{bmatrix} S & I \end{bmatrix}\mathcal{H}_\Theta \in c, \tag{15.22}$$

a relation that plays a key role in the representation of the solution, as given in Theorem 15.7.

Theorem 15.7 *The constrained interpolation problem with data $\{W,\xi,\eta\}$, such that the pair $\{W',\xi'\}$ is strictly reachable, has a strictly contractive solution if and only if the corresponding Pick matrix*

$$P = \Pi_0((Z' - W')^{-1}(\xi'\xi - \eta'\eta)(Z - W)^{-1}) \tag{15.23}$$

is strictly positive definite. All solutions are given by

$$S = \sigma_\Theta(S_L), \tag{15.24}$$

in which σ_Θ is a chain scattering transformation by the chain scattering matrix Θ, where Θ has the reachability data $\{W', \begin{bmatrix} \xi' & -\eta' \end{bmatrix}\}$, and S_L is a conformal but otherwise arbitrary causal contractive operator (in particular, S_L may be chosen zero).

Proof

(We follow the proof from [29, p. 240], with some adaptation.)

Sufficiency: $P \gg 0 \Rightarrow (\sigma_\Theta(S_L)\xi)^\wedge W = \eta$.
Since P is assumed to have a bounded inverse, we can define $P = r^{-1}r^{-\prime}$, with both r and r^{-1} bounded. Then $\left[\, r^{\langle 1\rangle}W'r^{-1} \;\middle|\; r^{\langle 1\rangle}\xi' \quad -r^{\langle 1\rangle}\eta' \,\right]$ is J-co-isometric and strictly positive. A J-unitary realization for Θ is obtained by completion to a full J-unitary matrix. (There is quite a bit of latitude for doing so; see Chapter 14). Next, $S = \sigma_\Theta(S_L)$ satisfies

$$\begin{bmatrix} S & -I \end{bmatrix} = \Phi^{-1}\begin{bmatrix} S_L & -I \end{bmatrix}\Theta, \tag{15.25}$$

in which $\Phi := -S_L\Theta_{1,2} + \Theta_{2,2}$ is outer and has a bounded inverse. This follows from the fact that (1) $\Theta_{2,2} = \Sigma_{2,2}^{-1}$ is outer with a bounded inverse (and so is $\Sigma_{2,2}$) and (2) by

Lemma 14.2 and the strict contractivity of $S_L\Sigma_{1,2}$, Φ is outer and invertible as well, together with $I - S_L\Theta_{1,2}\Theta_{2,2}^{-1} = I + S_L\Sigma_{1,2}$. The interpolation condition formulated as $\begin{bmatrix} S & -I \end{bmatrix} J\mathcal{H}_\Theta \in c$ is now evaluated as follows, making use of the mapping property of Θ from the reachability space to the observability space:

$$\begin{bmatrix} S & -I \end{bmatrix} J\mathcal{H}_\Theta = \Phi^{-1}\begin{bmatrix} S_L & -I \end{bmatrix}\Theta J\mathcal{H}_\Theta = \Phi^{-1}\begin{bmatrix} S_L & -I \end{bmatrix}\mathcal{H}_{o,\Theta} \in c, \tag{15.26}$$

showing that interpolation is achieved as soon as $S_L \in c$ and S_L contractive (since $\Theta_{1,2}\Theta_{2,2}^{-1}$ is strictly contractive, the resulting S will be as well).

Necessity: If $S = \sigma_\Theta(S_L)$ is interpolating, then S_L is causal contractive.
If the proposed $S = \sigma_\Theta(S_L)$, then Eq. 15.25 is satisfied, with $S \in c$ by hypothesis, and assumed as the interpolation condition, written as

$$\Pi_{\text{sp}}\begin{bmatrix} S & -I \end{bmatrix} J\mathcal{H}_\Theta = 0. \tag{15.27}$$

Let now $\begin{bmatrix} G_1 & -G_2 \end{bmatrix} := \begin{bmatrix} S & -I \end{bmatrix}\Theta^{-1}$; then $G \in c$, for

$$\Pi_{\text{sp}}(G\mathcal{L}_2) = \Pi_{\text{sp}}(\begin{bmatrix} S & -I \end{bmatrix}\Pi_{\text{sp}}(\Theta^{-1}\mathcal{L}_2)) = \Pi_{\text{sp}}\begin{bmatrix} S & -I \end{bmatrix} J\mathcal{H}_\Theta = 0, \tag{15.28}$$

since $S \in c$ by hypothesis. Since $\begin{bmatrix} G_1 & -G_2 \end{bmatrix} = \Phi^{-1}\begin{bmatrix} S_L & -I \end{bmatrix}$, we also have $S_L = G_2^{-1}G_1$, which shall be causal when $\Phi = G_2^{-1} = (I - S_L\Theta_{1,2}\Theta_{2,2}^{-1})\Theta_{2,2}$ is causal. This we now proceed to do, showing invertibility as a bounded operator first. Since Θ has a bounded inverse (namely $J\Theta' J$), there is an ϵ so that $\Theta'\Theta \geq \epsilon$ for the operator norm,[4] from which it follows that

$$G_1G_1' + G_2G_2' \geq \begin{bmatrix} S & -I \end{bmatrix} J\Theta'\Theta J\begin{bmatrix} S' \\ -I \end{bmatrix} \geq \epsilon(SS' + I) \geq \epsilon I. \tag{15.29}$$

In addition, $G_2G_2' \geq G_1G_1'$, since $G_1 = G_2S_L$, and S_L is contractive because Θ is J-unitary and because of the connection between S and S_L (as shown in Eq. 15.25). It follows that $G_2G_2' \geq \epsilon/2$ (an easy exercise! Hint: apply $x' \cdots x$, with x being any conformal vector of unit norm), and hence G_2^{-1} is Eq. (15.25) bounded. Causality of the inverse is obtained from the special form of G_2: from the definition of G, we have $G_2^{-1}\Theta_{2,2}^{-1} = I - S_L\Theta_{1,2}\Theta_{2,2}^{-1}$, and

$$\Theta_{2,2}G_2 = (I - S_L\Theta_{1,2}\Theta_{2,2}^{-1})^{-1} \in c, \tag{15.30}$$

from which it follows that $S_L\Theta_{1,2}\Theta_{2,2}^{-1} \in c$ by the next Lemma 15.8:

Lemma 15.8 *Let S be strictly contractive and $T = (I - S)^{-1}$ causal; then S is causal as well.*

[4] If a Hilbert space operator A has a bounded inverse, then there exist $\sqrt{\epsilon}$ with $\epsilon > 0$ such that $\forall(x \in \|x\|_2 = 1) : \|Ax\|_2 \geq \sqrt{\epsilon}$, namely $\sqrt{\epsilon} = \|A^{-1}\|^{-1}$, and hence $\forall(x \in \|x\|_2 = 1) : x'A'Ax \geq \epsilon$.

Proof of the Lemma

Let $Y := (I+S)(I-S)^{-1} = 2(I-S)^{-1} - I = 2T - I$; then since S is strictly contractive, $Y + Y' \gg 0$, for

$$Y + Y' = 2(I - S')^{-1}(I - S'S)(I - S)^{-1}, \tag{15.31}$$

and since Y is causal together with T, Y is strictly positive real (PR). It follows that Y is outer. For suppose there would exist an $x \in \mathcal{L}_2$, such that $x \perp Y\mathcal{L}_2$ in the Hilbert–Schmidt metric; then we would also have $(x, Yx)_{\mathrm{HS}} = 0$, and hence, taking adjoint and adding, $(x, (Y + Y')x)_{\mathrm{HS}} = 0$ and, because $Y + Y'$ is nonsingular, $x = 0$. (Notice that under the hypotheses of the lemma, Y is bounded as well.) Hence Y is outer ($\overline{Y\mathcal{L}_2} = \mathcal{L}_2$). It also follows immediately that $I + Y$ is bounded and has a causal and bounded inverse, and hence $S = I - 2(Y + I)^{-1}$ is causal. It follows that $S_L\Theta_{1,2} \in c$, since $\Theta_{2,2} \in c$, and finally, $G_2^{-1} = \Theta_{2,2} - S_L\Theta_{1,2} \in c$ as well as $S_L = G_2^{-1}G_1$. □

Extension*

Up to this point, we requested the interpolating S to be strictly contractive. However, this is not necessary: just S contractive, or equivalently $I - S'S \geq 0$ is sufficient, leading to the result $S = \sigma_\Theta(S_L)$, with S_L being just causal contractive, and the Pick matrix just positive definite with an inverse defined on a dense subset in the relevant $\mathcal{X}_2$. The corresponding Θ may be unbounded but remains causal and corresponding to an inner Σ with outer $\Sigma_{2,2}$:

$$\Theta = \begin{bmatrix} \Sigma_{1,1} - \Sigma_{1,2}\Sigma_{2,2}^{-1}\Sigma_{2,1} & \Sigma_{1,2}\Sigma_{2,2}^{-1} \\ -\Sigma_{2,2}^{-1}\Sigma_{2,1} & \Sigma_{2,2}^{-1} \end{bmatrix}. \tag{15.32}$$

In this expression, outer $\Sigma_{2,2}$ is necessarily square (equal input and output dimensions, but state dimensions can change) and has a bounded causal inverse. However, the case where $\Sigma_{2,2}^{-1}$ is unbounded, but $\Sigma_{2,2}$ has full dense range, that is, $\overline{\Sigma_{2,2}\mathcal{L}_2^{\mathbf{m}_2}} = \mathcal{L}_2^{\mathbf{m}_2}$, where $\mathbf{m}_2$ is its input/output dimension vector, is, as already mentioned, of great technical importance and needs some special consideration.[5] It turns out that most of the previous theory can be saved for this case. There remains a valid, although unbounded Θ, which can still be called J-inner, and which has a J-unitary representation, in which, however, the state transition operator A_Θ has $\ell_{A_\Theta} = 1$. This transition operator is derived from the original W and the Pick matrix, factored as $P = r'r$ leading to $A_\Theta = r^{-\prime}W'r^{\langle 1\rangle\prime}$ and $B_\Theta = r^{-\prime}\begin{bmatrix} \xi' & \eta' \end{bmatrix}$. We do not explore this theory further here; let a simple LTI example suffice to illustrate the issue.

[5] The technical importance is due to the fact that in the rational LTI case, the unboundedness stems from the choice of interpolation points on the unit circle, rather than in the open unit disc. This results in a corresponding Θ-matrix that has poles in these points (since an interpolation point a produces a pole at $1/a'$). The occurrence of such a pole effectively blocks the transmission of a signal component with that frequency through the chain connecting left input/output to right input/output, or, equivalently, S to S_L. a is therefore called a transmission zero. Such zeros are of great importance in selective filtering, in order to create a valid stop band, preventing one frequency band from interfering with another.

Take, for example, the simplest possible low-pass filter with $\Sigma_{2,2} = \frac{z+1}{2}$. We obtain

$$\Sigma = \begin{bmatrix} \frac{z+1}{2} & \frac{z-1}{2} \\ \frac{z-1}{2} & \frac{z+1}{2} \end{bmatrix}, \; \Theta = \begin{bmatrix} \frac{2z}{z+1} & \frac{z-1}{z+1} \\ -\frac{z-1}{z+1} & \frac{2}{z+1} \end{bmatrix}, \tag{15.33}$$

leading to the unitary and J-unitary respectively with $J = \left[\begin{array}{c|cc} 1 & & \\ \hline & 1 & \\ & & -1 \end{array}\right]$ realizations

$$\Sigma = \left[\begin{array}{c|cc} 0 & \frac{1}{\sqrt{2}} & \frac{1}{\sqrt{2}} \\ \hline \frac{1}{\sqrt{2}} & \frac{1}{2} & \frac{1}{2} \\ \frac{1}{\sqrt{2}} & -\frac{1}{2} & \frac{1}{2} \end{array}\right], \; \Theta = \left[\begin{array}{c|cc} -1 & \sqrt{2} & \sqrt{2} \\ \hline \sqrt{2} & 0 & -1 \\ -\sqrt{2} & 1 & 2 \end{array}\right]. \tag{15.34}$$

$\Sigma_{2,2}$ is obviously outer with the unbounded inverse $\Theta_{2,2}$, which has a pole at the border point $z = -1$. This pole is the "transmission zero," and is at the same time the pole of the chain scattering operator Θ. A_Θ is obviously not u.e.s., but all the quantities are well defined, and the interpolation formula gives all the passive circuits that have a transmission zero at $z = -1$. We find

$$S = -1 + (z+1)\frac{1 + S_L}{2 - (z-1)S_L}, \tag{15.35}$$

which looks like an interpolation formula with $S(-1) = -1$ for any contractive S_L. To prove this, one must show that $(1 + S_L)(2 - (z-1)S_L)^{-1}$ cannot cancel $(z+1)$ in the numerator; this follows from S_L and $2 - (z-1)S_L = \frac{1}{2}(1 - s)$, with $s = \frac{z-1}{2}S_L$ both being causal and strictly contractive, and a limiting argument when S_L is just causal contractive (the proof amounts to the previous treatment!).

This example clearly exhibits the borderline transmission zero as a pole of Θ, resulting in a full reflection -1 at the same frequency ($z = -1$ in this case). Θ is unbounded in this case, but the corresponding $\Sigma_{2,2}$ is causal, so Θ still qualifies as J-inner.

15.4 Notes

- The method we followed to treat the constrained interpolation problem, in particular Proposition 15.5 is due, as far as we know, to Harry Dym.
- *Double-sided interpolation and Sylvester's equation*. One may, of course, consider more complex interpolation problems, in particular double-sided interpolation. This type of interpolation leads to what is known as a *Sylvester equation*, which may easily be very ill conditioned and whose solution may therefore be difficult to compute. In [29, p. 250 ff], we propose a pretty elaborate theory to deal with double-sided interpolation and the role noncausal but inner chain scattering matrices play in it. In particular, we show how, under mild conditions, mixed causal and anti-causal J-inner matrices solve mixed interpolation problems formulated as: find a causal and strictly contractive S which, given right interpolation data $\{V, \xi, \eta\}$ and left interpolation data $\{W, \zeta, \iota\}$, satisfies the following interpolation relations:

$$\begin{cases} (S\xi - \eta)(Z - V)^{-1} & \in & c, \\ (Z - W)^{-1}(\zeta S - \iota) & \in & c. \end{cases} \tag{15.36}$$

It turns out that, under broad conditions, a mixed causal anti-causal J-inner chain scattering matrix of the type

$$\Theta = \begin{bmatrix} C_1 \\ C_2 \end{bmatrix} (Z - W)^{-1} \begin{bmatrix} \iota & -\zeta \end{bmatrix} + D + \begin{bmatrix} C_3' \\ C_4' \end{bmatrix} (Z' - V')^{-1} \begin{bmatrix} \xi' & \eta' \end{bmatrix} \tag{15.37}$$

solves the problem. Here the missing entries C and D follow from J-unitary completion on the related mixed causal anti-causal and normalized realization.

An essential condition for the existence of solutions to the mixed interpolation problem requires the invertibility of the often ill-conditioned Sylvester equation

$$\mathcal{D} \to \mathcal{D} : X \mapsto Y : XV - WX^{\langle -1 \rangle} = Y, \tag{15.38}$$

which plays the role of the Pick equations in this case. Ill conditioning may be expected, because the Sylvester equation cannot be solved by a numerically stable unilateral recursion, as is the case for the Pick equation, depending on the properties of the operators V and W (right interpolation is solved by a forward recursion, while left interpolation leads to a backward recursion; a Lyapunov–Stein equation is not, in general, stably invertible). Nonetheless, it is possible (under broad conditions) to convert a right interpolation problem into a left one using unitary embedding theory. This is the content of Theorem 9.15 in [29].

16 Constrained Model Reduction

How can a causal and uniformly exponentially stable system be approximated by a system of the same class and of the lowest possible degree, given an acceptable error expressed in a strong norm? The answer is: by using a generalized form of interpolation, originally proposed (independently) by Schur and Tagaki, later worked out by Adamyan, Arov and Krein (AAK) for the matrix function (LTI) case and shown thereafter to provide the solution of the constrained model reduction problem for the (strong) so-called Hankel norm. This chapter shows that the extension of the constrained interpolation problem first proposed by Schur and Takagi using complex function calculus also works in the LTV case, now using time-variant state-space models and scattering theory.

Menu

Hors d'oeuvre
Constrained Interpolation: A Summary

First course
Solving the Model Reduction Problem via Schur–Takagi Interpolation

Second course
Computing Realizations

Dessert
Subspace Tracking: A Fully Worked Out Example

16.1 Introduction

In Chapter 15 on constrained interpolation, we saw how a chain scattering matrix solves a constrained interpolation problem, that is, a problem in which a bounded and causal transfer function is asked to meet specified data. The required causality and boundedness of the solution force a condition on the data, namely the positive definiteness of the related Pick operator. In classical interpolation theory for functions on the complex plane, causality translates into *analyticity in the unit disc of the complex plane* and boundedness into "belonging to the Hardy space $\mathcal{H}_\infty$, i.e., functions analytic and bounded in norm in the open unit disc." However, the conditions on constrained interpolation can be relaxed, by allowing the interpolating function to have a limited number of poles in the open unit disk, while remaining bounded on the unit circle. This is known as "Schur–Takagi interpolation." Translated to our context of bounded LTV operators, the relaxation of the causality condition leads to transfer functions that are not strictly causal, but have a limited (low-degree) anti-causal part.

It turns out that this strategy allows one to solve a pretty general constrained model reduction problem, that is, whereby a causal and bounded transfer function is approximated in a strong norm by a causal and bounded transfer function of low degree. To achieve this result, it is necessary to make a translation between the model reduction problem and the generalized Schur–Takagi problem. In a broad outline and for the case of constrained model reduction, the interpolation is done on the *approximation error*, while a remaining singular part of "Schur–Takagi-type" interpolation provides for the approximation that cancels that part of the original that does not meet the error constraint. That this works is due to the fact that the norm constraint is not on the transfer functions themselves but on their difference. To achieve the transition between the model reduction problem and the (generalized) interpolation problem, a link has to be provided between the norm used for model reduction and the norm used for interpolation.

Before moving to the extended interpolation problem of Schur–Takagi type, let us summarize some salient features of the interpolation strategy developed in Chapter 15. We bring the inspirational ingredients together in Proposition 16.1.

Proposition 16.1 1. *Let* $\Theta = \begin{bmatrix} \Theta_{1,1} & \Theta_{1,2} \\ \Theta_{2,1} & \Theta_{2,2} \end{bmatrix}$ *be a causal, bounded and J-unitary locally finite chain scattering matrix with $\Theta_{2,2}$ outer and J-unitary realization such that*

$$\Theta \sim \left[\begin{array}{c|cc} \alpha & \beta_1 & \beta_2 \\ \hline \gamma_1 & \delta_{1,1} & \delta_{1,2} \\ \gamma_2 & \delta_{2,1} & \delta_{2,2} \end{array}\right]. \tag{16.1}$$

Let $\Sigma_{2,1} = -\Theta_{2,2}^{-1}\Theta_{2,1} = -\Theta_{1,2}'\Theta_{1,1}^{-\prime} \in c$ be the zero-load input scattering matrix; then $\Sigma_{2,1}$ solves the interpolation problem

$$[\Sigma_{2,1}\beta_1' - \beta_2'](Z - \alpha')^{-1} \in c. \tag{16.2}$$

2. *Let $T = D + C(I - ZA)^{-1}ZB$ be a strictly observable transfer function in the output normal form ($A'A + C'C = I$) with the external left co-prime factorization $T = -U\Delta'$, $\Delta \in c$ and $U = D_U + C(I - ZA)^{-1}ZB_U$ inner, and suppose Θ is a bounded J-inner chain-transfer function with* $\mathcal{H}_\Theta = \begin{bmatrix} B_U' \\ B' \end{bmatrix}(Z - A')^{-1}$. *Then*

$$\begin{bmatrix} \overline{a} \\ -\overline{b} \end{bmatrix} := \Theta J \begin{bmatrix} U' \\ T' \end{bmatrix} \in c. \tag{16.3}$$

Let $\Sigma_{2,1} := -\Theta_{2,2}^{-1}\Theta_{2.1}$. Then

$$\Delta - \Sigma_{2,1} = -\Theta_{2,2}^{-1}\overline{b}U, \tag{16.4}$$

that is, $\Delta = \Sigma_{2,1} \bmod U$.

3. *Let*

$$S = \Sigma_{2,1} + \Sigma_{2,2} S_L (I - \Sigma_{1,2} S_L)^{-1} \Sigma_{1,1} \tag{16.5}$$

be a contractive input scattering matrix obtained by loading Θ *with a contractive* $S_L \in c$*; then also*

$$\Delta - S = rU, \tag{16.6}$$

for $r = (S_L \Theta_{1,2} - \Theta_{2,2})^{-1}(S_L \overline{a} + \overline{b}) \in c$.

Proof

1. We have $\mathcal{H}_\Theta = \begin{bmatrix} \beta_1' \\ \beta_2' \end{bmatrix} (Z - \alpha')^{-1} \mathcal{D}_2$ by the definition of $\mathcal{H}$ as the range of Θ' and $\Theta J \mathcal{H}_\Theta \in c$ from Proposition 14.6. This shows the interpolation defined by Θ.
2. Next, Eq. (16.3) is valid because $\Theta J \mathcal{H}_\Theta \in c$, and working out the second block row, we find $\Theta_{2,1} U' - \Theta_{2,2} T' = -\overline{b}$ and hence the result.
3. We have from the input–output relation of the Θ matrix

$$\begin{bmatrix} S_L & -I \end{bmatrix} \Theta = Y \begin{bmatrix} -S & I \end{bmatrix}, \tag{16.7}$$

with $Y := (S_L \Theta_{1,2} - \Theta_{2,2}) = -(I - S_L \Theta_{1,2} \Theta_{2,2}^{-1}) \Theta_{2,2}$. Since Θ is assumed strictly J-inner, we have $\Theta_{2,2}^{-1} \in c$ as well as $(I - S_L \Theta_{1,2} \Theta_{2,2}^{-1})^{-1} \in c$ because $\Theta_{1,2} \Theta_{2,2}^{-1} \in c$ is strictly contractive. Hence,

$$\begin{bmatrix} -S & I \end{bmatrix} \begin{bmatrix} U' \\ T' \end{bmatrix} = Y^{-1} \begin{bmatrix} S_L & -I \end{bmatrix} \begin{bmatrix} \overline{a} \\ -\overline{b} \end{bmatrix} = Y^{-1}(S_L \overline{a} + \overline{b}). \tag{16.8}$$

□

We see that the constrained interpolation problem defined by $(SB_U' - B')(Z - A')^{-1} \in c$ and $S \in c$ contractive results in a projective equality

$$\Pi_{\mathcal{H}_{o,U}} S = \Pi_{\mathcal{H}_{o,U}} \Delta, \tag{16.9}$$

in which Δ and U are defined by the relation $\Delta = -T'U \in c$, with T and U left inner coprime. Actually, the properties mentioned (interpolation and projective equality) are equivalent as shall be seen further in this chapter. The condition for the existence of the solution of the interpolation problem is the existence of the J-inner Θ, which, in turn, is the strict positivity of the Gramian

$$\Lambda^{\langle -1 \rangle} = A \Lambda A' + B_U B_U' - BB'. \tag{16.10}$$

This intimate relation between "approximation through projection" and interpolation is how the Schur–Takagi interpolation theory is connected to the constrained norm approximation. We shall see that this is achieved by letting $\Theta_{2,2}^{-1}$ to become partly anti-causal, while staying contractive, keeping Θ causal and allowing the Gramian Λ to be nonsingular indefinite. It turns out that the model reduction problem produces an interpolation problem of Schur–Takagi type on the derived *error* $\Delta - S$. How this works is the central topic of this chapter.

16.2 Model Reduction

Let T be a strictly lower (causal) and bounded operator of dimensions $\mathbf{n}\times\mathbf{m}$ with state-space realization $\{A, B, C, 0\}$, in which A is assumed u.e.s., and uniformly observable and controllable (i.e., with bounded Hankel operator). Let moreover Γ be a diagonal, Hermitian operator with bounded inverse (Γ may be taken small and constant – like the traditional ϵ, but we do not have to assume time invariance here: the required accuracy is allowed to change with the index). The goal of constrained model reduction is to find an approximating, low degree and causal $\widehat{T}$ such that

$$\|(T - \widehat{T})\Gamma^{-1}\|_{?} \le I \tag{16.11}$$

for some adequate strong norm to be determined.

From classical LTI theory, we know that the regular operator norm is too strong (see also the discussion on this). It turns out that a somewhat weaker but still pretty strong norm, known as the Hankel norm, will be adequate. Given a strictly lower operator such as T (or $\widehat{T}$), the *Hankel norm* on T is defined as

$$\|T\|_H := \|H_T\|, \tag{16.12}$$

where H_T is the Hankel operator associated with T and $\|\cdot\|$ is its regular operator norm.

An important observation is the connection between the operator norm and the Hankel norm as follows. Let T_r be a mixed operator with a bounded strictly causal part $\widehat{T}$; then the following holds:

Lemma 16.2

$$\|(T - \widehat{T})\Gamma^{-1}\|_H \le \|(T - T_r)\Gamma^{-1}\|. \tag{16.13}$$

Proof

We have (with $\|\cdot\|_2$ being the usual Hilbert–Schmidt or Frobenius norm)

$$\begin{aligned} \|(T - \widehat{T})\Gamma^{-1}\|_H &= \sup_{u\in\mathcal{U}_2^-, \|u\|_2\le 1}\|\Pi_f((T - \widehat{T})\Gamma^{-1}u)\|_2 \\ &= \sup_{u\in\mathcal{U}_2^-, \|u\|_2\le 1}\|\Pi_f((T - T_r)\Gamma^{-1}u)\|_2 \\ &\le \|(T - T_r)\Gamma^{-1}\|. \end{aligned} \tag{16.14}$$

□

Hankel Norm Approximation Procedure

We first formulate the basic strategy. Assume we can determine a factorization $\Theta\begin{bmatrix} U' \\ -\Gamma^{-1}T' \end{bmatrix} = \begin{bmatrix} \overline{a} \\ -\overline{b} \end{bmatrix}$, in which T is strictly causal, $T = U\Delta'$ is a left coprime external factorization, Θ is causal J-unitary, and both $\overline{a}$ and $\overline{b}$ are causal. Moreover, T will be assumed to have a realization that is both strictly reachable and observable, with A u.e.s.

Proposition 16.3 *Let $T = U\Delta'$ be a (left-coprime) external factorization of T such that U is inner and Δ causal, and let Θ be a causal bounded J-unitary operator such that*

$$\Theta \begin{bmatrix} U' \\ -\Gamma^{-1}T' \end{bmatrix} = \begin{bmatrix} \overline{a} \\ -\overline{b} \end{bmatrix} \in c \tag{16.15}$$

(i.e., with $\overline{a}$ and $\overline{b}$ causal); let $T_r' := \Gamma\Theta_{2,2}^{-1}\overline{b}$ and $\widehat{T}' := \Pi_{sp}T_r'$. Then,

$$\|(T - \widehat{T})\Gamma^{-1}\|_H \leq \|(T - T_r)\Gamma^{-1}\| \leq 1. \tag{16.16}$$

Proof

Multiplying out the second block row, we find that $\Theta_{2,1}U' - \Theta_{2,2}\Gamma^{-1}T' = -\overline{b}$. $\Theta_{2,2}^{-1}$ is a contraction by the J-unitarity of Θ, and so is $\Sigma_{2,1} = -\Theta_{2,2}^{-1}\Theta_{2,1}$ as well. Dividing by $\Theta_{2,2}$ and introducing $T_r := (\Gamma\Theta_{2,2}^{-1}\overline{b})'$ gives

$$\Gamma^{-1}(T' - T_r') = \Sigma_{2,1}U'. \tag{16.17}$$

With $\widehat{T} := \Pi_f^+ T_r$, the result follows by Lemma 16.2, since $\Sigma_{2,1}U'$ is contractive. □

Supposing the conditions of the theorem fulfilled, $\widehat{T}$ appears to be an approximation to T with an error whose norm is limited by Γ. (Suppose $\Gamma = \epsilon$ constant; then the error is smaller than ϵ.) But will it be an approximation of low degree? We shall analyze this question in detail, but let it be remarked here that since $\widehat{T} = \Pi_f^+(\overline{b}'\Theta_{2,2}^{-\prime}\Gamma)$ (with Π_f^+ the projection on the strict future), and $\overline{b}'$ is anti-causal, the degree of $\widehat{T}$ is limited by the degree of the causal part of $\Theta_{2,2}^{-\prime}$, which equals the degree of the anti-causal part of $\Theta_{2,2}^{-1}$, and this may indeed turn out to be small, *provided $\overline{b}$ is causal*. We shall show that this degree has a direct relation to significant singular values of the Hankel operator of T. This will follow from an explicit construction of Θ based on the realizations of T and U.

The Construction of Θ

The external factorization $T = U\Delta'$ is simplest when T is given in output canonical form. U characterizes the natural output state space $\mathcal{H}_{o,T}$ of T. This means that U borrows the isometric $\{A, C\}$ from T, and a realization for U then follows by the orthogonal completion of $\begin{bmatrix} A \\ C \end{bmatrix}$ with $\begin{bmatrix} B_U \\ D_U \end{bmatrix}$. *Let us therefore assume that we use a realization $\{A, B, C\}$ for T in output normal form (and $D = 0$).*

Θ is required to map $\begin{bmatrix} U' \\ -\Gamma^{-1}T' \end{bmatrix}$ to causality. We know from Proposition 14.6 that Θ maps J times its reachability space to causality: $\Theta J\mathcal{H}_\Theta \in c$. A sufficient condition for $\overline{a}$ and $\overline{b}$ to be causal is that the reachability space of $\begin{bmatrix} U & T\Gamma^{-1} \end{bmatrix}$ is contained in $\mathcal{H}_\Theta$ or, to put it differently, the reachability data for Θ have to correspond minimally to that of $\begin{bmatrix} U & T\Gamma^{-1} \end{bmatrix}$, that is, to $\begin{bmatrix} A \mid B_U & B\Gamma^{-1} \end{bmatrix}$. (The condition turns out to be necessary as well; this statement requires proof, but remark that if $u \in \mathcal{U}_2^-$ is not contained in $\mathcal{H}_\Theta$, then ΘJu will have a remaining nonzero component in $\mathcal{U}_2^-$.)

Going for minimality, such a (bounded, causal and J-unitary) Θ will exist when the Lyapunov–Stein equation

$$\Lambda^{\langle -1\rangle} = B_U B_U' - B\Gamma^{-2}B' + A\Lambda A' \tag{16.18}$$

yields a strictly nonsingular solution Λ, in which case, a J-unitary realization $\boldsymbol{\Theta}$ can be built corresponding to a bounded, J-unitary Θ (see further, or Section 14.5, for the construction).

The same equation can be expressed in terms of the original data. Let N be the reachability Gramian for $T\Gamma^{-1}$; then N satisfies the Lyapunov–Stein equation

$$N^{\langle -1\rangle} = B\Gamma^{-2}B' + ANA', \tag{16.19}$$

and putting $M := \Lambda + N$, we see that M satisfies

$$M^{\langle -1\rangle} = B_U B_U' + AMA'. \tag{16.20}$$

Since $\begin{bmatrix} A & B_U \end{bmatrix}$ are co-isometric, we find a unique solution of this Lyapunov–Stein equation $M = I$, and hence

$$\Lambda = I - N. \tag{16.21}$$

Given the hypotheses for T, $\{A, B\Gamma^{-1}\}$ is strictly reachable, and hence the corresponding reachability Gramian N is strictly nonsingular. (One notices that T is strictly reachable if and only if $T\Gamma^{-1}$ is, provided Γ is bounded with bounded inverse; this requires some proof – use the expression $\mathcal{R}\mathcal{R}'$ for the reachability Gramian.) The issue is whether $\Lambda = I - N$ will be nonsingular, and, more generally, what its inertia will be. We now show that a necessary and sufficient condition for the existence of a bounded Θ is that Λ be strictly nonsingular, meaning that there is an ϵ and an interval $(1-\epsilon, 1+\epsilon)$ around 1 that contains no eigenvalue of any N_k:

Proposition 16.4 Λ *is strictly nonsingular, if and only if there is an* $\epsilon > 0$ *(independent of* k*) such that all singular values of each* H_k *lie outside the interval* $(1-\epsilon, 1+\epsilon)$.

Proof

The singular values squared of each H_k are, by definition, the eigenvalues of N_k. On the other hand, Λ has a bounded inverse if and only if the eigenvalues of each $I - N_k$ are bounded uniformly over k as well as uniformly bounded away from zero over k, that is, $\forall \sigma_k : |1 - \sigma_k^2| \geq \epsilon$ for some fixed ϵ, any eigenvalue σ_k^2 of N_k and any k. In fact, $\|(I - N_k)^{-1}\| = \sup_{\sigma_k} |1 - \sigma_k^2|^{-1}$ because $(I - N)$ has a bounded inverse if and only if each $(I - N_k)$ is nonsingular and there is a uniform bound on $(I - N_k)^{-1}$, from which follows $\|(I - N)^{-1}\| = \sup_k \|(I - N_k)^{-1}\| \leq \epsilon^{-1}$. □

Remark: The singular values of the local H_k bear no relation to eventual singular values of the global H. This is as expected: global quantities (eigen-values or singular values) are only relevant in a time-invariant context, because they force equal behavior throughout. Note that interpolation theory only involves local quantities. Boundedness is, of course, a global property, and it plays a role here, but the requirement that Θ is bounded is somehow artificial (we have already noted that the theory can be valid when Θ is unbounded); the boundedness condition makes the treatment considerably easier. We shall not consider the unbounded Hankel norm model reduction case in this book, as it leads to more advanced Hilbert space considerations, and is not relevant for the finitely indexed case.

Assume now that Λ has a bounded inverse, and let $X = \mathrm{diag}_{k=:} X_k$ be defined by $\Lambda_k = X_k J_{\mathcal{B},k} X_k'$, with $J_{\mathcal{B},k} = \begin{bmatrix} I_{\mathcal{B}_+,k} & \\ & -I_{\mathcal{B}_-,k} \end{bmatrix}$. We may define $J_{\mathcal{B}} = \mathrm{diag}_{k=:} \begin{bmatrix} I_{\mathcal{B}_+,k} & \\ & -I_{\mathcal{B}_-,k} \end{bmatrix}$ and write symbolically $\boldsymbol{\eta} = \boldsymbol{\eta}^+ + \boldsymbol{\eta}^-$, with $\eta_k = \eta_k^+ + \eta_k^-$ defining what we call Λ'*s global inertia*. Consider

$$\begin{bmatrix} \alpha \mid \beta \end{bmatrix} = (X^{\langle -1 \rangle})^{-1} \begin{bmatrix} AX \mid B_U \quad B\Gamma^{-1} \end{bmatrix}. \tag{16.22}$$

Then from the Lyapunov–Stein equation for Λ it follows that $\begin{bmatrix} \alpha \mid \beta \end{bmatrix}$ is $J_{\mathcal{B}}$-co-isometric. The theory in Chapter 14 allows us to augment $\begin{bmatrix} \alpha \mid \beta \end{bmatrix}$ to a full $J_{\mathcal{B}}$-unitary matrix (which will define a realization of a causal, bounded, $J_{\mathcal{B}}$-unitary Θ). Let the original dimensions of $T = U\Delta'$ be $\mathbf{n}\times\mathbf{m}$, with $\boldsymbol{\eta}$ as state-space dimensions (as follows from Eq. (16.18)). The dimensions of U are then $\mathbf{n}\times\mathbf{m}_U$, with $\mathbf{m}_{U,k} := \eta_{k+1} - \eta_k + n_k$. The dimensions of $\begin{bmatrix} \alpha_{:,1} & \alpha_{:,2} \mid \beta_{:,1} & \beta_{:,2} \end{bmatrix}_k$ are $\eta_{k+1} \times (\eta_{+,k} + \eta_{-,k} + m_{U,k} + m_k)$ with incoming $\eta_{+,k} + m_{U,k}$ having + signature and $\eta_{-,k} + m_k$ having − signature, while there are $\eta_{+,k+1}$ positive outgoing arrows in the state and $\eta_{-,k+1}$ negative outgoing arrows (equivalent to positive ingoing arrows; see Fig. 16.1).

Note: By convention, an arrow, which carries data (say a) and is adorned with a negative sign, carries the energy $\|a\|_{\mathrm{HS}}^2$ in the direction opposite to the arrow, subtracts the same amount from the terminal to which the point of the arrow connects and inputs it to the terminal connected to the tail. It is equivalent to an opposite arrow with the same data a and a positive sign.

A local, J-unitary, $\boldsymbol{\Theta}_k$ will then be obtained by the completion of the row basis consisting of η_{k+1}^+ strictly positive and η_{k+1}^- strictly negative, J-orthonormal basis vector. To achieve the same count of positive and negative outgoing vectors as incoming ones, $\eta_k^+ + m_{U,k} - \eta_{k+1}^+$ new outgoing +vectors (called $a_{2,k}$) and $\eta_k^- + m_k - \eta_{k+1}^-$ new outgoing −vectors (called $b_{2,k}$) have to be generated, together with the complementation of the J-orthonormal basis. Figure 16.1 shows the counts. The treatment of J-unitary matrices given in Section 14.5 indicates how the complementation can be done (e.g., recursively, one new base vector at a time, or globally, via the construction of an orthogonal complement).

The result is a full realization

$$\boldsymbol{\Theta} = \left[\begin{array}{cc|cc} \alpha_{1,1} & \alpha_{1,2} & \beta_{1,1} & \beta_{1,2} \\ \alpha_{2,1} & \alpha_{2,2} & \beta_{2,1} & \beta_{2,2} \\ \hline \gamma_{1,1} & \gamma_{1,2} & \delta_{1,1} & \delta_{1,2} \\ \gamma_{2,1} & \gamma_{2,2} & \delta_{2,1} & \delta_{2,2} \end{array}\right], \tag{16.23}$$

in which:

1. α is u.e.s. provided Λ has a bounded inverse (which we have assumed), and the other block diagonals β, γ and δ are bounded as well (since the originals

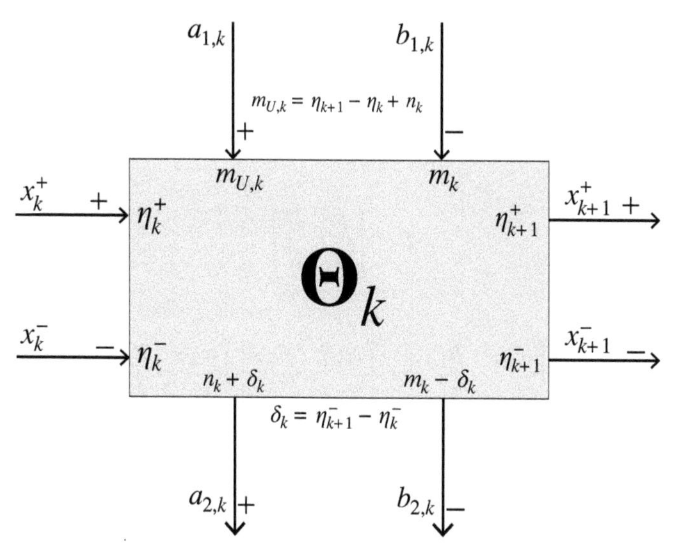

Figure 16.1 The completion of Θ. (Note: δ_k in this figure is a dimension, unrelated to the operator $\delta_{i,j}$ in $\mathbf{\Theta}$.)

were bounded, and thanks to the further assumptions of strict reachability and observability).

2. Both the input and the output signatures are $[+,-,+,-]$ with dimensions as detailed before; see Fig. 16.1.

It follows that $\Theta = \delta + \gamma(I - Z\alpha)^{-1}Z\beta$ is not only causal and bounded but also globally J-unitary (the same brute force proof as for unitary but now using the u.e.s. property together with the local J-unitarity). A further consequence is the existence of $\Sigma_{2,2} = \Theta_{2,2}^{-1}$ globally, as a contractive, but generally not causal operator. (Notice that the local $\mathbf{\Theta}_{k;2,2}$ is not necessarily square.) Together with $\Theta_{2,2}^{-1}$, the corresponding

$$\Sigma = \begin{bmatrix} \Theta_{1,1} - \Theta_{1,2}\Theta_{2,2}^{-1}\Theta_{2,1} & \Theta_{1,2}\Theta_{2,2}^{-1} \\ -\Theta_{2,2}^{-1}\Theta_{2,1} & \Theta_{2,2}^{-1} \end{bmatrix} \tag{16.24}$$

is unitary (and not causal), with its noncausal dynamics being determined by $\Theta_{2,2}^{-1}$, since the latter is the only source of noncausality in the formula for Σ, and this property propagates to $T_r' = \Gamma\Theta_{2,2}^{-1}\overline{b}$ and $\widehat{T}' = \Pi_{\text{sp}}T_r'$. (As mentioned, $\mathbf{\Theta}_{k;2,2}$ does not have to be square nor invertible, and Σ will necessarily have a mixed causal–anti-causal realization, to be explored next.)

Remark, for that matter, that $\begin{bmatrix} \alpha_{1,1} & \beta_{1,1} \\ \gamma_{1,1} & \delta_{1,1} \end{bmatrix}$ and $\begin{bmatrix} \alpha_{2,2} & \beta_{2,2} \\ \gamma_{2,2} & \delta_{2,2} \end{bmatrix}$ have strictly contractive inverses, and that, the same as before, $\begin{bmatrix} \alpha_{2,2} & \beta_{2,2} \\ \gamma_{2,2} & \delta_{2,2} \end{bmatrix}^{-1} \begin{bmatrix} \alpha_{2,1} & \beta_{2,1} \\ \gamma_{2,1} & \delta_{2,1} \end{bmatrix}$ and $\begin{bmatrix} \alpha_{1,1} & \beta_{1,1} \\ \gamma_{1,1} & \delta_{1,1} \end{bmatrix}^{-1} \begin{bmatrix} \alpha_{1,2} & \beta_{1,2} \\ \gamma_{1,2} & \delta_{1,12} \end{bmatrix}$ are strictly contractive (and dually for right inverses). This allows us to determine a unitary mixed causal–anti-causal unitary realization $\mathbf{\Sigma}$ for Σ:

$$\left[\begin{array}{cc|cc} F_{1,1} & F_{1,2} & G_{1,1} & G_{1,2} \\ F_{2,1} & F_{2,2} & G_{2,1} & G_{2,2} \\ \hline H_{1,1} & H_{1,2} & K_{1,1} & K_{1,2} \\ H_{2,1} & H_{2,2} & K_{2,1} & K_{2,2} \end{array}\right]\left[\begin{array}{c} x^+ \\ Z^{-1}x^- \\ \hline a_1 \\ b_2 \end{array}\right] = \left[\begin{array}{c} Z^{-1}x^+ \\ x^- \\ \hline a_2 \\ b_1 \end{array}\right] \tag{16.25}$$

with

$$\mathbf{\Sigma} = \begin{bmatrix} F & G \\ H & K \end{bmatrix} \tag{16.26}$$

and

$$\begin{aligned}
\begin{bmatrix} F_{1,1} & G_{1,1} \\ H_{1,1} & K_{1,1} \end{bmatrix} &= \begin{bmatrix} \alpha_{1,1} & \beta_{1,1} \\ \gamma_{1,1} & \delta_{1,1} \end{bmatrix} - \begin{bmatrix} \alpha_{1,2} & \beta_{1,2} \\ \gamma_{1,2} & \delta_{1,2} \end{bmatrix}\begin{bmatrix} \alpha_{2,2} & \beta_{2,2} \\ \gamma_{2,2} & \delta_{2,2} \end{bmatrix}^{-1}\begin{bmatrix} \alpha_{2,1} & \beta_{2,1} \\ \gamma_{2,1} & \delta_{2,1} \end{bmatrix} \\
\begin{bmatrix} F_{2,1} & G_{2,1} \\ H_{2,1} & K_{2,1} \end{bmatrix} &= -\begin{bmatrix} \alpha_{2,2} & \beta_{2,2} \\ \gamma_{2,2} & \delta_{2,2} \end{bmatrix}^{-1}\begin{bmatrix} \alpha_{2,1} & \beta_{2,1} \\ \gamma_{2,1} & \delta_{2,1} \end{bmatrix} \\
\begin{bmatrix} F_{1,2} & G_{1,2} \\ H_{1,2} & K_{1,2} \end{bmatrix} &= \begin{bmatrix} \alpha_{1,2} & \beta_{1,2} \\ \gamma_{1,2} & \delta_{1,2} \end{bmatrix}\begin{bmatrix} \alpha_{2,2} & \beta_{2,2} \\ \gamma_{2,2} & \delta_{2,2} \end{bmatrix}^{-1} \\
\begin{bmatrix} F_{2,2} & G_{2,2} \\ H_{2,2} & K_{2,2} \end{bmatrix} &= \begin{bmatrix} \alpha_{2,2} & \beta_{2,2} \\ \gamma_{2,2} & \delta_{2,2} \end{bmatrix}^{-1},
\end{aligned} \tag{16.27}$$

in which all $\begin{bmatrix} F_{i,j} & G_{i,j} \\ H_{i,j} & K_{i,j} \end{bmatrix}$ are strictly contractive.

As done before, we can decompose Θ into a past operator Θ_p, which maps its global strict past inputs $\{a_{1,p}, b_{1,p}\} \in \mathcal{U}_2^-$ to its current state

$$\{x^+_{[0]}, x^-_{[0]}\} \in \mathcal{D}_2, \tag{16.28}$$

and strict past outputs $\{a_{2,p}, b_{2,p}\} \in \mathcal{U}_2^-$ (Θ_p includes the reachability operator as a suboperator), and a future operator that maps the current state $\{x^+_{[0]}, x^-_{[0]}\} \in \mathcal{D}_2$ and future inputs $\{a_{1,f}, b_{1,f}\} \in \mathcal{L}_2$ to the future outputs $\{a_{2,f}, b_{2,f}\} \in \mathcal{L}_2$; see Fig. 16.2. Both Θ_p and Θ_f are J-unitary with appropriate J's reflecting the signatures of inputs, outputs and states.

A similar decomposition can be done for Σ as well, leading to unitary suboperators, which we can call Σ_p and Σ_f again, but there is a major distinction. As Σ is no longer

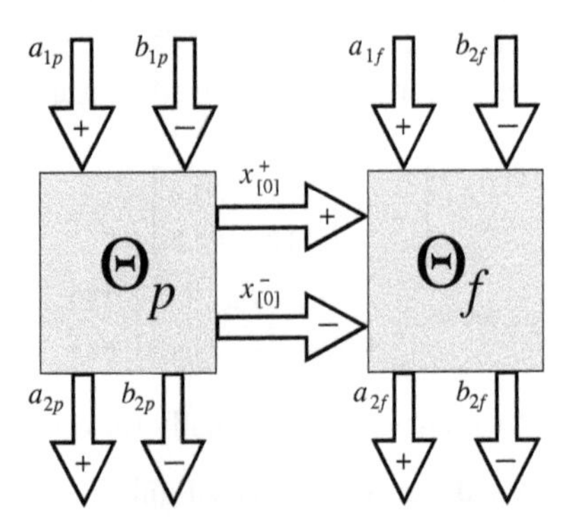

Figure 16.2 Θ_p (past) and Θ_f (future).

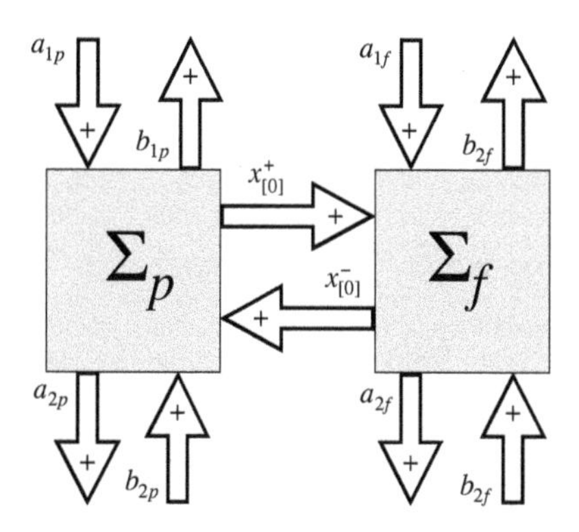

Figure 16.3 Σ_p (past) and Σ_f (future).

causal, the mixed state will produce state inputs and outputs for both the past processor and the future processor, as shown in Fig. 16.3. Notice that the realization shown for Σ is not computable in general, because there are internal loops that have to be resolved (to compute $x^-_{[0]}$, you need $x^+_{[0]}$ and vice versa). However, the graph defines correct algebraic relations, and how the loops can be made computable will be shown in Section 16.3.

These partial operators can be viewed as full scattering operators in their own right, for which the time-variant analysis remains valid. The next section is devoted to a further analysis of the structure of Σ, the determination of a realization for $\widehat{T}$ and the derivation of all possible solutions that meet the norm constraint on the error.

16.3 State-Space Analysis of the Hankel Norm Approximant

Key to the state-space analysis is the strict anti-causal part of $\Theta^{-1}_{2,2}$, since

$$\widehat{T}' = \Pi_{\text{sp}}(\Gamma\Theta^{-1}_{2,2}\overline{b}) = \Pi_{\text{sp}}(\Gamma(\Pi_{\text{sp}}\Theta^{-1}_{2,2})\overline{b}). \tag{16.29}$$

Consider $\mathcal{H}(\Theta^{-\prime}_{2,2}) := \Pi_{\text{sp}}(\Theta^{-1}_{2,2}\mathcal{L}_2)$. $\mathcal{H}$ is a regular natural input state space, and hence has a representation $\widehat{B}'Z'(I-\widehat{A}'Z')^{-1}\mathcal{D}_2$ with $\widehat{A}$ u.e.s. (assuming, in the infinitely indexed case, that this anti-causal part is bounded[1]). It then follows also that Θ^{-1}_{22} can be decomposed in a bounded anti-causal part and a strictly causal part $C_r(I-Z\widehat{A})^{-1}Z\widehat{B}$ for some bounded C_r to be determined.

A central property of the relation between $\widehat{T}$ and $\Theta^{-\prime}_{2,2}$ is

[1] Let T be a quasi-separable causal operator, and consider how to determine its inverse, assumed to exist as a bounded operator. T^{-1} can be found by performing an inner–outer factorization on $T := UT_o$, in which $T_o \sim \begin{bmatrix} A_o & B_o \\ C_o & D_o \end{bmatrix}$ will have a bounded causal $T_o^{-1} \sim \begin{bmatrix} A_o - B_oD_o^{-1}C_o & B_oD_o^{-1} \\ -D_o^{-1}C_o & D_o^{-1} \end{bmatrix}$ and $T^{-1} = T_o^{-1}U'$ *in case U is indeed unitary and hence inner*, which is not necessarily the case. In fact, $U^{-1} \neq U'$ may be unbounded in the general case, even though U has a unitary realization (see the example in the "Notes" of Chapter 11), but this case cannot occur in the finitely indexed case.

Proposition 16.5 *Let $\mathcal{H}(\widehat{T}\Gamma^{-1})$ be the natural input state space of $\widehat{T}\Gamma^{-1}$. Then,*

$$\mathcal{H}(\widehat{T}\Gamma^{-1}) \subset \mathcal{H}(\Theta_{2,2}^{-\prime}) = \mathcal{H}(\Sigma_{2,2}^{\prime}). \tag{16.30}$$

Proof

Recalling the definitions $T_r := (\Gamma\Theta_{2,2}^{-1}\overline{b})'$ and $\widehat{T} = \Pi_{\mathrm{f}}^{+}T_r$, we have

$$\begin{aligned}
\mathcal{H}(\widehat{T}\Gamma^{-1}) &= \Pi_{\mathrm{sp}}(\Gamma^{-1}\widehat{T}'\mathcal{L}_2)\\
&= \Pi_{\mathrm{sp}}(\Gamma^{-1}T_r'\mathcal{L}_2)\\
&= \Pi_{\mathrm{sp}}(\Theta_{2,2}^{-1}\overline{b}\mathcal{L}_2)\\
&\subset \Pi_{\mathrm{sp}}(\Theta_{2,2}^{-1}\mathcal{L}_2) \qquad \text{since } \overline{b} \in c\\
&= \mathcal{H}(\Theta_{2,2}^{-\prime}) = \mathcal{H}(\Sigma_{2,2}^{\prime}).
\end{aligned} \tag{16.31}$$

□

Except for C_r, $\widehat{A}$ and $\widehat{B}$ already define part of a realization for $\widehat{T}$, and $\widehat{C}$ follows from $\Gamma^{-1}\widehat{T}' = \Pi_{\mathrm{sp}}(\Theta_{22}^{-1}\overline{b})$. That $\widehat{T}$ inherits $\widehat{A}$ and $\widehat{B}$ follows from the required causality of $\overline{b}$. Indeed, Proposition 16.5, shows that only the strict anti-causal part of Θ_{22}^{-1} plays a role in determining $\widehat{T}$. This anti-causal part can be obtained from a further state-space analysis of Θ, Σ and $\overline{b}$. It is, of course, possible to obtain state-space realizations of the various operators needed by "brute force," using standard inner–outer, external factorization and multiplication methods, but it turns out that much of the necessary algebraic work has already been done and further derivations can be made fairly directly from the quantities obtained in the determination of Θ. This is what we proceed to show now.

Key to the analysis are the strict past–future decompositions for the J-unitary but causal Θ:

$$\Theta_p \begin{bmatrix} a_{1p} \\ b_{1p} \end{bmatrix} = \begin{bmatrix} x_{[0]}^{+} \\ x_{[0]}^{-} \\ \hline a_{2p} \\ b_{2p} \end{bmatrix}, \quad \Theta_f \begin{bmatrix} x_{[0]}^{+} \\ x_{[0]}^{-} \\ \hline a_{1f} \\ b_{1f} \end{bmatrix} = \begin{bmatrix} a_{2f} \\ b_{2f} \end{bmatrix} \tag{16.32}$$

and the corresponding unitary Σ of mixed causality (see Fig. 16.3):

$$\Sigma_p \begin{bmatrix} x_{[0]}^{-} \\ a_{1p} \\ b_{2p} \end{bmatrix} = \begin{bmatrix} x_{[0]}^{+} \\ a_{2p} \\ b_{1p} \end{bmatrix}, \quad \Sigma_f \begin{bmatrix} x_{[0]}^{+} \\ a_{1f} \\ b_{2f} \end{bmatrix} = \begin{bmatrix} x_{[0]}^{-} \\ a_{2f} \\ b_{1f} \end{bmatrix}. \tag{16.33}$$

Taking all signs properly into account, Θ_p and Θ_f are J-unitary, while Σ_p and Σ_f are correspondingly unitary. Assuming Θ to be bounded (and all the partial maps as well) and all inputs to be in corresponding spaces $\mathcal{X}_2$, we have $x^{+} \in \mathcal{X}_2^{\boldsymbol{\eta}^{+}}$ and $x^{-} \in \mathcal{X}_2^{\boldsymbol{\eta}^{-}}$. $x_{[0]}^{+}$ is then the main diagonal of x^{+}, and likewise $x_{[0]}^{-}$. Together they represent the present state produced by strict past inputs at each k. (See the construction of the Hilbert–Smith or Frobenius space $\mathcal{X}_2$ in Chapter 7; $u_{[k]}$ is defined as the main block diagonal $u_{[k]} = \Pi_0(u(Z')^k)$ corresponding to the diagonal representation $u = \sum u_{[k]}Z^k$.)

Of major interest for the model reduction theory is the anti-causal part of $\Theta_{2,2}^{-1} = \Sigma_{2,2}$, which is the partial map $b_2 \mapsto b_1$, assuming $a_1 = 0$. The Hankel operator of the strictly causal part of $\Sigma_{2,2}^{\prime}$ is $\widehat{H}$ with (see Figs. 16.1 and 16.4; here δ is a dimension!)

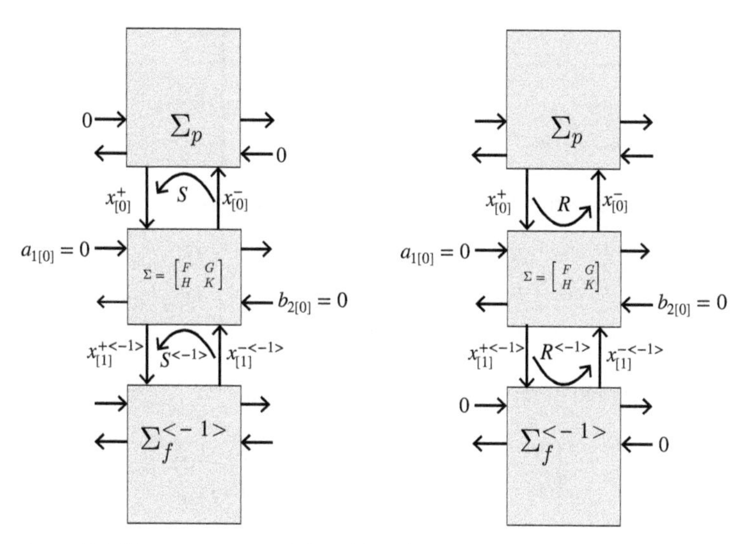

Figure 16.4 The propagation of S and R, using only the contributions of the states.

$$\widehat{H}': (\mathcal{L}_2^{\mathbf{m}-\delta} \to \mathcal{U}_2^{\mathbf{m},-})|_{a_1=0} : b_{2f} \mapsto b_{1p}, \tag{16.34}$$

whose range is the space $\mathcal{H}(\Sigma'_{2,2})$. $x^-_{[0]}$ acts as intermediary, and we have (always under the condition $a_1 = 0$) $\widehat{H}' = \tau\sigma$, with

$$\begin{aligned} \sigma: &\quad b_{2f} \mapsto x^-_{[0]}, \\ \tau: &\quad x^-_{[0]} \mapsto b_{1p}. \end{aligned} \tag{16.35}$$

These two maps are evidently contractive, as submaps of a unitary operator, and it can be shown that σ is onto with bounded pseudo-inverse, and τ is one-to-one with a bounded inverse on its range, from which it follows that $\widehat{H}$ has a full rank $\boldsymbol{\eta}^-$. (Proofs are in [32], but the properties may not be surprising, in view of the assumed strong boundedness of Θ.)

The connection between strict past and future in Σ is characterized by two diagonal operators S and R defined as

$$\begin{aligned} Sx^-_{[0]} &= x^+_{[0]} \qquad \text{when } a_{1p} = 0 \text{ and } b_{2p} = 0, \\ Rx^+_{[0]} &= x^-_{[0]} \qquad \text{when } a_{1f} = 0 \text{ and } b_{2f} = 0. \end{aligned} \tag{16.36}$$

See Fig. 16.4 from which a realization for the upper part of $\Sigma'_{2,2}$ can be determined:

Proposition 16.6 *The operators S and R are given by the recursive relations*

$$\begin{aligned} S^{\langle -1\rangle} &= F_{1,2} + F_{1,1}S(I - F_{2,1}S)^{-1}F_{2,2}, \\ R &= F_{2,1} + F_{2,2}R^{\langle -1\rangle}(I - F_{1,2}R^{\langle -1\rangle})^{-1}F_{1,1}. \end{aligned} \tag{16.37}$$

Proof

The state transition in $\boldsymbol{\Sigma}$ is globally given by how the 0th diagonals in the inputs are transformed to the 0th diagonals in the outputs, involving states in the 0th and first diagonals (see Fig. 16.4):

$$\Sigma\left[\begin{array}{c} x^{+}_{[0]} \\ x^{-\langle -1\rangle}_{[1]} \\ \hline a_{1[0]} \\ b_{2[0]} \end{array}\right] = \left[\begin{array}{c} x^{+\langle -1\rangle}_{[1]} \\ x^{-}_{[0]} \\ \hline a_{2[0]} \\ b_{1[0]} \end{array}\right]. \tag{16.38}$$

With $a_1 = 0$ and inserting the partitioning of Σ, this gives

$$\begin{cases} x^{+\langle -1\rangle}_{[1]} &= F_{1,1}x^{+}_{[0]} + F_{1,2}x^{-\langle -1\rangle}_{[1]} + G_{1,2}b_{2[0]}, \\ x^{-}_{[0]} &= F_{2,1}x^{+}_{[0]} + F_{2,2}x^{-\langle -1\rangle}_{[1]} + G_{2,2}b_{2[0]}, \\ b_{1[0]} &= H_{2,1}x^{+}_{[0]} + H_{2,2}x^{-\langle -1\rangle}_{[1]} + K_{1,2}b_{2[0]}. \end{cases} \tag{16.39}$$

Putting $b_{2[0]} = 0$ and observing that $Sx^{-}_{[0]} = x^{+}_{[0]} \Leftrightarrow S^{\langle -1\rangle}x^{-\langle -1\rangle}_{[1]} = x^{+\langle -1\rangle}_{[1]}$ gives

$$\begin{cases} S^{\langle -1\rangle}x^{-\langle -1\rangle}_{[1]} &= F_{1,1}Sx^{-}_{[0]} + F_{1,2}x^{-\langle -1\rangle}_{[1]}, \\ x^{-}_{[0]} &= F_{2,1}Sx^{-}_{[0]} + F_{2,2}x^{-\langle -1\rangle}_{[1]}. \end{cases} \tag{16.40}$$

The second equation gives $x^{-}_{[0]} = (I - F_{2,1}S)^{-1}x^{-\langle -1\rangle}_{[1]}$. Filling this in the first equation gives the expression for $S^{\langle -1\rangle}x^{-\langle -1\rangle}_{[1]}$:

$$S^{\langle -1\rangle}x^{-\langle -1\rangle}_{[1]} = \left(F_{1,1}S(I - F_{2,1}S)^{-1} + F_{1,2}\right)x^{-\langle -1\rangle}_{[1]}, \tag{16.41}$$

and because of the full range, the expression for $S^{\langle -1\rangle}$ and hence claimed expression for S. For R, a similar reasoning applies. □

Next, we look for a realization for the strictly lower part of $\Sigma'_{2,2} = \Theta^{-\prime}_{2,2}$.

Proposition 16.7 *A realization* $\{\widehat{A}, \widehat{B}, C_r\}$ *for the strict causal (lower) part of* $\Sigma'_{2,2}$ *in terms of S and R is*

$$\begin{aligned} \widehat{A} &= \left((I - F_{2,1}S)^{-1}F_{2,2}\right)', \\ \widehat{B} &= \left(H_{2,2} + H_{1,2}S(I - F_{2,1}S)^{-1}F_{2,2}\right)', \\ C_r &= \left(G_{2,2} + F_{2,2}R^{\langle -1\rangle}(I - F_{1,2}R^{\langle -1\rangle})^{-1}G_{1,2}\right)'(I - RS)^{-\prime}. \end{aligned} \tag{16.42}$$

Proof

Since the Hankel $\widehat{H}$ related to the strictly causal part $\Sigma'_{2,2}$ maps b_{2f} to b_{1p}, with $b_{2p} = 0$ by definition, and, as previously observed, $\widehat{H}$ decomposes as $\widehat{H} = \tau\sigma$ with the map $\sigma\colon b_{2f} \mapsto x^{-}_{[0]}$ onto and $\tau\colon x^{-}_{[0]} \mapsto b_{1p}$ one-to-one, $x^{-}_{[0]}$ is a valid minimal state for the strictly anticausal part of $\Sigma_{2,2}$, with σ being the corresponding (anti-causal) reachability map and τ the observability map. The resulting realization defines the anti-causal global state equations

$$\begin{cases} x^{-}_{[0]} &= \widehat{A}'x^{-\langle -1\rangle}_{[1]} + C_r'b_{2[0]}, \\ b_{1[0]} &= \widehat{B}'x^{-\langle -1\rangle}_{[1]}, \end{cases} \tag{16.43}$$

in which $\widehat{A}'$ maps $x^{-\langle -1\rangle}_{[1]}$ conditioned on $b_{2[0]} = 0$ as well as $b_{2p} = 0$. In this situation, $x^{+}_{[0]} = Sx^{-}_{[0]}$. Filling this in the second equation of (16.39) gives

$$x^{-}_{[0]} = (I - F_{2,1}S)^{-1}F_{2,2}x^{-\langle -1\rangle}_{[1]}, \tag{16.44}$$

and hence $\widehat{A}' = (I - F_{2,1}S)^{-1}F_{2,2}$. $\widehat{B}'$ is obtained similarly from the third equation in (16.39) under the same conditions $a_1 = 0$ and $b_{2[0]} = 0$:

$$b_{1[0]} = H_{2,1}Sx^{-}_{[0]} + H_{2,2}x^{-\langle -1\rangle}_{[1]} = \left(H_{2,1}S(I - F_{2,1}S)^{-1}F_{2,2} + H_{2,2}\right) x^{-\langle -1\rangle}_{[1]}. \quad (16.45)$$

Hence, $\widehat{B}' = H_{2,1}S(I - F_{2,1}S)^{-1}F_{2,2} + H_{2,2}$. As to C_r', it defines the relation between $b_{2[0]} \mapsto x_{-[0]}$ under the circumstances $a_1 = 0$, $b_{2p} = 0$, as well as $b_{2f}^{\langle -1\rangle} = 0$, so that $Sx^{-}_{[0]} = x^{+}_{[0]}$ and $x^{-\langle -1\rangle}_{[1]} = R^{\langle -1\rangle}x^{+}_{[1]}$. Inserting this in the first equation of (16.39) gives $x^{-\langle -1\rangle}_{[1]}$ in function of $x^{+}_{[0]}$ and $b_{2[0]}$:

$$x^{+\langle -1\rangle}_{[1]} = (I - F_{1,2}R^{\langle -1\rangle})^{-1}F_{1,1}Sx^{-}_{[0]} + (I - F_{1,2}R^{\langle -1\rangle})^{-1}G_{1,2}b_{2[0]}. \quad (16.46)$$

The second equation of (16.39) then produces, together with $x^{-\langle -1\rangle}_{[1]} = R^{\langle -1\rangle}x_1^{+\langle -1\rangle}$, and using the recursion for R:

$$\begin{aligned} x^{-}_{[0]} &= \left(F_{2,1} + F_{2,2}R^{\langle -1\rangle}(I - F_{1,2}R^{\langle -1\rangle})^{-1}F_{1,1}\right) Sx^{-}_{[0]} \\ &\quad + \left(F_{2,2}R^{\langle -1\rangle}(I - F_{1,2}R^{\langle -1\rangle})^{-1}G_{1,2} + G_{2,2}\right) b_{2[0]} \\ &= RSx^{-}_{[0]} + \left(F_{2,2}R^{\langle -1\rangle}(I - F_{1,2}R^{\langle -1\rangle})^{-1}G_{1,2} + G_{2,2}\right) b_{2[0]}, \end{aligned} \quad (16.47)$$

and finally,

$$C_r' = (I - RS)^{-1}\left(F_{2,2}R^{\langle -1\rangle}(I - F_{1,2}R^{\langle -1\rangle})^{-1}G_{1,2} + G_{2,2}\right), \quad (16.48)$$

in which all inverses exist because R, S and F are strict contractions. □

The Realization of $\widehat{T}$

The realization for $\widehat{T}$ combines the realization of the anti-causal part of $\Sigma_{2,2}$ with a realization for the causal $\overline{b}$. A realization for $\overline{b}$ is readily obtained from the defining equation $\Theta J_1 \begin{bmatrix} U' \\ \Gamma^{-1}T' \end{bmatrix} = \begin{bmatrix} \overline{a} \\ -\overline{b} \end{bmatrix} \in c$. Reverting the normalized realization back to the originally defining data, we may put

$$\Theta = \begin{bmatrix} d_{1,1} & d_{1,2} \\ d_{2,1} & d_{2,2} \end{bmatrix} + \begin{bmatrix} c_1 \\ c_2 \end{bmatrix} (I - ZA)^{-1} \begin{bmatrix} B_U & \Gamma^{-1}B \end{bmatrix} \quad (16.49)$$

with $d := \delta$ and

$$\begin{bmatrix} c_1 \\ c_2 \end{bmatrix} = \begin{bmatrix} \gamma_1 \\ \gamma_2 \end{bmatrix} X^{-1}, \begin{bmatrix} B_U & B\Gamma^{-1} \end{bmatrix} = X^{\langle -1\rangle} \begin{bmatrix} \beta_1 & \beta_2 \end{bmatrix}. \quad (16.50)$$

Hence,

$$\begin{aligned} \Theta \begin{bmatrix} B_U' \\ -\Gamma^{-1}B' \end{bmatrix} Z'(I - A'Z')^{-1}C' &= \Theta J_1\beta' Z'(I - \alpha' Z')^{-1}X'C' \\ &= \gamma(I - Z\alpha)^{-1}J_B X'C' \\ &= \begin{bmatrix} c_1 \\ c_2 \end{bmatrix} (I - ZA)^{-1}\Lambda C'. \end{aligned} \quad (16.51)$$

Hence,

$$\begin{aligned}\Theta\begin{bmatrix} U' \\ -\Gamma^{-1}T' \end{bmatrix} &= \Theta\begin{bmatrix} D_U' \\ 0 \end{bmatrix} + \begin{bmatrix} c_1 \\ c_2 \end{bmatrix}(I - ZA)^{-1}\Lambda C' \\ &= \begin{bmatrix} d_{1,1}D_U' + c_1 C' \\ d_{2,1}D_U' + c_2 C' \end{bmatrix} - \begin{bmatrix} c_1 \\ c_2 \end{bmatrix}(I - ZA)^{-1}MC',\end{aligned} \tag{16.52}$$

in which derivation use has been made of $(I - ZA)^{-1} = I + (I - ZA)^{-1}ZA$, $AC' + B_U D_U' = 0$ because T has been put in the output normal form, and $\Lambda = I - M$. Defining

$$D_{\overline{b}} := -(d_{2,1}D_U' + c_2(I - M)C'), \tag{16.53}$$

it follows that

$$\overline{b} = D_{\overline{b}} + c_2(I - ZA)^{-1}MC', \tag{16.54}$$

and we can state

Proposition 16.8 *Under the stated conditions on T and the Hankel operator of $T\Gamma^{-1}$, a realization for a minimal Hankel norm approximation $\widehat{T}$ is given by $\{\widehat{A}, \widehat{B}\Gamma, \widehat{C}, 0\}$, in which $\widehat{A}$ and $\widehat{B}$ are inherited from Σ_{22} (Eq. (16.42)),*

$$\widehat{C} = (c_2C' - d_{2,1}D_U')'C_r + CMY', \tag{16.55}$$

and Y satisfies the Lyapunov–Stein equation

$$Y = C_r'c_2 + A'Y^{\langle -1\rangle}\widehat{A}. \tag{16.56}$$

Proof

We have

$$\begin{aligned}\Gamma^{-1}\widehat{T}' &= \Pi_{\text{sp}}(\Gamma^{-1}T_a') = \Pi_{\text{sp}}(\Theta_{22}^{-1}b) \\ &= \Pi_{\text{sp}}\left(\widehat{B}'Z'(I - \widehat{A}'Z')^{-1}C_r'(D_b + c_2Z(I - AZ)^{-1}AMC')\right).\end{aligned} \tag{16.57}$$

The mixed term $\widehat{B}'Z'(I - \widehat{A}'Z')^{-1}C_r'c_2Z(I - AZ)^{-1}AMC'$ splits, as usual, into a strictly anti-causal part $\widehat{B}'Z'(I - \widehat{A}'Z')^{-1}\widehat{A}'Y^{\langle -1\rangle}AMC'$ and a causal part, in which $Y = C_r'c_2 + A'Y^{\langle -1\rangle}\widehat{A}$. The causal part gets canceled by Π_{sp}. Hence,

$$\Gamma^{-1}\widehat{T}' = \widehat{B}'Z'(I - \widehat{A}'Z')^{-1}(C_r'D_b + \widehat{A}'Y^{\langle -1\rangle}AMC'). \tag{16.58}$$

□

Parametrization of All Causal Approximations

In this subsection, we give a short summary of the theory that produces a parametrization of all Hankel norm approximations that have degrees at most equal to the dimensions of $\mathcal{H}_{\Sigma_{2,2}'}$ (what has been called the "sdim" or sequence dimension, meaning the series of dimensions, one for each index point). Or, to put it differently: What are all the possible, partially noncausal, extensions T_e of $\widehat{T}$ that satisfy $\|(T - T_e)\Gamma^{-1}\| \leq 1$?

The full theory is given in chapter 10 of [29] and will not be repeated here; we give only the main results and refer to that chapter for further information. Let us just mention that the theory is an extension of the parametrization results of Chapter 15, and boils down to showing that (1) any contractive load $S_L : b_2 = S_L a_2$ of Θ produces an interpolation, but (2) only causal contractive loads produce interpolations with the minimal degree sdim$\mathcal{H}_{\Sigma'_{2,2}}$. This is due to the fact that any noncausality in S_L is propagated to an increase in noncausal degree of the input scattering S, which, expressed in terms of the load S_L, is given by

$$\begin{aligned} S(S_L) &= (-S_L\Theta_{1,2} + \Theta_{2,2})^{-1})(S_L\Theta_{1,1} - \Theta_{2,1}) \\ &= \Sigma_{2,1} + \Sigma_{2,2}(I - S_L\Sigma_{1,2})^{-1}S_L\Sigma_{1,1}, \end{aligned} \tag{16.59}$$

and S is seen to reduce to $\Sigma_{2,1}$ when $S_L = 0$. The starting point is the input–output relation of the Θ matrix loaded with S_L: $\begin{bmatrix} S_L & -I \end{bmatrix} \Theta = Y \begin{bmatrix} -S & I \end{bmatrix}$ with[2] $Y = S_L\Theta_{12} - \Theta_{22}$. S_L is strictly contrative, so is S. In the case where Θ_{22} is not assumed to be outer, S is not necessarily causal even when S_L is. However, this does not prevent the parametrization to go through; $Y = -(I - S_L\Theta_{12}\Theta_{22}^{-1})\Theta_{22}$ has a bounded inverse because $\Theta_{12}\Theta_{22}^{-1}$ is strictly contractive (as well as S), and defining $\Gamma^{-1}T_r' := -Y^{-1}(S_L\overline{a} + \overline{b})$, we find

$$T' - T_r' = -\Gamma S U', \tag{16.60}$$

with SU' strictly contractive as before. This T_r' can, of course, have a much larger degree than in the previous simple solution, but, importantly, *its anti-causal part turns out to have the same degree as before, namely* sdim$\mathcal{H}_{\Theta_{2,2}^{-'}}$, so that $\widehat{T} := \Pi_{\mathrm{f}^+}T_r$ (projection on the strict future) is of the same degree sdim$\mathcal{H}_{\Theta_{2,2}^{-'}}$ and hence a valid approximation of lowest possible degree that meets the norm constraint.

Nehari Parametrization

The *Nehari parametrization problem* – assume T strictly contractive, what are all causal T_a such that $\|T - T_a\| \leq 1$? – is a special case of the extension problem considered in this chapter: in the Nehari case, the original T is strictly contractive, $\widehat{T} = T$, Θ is J-inner and the approximation $\widehat{T} = 0$ because T already meets the constraint (we take $\Gamma = I$ here). Let

$$\Theta \begin{bmatrix} U' \\ -T' \end{bmatrix} = \begin{bmatrix} \overline{a} \\ -\overline{b} \end{bmatrix} \in c \tag{16.61}$$

as before; then

$$\begin{aligned} \begin{bmatrix} S_L & -I \end{bmatrix} \Theta \begin{bmatrix} U' \\ -T' \end{bmatrix} &= (S_L\Theta_{1,1} - \Theta_{2,1})U' - (S_L\Theta_{1,2} - \Theta_{2,2})T' \\ &= S_L\overline{a} + \overline{b}. \end{aligned} \tag{16.62}$$

$\Theta_{2,2}^{-1}$ is now causal and strictly contractive (assuming that $M = \Lambda - I$ has a bounded inverse); hence $(S_L\Theta_{2,1} - \Theta_{2,2}) = -(I - S_L\Theta_{2,1}\Theta_{2,2}^{-1})\Theta_{2,2}$ also has a bounded causal inverse, and we obtain, with $T_a(S_L)' = -(S_L\Theta_{1,2} - \Theta_{2,2})^{-1}(S_L\overline{a} + \overline{b}) \in c$,

[2] This follows directly from $\Theta \begin{bmatrix} I \\ S \end{bmatrix} = \begin{bmatrix} I \\ S_L \end{bmatrix} a_2$, in which a_2 is the output pertaining to the unit input $a_1 = I$.

$$T' - T_a(S_L)' = -S(S_L)U'. \tag{16.63}$$

Causal and contractive S_Ls then parametrize all T_a such that $\|T - T_a\| \leq 1$, with $T_a(S_L) = (S_L\Theta_{2,1} - \Theta_{2,2})^{-1}(S_L\overline{a} + \overline{b})$, in which T_a is anti-causal, and $T - T_a(S_L)$ produce all Nehari extensions whose causal part equals T. (The proof of this special case is similar to the parametrization for constrained interpolation given in Chapter 15.) All such S's are called *Nehari extensions*, and the result is that Nehari's theorem, originally proven in the context of Hardy-space extensions, is valid in the LTV context as well:

$$\|T\|_H = \inf_{S_L:\|S_L\|\leq 1 \& S_L \in \mathcal{C}} (T - T_a(S_L)), \tag{16.64}$$

that is, causal contractive S_L parametrize all solutions. All causal contractive S_L are then further parametrized by Schur coefficients; see Chapter 13 in this respect.

16.4 Example: Matrix Approximation and Subspace Tracking

Possibly the most direct and simplest example of an application of the theory of this chapter is the case of *low-rank approximations of a single matrix M, which meet a Euclidean norm constraint*. This case can be treated directly without recourse to the general theory, as is done in [77], but in this section, we work all the details of the general theory out on this simple example, not only because it is a good example, but also because it reveals the general mechanics in a very illustrative skeletal way. An additional benefit is that the example checks the general theory for robustness and consistency.

To fit this case into the general theory, we define the operator T embedding the matrix M as its $(0,-1)$ entry, with all other entries empty:

$$T := \begin{bmatrix} \cdot & & & \\ \cdot & - & & \\ | & M & \boxed{|} & \\ \cdot & - & \cdot & \cdot \end{bmatrix}, \tag{16.65}$$

with the goal to find low-rank approximations $\widehat{T}$ for T meeting

$$\|(T - \widehat{T})\Gamma^{-1}\| \leq 1, \tag{16.66}$$

with Γ a nonsingular Hermitian matrix. In practice, one often takes $\Gamma = \epsilon I$ for some small positive number ϵ. Without loss of generality, we can just take $\Gamma = I$, which simplifies formulas (just put $T\Gamma^{-1}$ for T).

Let $M = QR$ be a minimal QR factorization of M, with Q isometric and R in the upper echelon form; then T has the minimal realization shown in Fig. 16.5, consisting of two nonempty stages at -1 and 0. Remark that the realization given is in the output

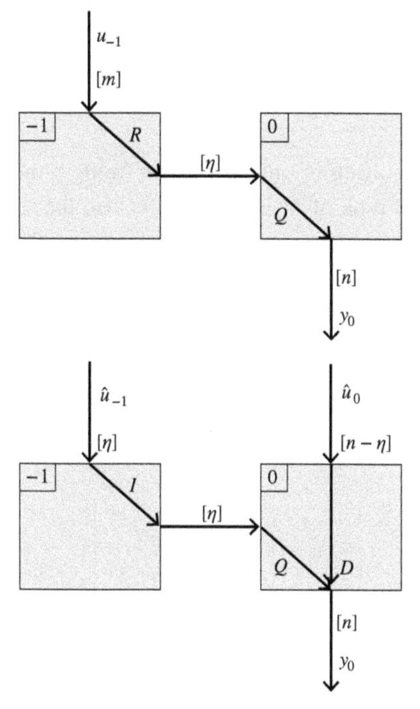

Figure 16.5 Minimal realizations of T and U. Dimensions are indicated between the square brackets.

normal form, with $C_0 = Q$ and $C_{-1} = \cdot$, with the realizations of T at -1: $\begin{bmatrix} | & R \\ \cdot & - \end{bmatrix}$ and at 0: $\begin{bmatrix} - & \cdot \\ Q & | \end{bmatrix}$.

Let now $T = U\Delta'$ be a coprime factorization of T in the output normal form. U is found by unitary completion of the $\begin{bmatrix} A \\ C \end{bmatrix}$ data at -1: $\begin{bmatrix} | & I \\ \cdot & - \end{bmatrix}$ and at 0: $\begin{bmatrix} - & - \\ Q & D \end{bmatrix}$ for D such that $\begin{bmatrix} Q & D \end{bmatrix}$ is unitary (see Fig. 16.5). Notice that the requirement of unitarity determines the column (input) dimensions of the realization of U. Hence (neglecting empty entries),

$$\begin{bmatrix} U \mid T \end{bmatrix} = \begin{bmatrix} Q & \boxed{D} \mid QR & \boxed{|} \end{bmatrix}. \tag{16.67}$$

So we obtain realizations for the reachability data of a causal J-unitary Θ such that $\Theta J \begin{bmatrix} U' \\ T' \end{bmatrix} \in c$, namely at -1: $\begin{bmatrix} A_{-1} \mid B_{U,-1} & B_{-1} \end{bmatrix} = \begin{bmatrix} | \mid I & R \end{bmatrix}$ and at 0: $\begin{bmatrix} A_0 \mid B_{U,0} & B_0 \end{bmatrix} = \begin{bmatrix} - \mid - & \cdot \end{bmatrix}$, in which all the column dimensions are already determined. The completion of Θ follows a pattern that is similar to the completion of U but with the extra requirement that the result must be a J-unitary matrix. Necessary for this to be possible is the nonsingularity of $I - RR'$, because this allows the completion of the realization at stage -1 of Θ. (The completion of the realization at 0 is trivially possible in this case with any compatible J-unitary matrix, since no further condition is imposed at 0, and all the other stages are necessarily empty.) Accordingly, the inertia of $I - RR'$ determines the rank of the approximation. Since the results are data dependent, the best we can do in this exposition is to tally the various dimensions that may arise during the procedure.

The dimensions of M are $n \times m$, resulting in only $\mathbf{m}_{-1} = m$ being nonempty, and similarly $\mathbf{n}_0 = n$. Let us assume that the rank of M is η, which is also the state dimension η_0 in Fig. 16.5. Furthermore, we take the signature of the nonsingular $I - RR'$, which is of dimension $\eta \times \eta$ with $\eta = [\eta^+, \eta^-]$, Θ is found next through the completion of its reachability data, globally characterized as $\left[\begin{array}{c|cc} A & B_U & B \end{array}\right]$. The realizations for T, and U given produce the only significant components of the realization for Θ: θ_{-1} at index -1 and θ_0 at index 0. Let us now define $I - RR' = N \begin{bmatrix} I_{\eta^+} & \\ & -I_{\eta^-} \end{bmatrix} N'$; then θ_{-1} and θ_0 are obtained by J-unitary completion as follows:

$$\theta_{-1} = \left[\begin{array}{c|cc} | & N^{-1} & N^{-1}R \\ \hline | & ? & ? \\ | & ? & ? \end{array}\right], \quad \theta_0 = \left[\begin{array}{cc|c} - & - & - \ \cdot \\ \hline ? & ? & ? \ \cdot \\ ? & ? & ? \ \cdot \end{array}\right]. \tag{16.68}$$

The column dimensions of θ_{-1} and θ_0 are determined respectively by the column dimensions of U and T and are $(\cdot \mid \eta, m)$ and $(\eta^+, \eta^- \mid n - \eta, \cdot)$ with $(\cdot \mid +, -)$ and $(+, - \mid +, \cdot)$ signs respectively. The row dimensions of θ_{-1} and θ_0 are obtained by requiring them to be square matrices and requiring the sums of plus dimensions equal the sums of minus dimensions as well. Choosing a natural order, they necessarily become $(\eta^+, \eta^- \mid \eta^-, m - \eta^-)$ for θ_{-1} and $(\cdot \mid n - \eta^-, \eta^-)$ for θ_0.

Using ${}^k\theta = \begin{bmatrix} {}^k\alpha & {}^k\beta \\ {}^k\gamma & {}^k\delta \end{bmatrix}$ for J-unitary realizations of Θ, we have been given already

$$\begin{bmatrix} N^{-1} & N^{-1}R \end{bmatrix} = \left[\begin{array}{c|cc} | & {}^{-1}\beta_{11} & {}^{-1}\beta_{12} \\ | & {}^{-1}\beta_{21} & {}^{-1}\beta_{22} \end{array}\right] \tag{16.69}$$

of dimensions $(\eta^+, \eta^-) \times (0 \mid \eta, m)$, and by J-unitary completion, we deduce

$${}^{-1}\theta = \left[\begin{array}{c|cc} | & {}^{-1}\beta_{11} & {}^{-1}\beta_{12} \\ | & {}^{-1}\beta_{21} & {}^{-1}\beta_{22} \\ \hline | & {}^{-1}\delta_{11} & {}^{-1}\delta_{12} \\ | & {}^{-1}\delta_{21} & {}^{-1}\delta_{22} \end{array}\right], \quad {}^{0}\theta = \left[\begin{array}{cc|cc} - & - & - & \cdot \\ \hline {}^0\gamma_{11} & {}^0\gamma_{12} & {}^0\delta_{11} & | \\ {}^0\gamma_{21} & {}^0\gamma_{22} & {}^0\delta_{21} & | \end{array}\right], \tag{16.70}$$

with dimensions and signs as given before (see Fig. 16.6). All these quantities are new and computed through completion in the case of ${}^{-1}\theta$, while in the case of ${}^0\theta$, an otherwise arbitrary J-unitary matrix of the correct dimensions can be filled in. This could be, in particular, with $u := \begin{bmatrix} u_1 & u_2 \end{bmatrix}$ any compatible but otherwise arbitrary unitary matrix:

$${}^0\theta = \left[\begin{array}{cc|cc} - & - & - & \cdot \\ \hline u_1 & 0 & u_2 & | \\ 0 & I & 0 & | \end{array}\right] \tag{16.71}$$

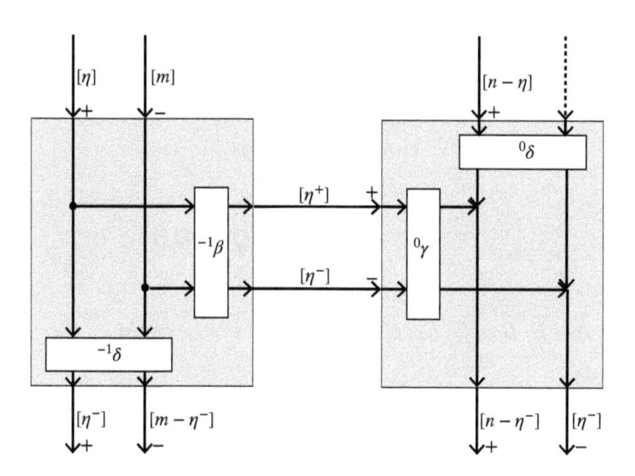

Figure 16.6 Realization of Θ.

of dimensions $(0 \mid n-\eta_-, \eta^-) \times (\eta^+, \eta_- \mid n-\eta, 0)$ and signs $(\cdot \mid +, -) \times (+, - \mid +, \cdot)$, where $\gamma_{1,2} = 0$ and $\gamma_{2,2} = I$. Hence, using progressing time indices as the main order

$$\overline{\Theta} = \begin{bmatrix} {}^{-1}\delta & \\ {}^{0}\gamma({}^{-1}\beta) & \boxed{{}^{0}\delta} \end{bmatrix}, \tag{16.72}$$

and

$$\begin{bmatrix} \overline{a} \\ -\overline{b} \end{bmatrix} = \Theta J \begin{bmatrix} U' \\ T' \end{bmatrix} \sim \overline{\Theta} J \begin{bmatrix} Q' \\ R'Q' \\ \hline \boxed{\begin{matrix} D' \\ - \end{matrix}} \end{bmatrix} = \begin{bmatrix} 0 \\ 0 \\ \hline \boxed{\begin{matrix} \overline{A} \\ -\overline{B} \end{matrix}} \end{bmatrix}, \tag{16.73}$$

in which $J := \begin{bmatrix} J_{-1} & \\ & \boxed{J_0} \end{bmatrix}$ is the previously determined sign matrix corresponding to the inputs in $\begin{bmatrix} U & T \end{bmatrix}$ and the realizations derived for Θ (see Fig. 16.6).[3] The outputs of Θ have the corresponding sign matrix $j := \begin{bmatrix} j_{-1} & \\ & \boxed{j_0} \end{bmatrix}$. As ${}^{-1}\beta J_{-1} \begin{bmatrix} Q' \\ R'Q' \end{bmatrix} =$

[3] Consider, for example, the join $\begin{bmatrix} S & T \end{bmatrix}$, with S and T being compatible transfer operators; then the time-index-based order represents it as $\begin{bmatrix} \ddots & \vdots & \iddots \\ \cdots & S_{i,j}\, T_{i,j} & \cdots \\ \iddots & \vdots & \ddots \end{bmatrix}$, instead of stacking S and T separately: $\left[\begin{array}{ccc|ccc} \ddots & \vdots & \iddots & \ddots & \vdots & \iddots \\ \cdots & S_{i,j} & \cdots & \cdots & T_{i,j} & \cdots \\ \iddots & \vdots & \ddots & \iddots & \vdots & \ddots \end{array}\right]$; these representations are permutations of each other, but the index-based order is more natural. Consequently, the corresponding signs have to be carefully tallied as per the time index.

$N'Q'$, we end up finding

$$\begin{bmatrix} \overline{A} \\ -\overline{B} \end{bmatrix} = \begin{bmatrix} {}^0\gamma N' & {}^0\delta \end{bmatrix} \begin{bmatrix} Q' \\ D' \end{bmatrix}, \tag{16.74}$$

and finally, since $\begin{bmatrix} {}^0\gamma' \\ {}^0\delta' \end{bmatrix} j_0 \begin{bmatrix} {}^0\gamma & {}^0\delta \end{bmatrix} = J_0 = I$ and $QQ' + DD' = I$,

$$\overline{A}'\overline{A} - \overline{B}'\overline{B} = Q(I - RR')Q' + DD' = I - MM' \tag{16.75}$$

as expected. $\begin{bmatrix} \overline{A} \\ -\overline{B} \end{bmatrix}$ has a fine structure that has to be made more explicit, due to the fact that N^{-1} converts $\begin{bmatrix} I & R \end{bmatrix}$ into a J-orthogonal basis with $J = \begin{bmatrix} 1_{\eta^+} & \\ & I_{-\eta^-} \end{bmatrix}$, and hence can be decomposed as $N^{-1} = \begin{bmatrix} \nu_1 \\ \nu_2 \end{bmatrix}$ with $N^{-1}\begin{bmatrix} I & R \end{bmatrix}$ of signature $(\eta_+, -\eta_-) \times (\eta, -m)$. Correspondingly, we shall have $N = \begin{bmatrix} N_1 & N_2 \end{bmatrix}$ of dimensions $\eta \times (\eta^+, \eta^-)$. Hence, we find

$$\begin{bmatrix} \overline{A} \\ -\overline{B} \end{bmatrix} = \begin{bmatrix} \gamma_{11}N_1' + \gamma_{12}N_2' & {}^0\delta_{11} \\ \gamma_{21}N_1' + \gamma_{22}N_2' & {}^0\delta_{21} \end{bmatrix} \begin{bmatrix} Q' \\ D' \end{bmatrix}. \tag{16.76}$$

The Computation of Θ_{22}^{-1}

The issue here is that Θ_{22}^{-1} is of mixed causal–anti-causal character. In this case, this can be directly dealt with, as Θ_{22} has a relatively simple form, $\begin{bmatrix} x_{-1,-1} \\ x_{0,-1} \end{bmatrix}$, so that its inverse, which is assumed to exist, will necessarily be of the form $\begin{bmatrix} y_{-1,-1} & y_{-1,0} \end{bmatrix}$, in which y is the straight matrix inverse of x. (Remark that both x and y without subdivision have to be square.)

For the special choice $\gamma_{22} = I, \gamma_{21} = 0, \gamma_{12} = 0$ and $\begin{bmatrix} \gamma_{11} \mid {}^0\delta_{11} \end{bmatrix}$ unitary, we have

$$\Theta_{22} = \begin{bmatrix} {}^{-1}\delta_{22} & | \\ {}^{-1}\beta_{22} & \boxed{|} \end{bmatrix}, \quad \Theta_{22}^{-1} = \begin{bmatrix} k_2 & h \\ - & \boxed{-} \end{bmatrix} \tag{16.77}$$

(for the motivation of the choice of the symbols h and k; see a bit further) and

$$\begin{bmatrix} \overline{A} \\ -\overline{B} \end{bmatrix} = \begin{bmatrix} \gamma_{11}N_1'Q' + {}^0\delta_{11}'D' \\ N_2'Q' \end{bmatrix}. \tag{16.78}$$

It follows (as in the general case) that $\Theta_{21}U' - \Theta_{22}T' = \begin{bmatrix} 0 \\ \boxed{-\overline{B}} \end{bmatrix}$. Hence, defining $T_r' = -\Theta_{22}^{-1}\overline{b}$, we obtain $T' - T_r' = -\Theta_{22}^{-1}\Theta_{21}U'$ contractive (as was our goal). In the special case $\gamma_{22} = I$, this then gives (simply)

$$T_r' = \begin{bmatrix} -h\overline{B} \\ \boxed{-} \end{bmatrix} = \begin{bmatrix} hN_2'Q' \\ \boxed{-} \end{bmatrix} \tag{16.79}$$

and (with $\widehat{M}$ for the low-rank approximation of M)

$$\widehat{T} = \begin{bmatrix} QN_2h' & \boxed{|} \end{bmatrix}, \quad \widehat{M} = QN_2h', \tag{16.80}$$

with $\widehat{M}$ of rank η^-.

Although we have already found the full solution of the problem through the use of a general J-unitary ${}^0\theta$, the full solution derived in the general theory consists in selecting one specific Θ that meets the interpolation condition and then providing it with a causal and strictly contractive but otherwise arbitrary load $S_L: a_2 \mapsto b_2 = S_L a_2$. In this case, $a_2 = \begin{bmatrix} {}^{-1}a_2 \\ \boxed{{}^0a_2} \end{bmatrix}$ and similarly $b_2 = \begin{bmatrix} {}^{-1}b_2 \\ \boxed{{}^0b_2} \end{bmatrix}$, S_L, must be causal and hence have the form

$$S_L = \begin{bmatrix} S_{L;-1,-1} & 0 \\ S_{L;0,-1} & \boxed{S_{L;0,0}} \end{bmatrix}. \tag{16.81}$$

S_L must be strictly contractive as well. It has dimensions $(m-\eta^-,\eta^-)\times(\eta^-,m-\eta^-)$. The general solution $\widehat{T} = \Pi_{\mathrm{f}^+}[(\overline{a}'S_L' + \overline{b}')Y^{-\prime}]$ reduces in our special case to: ${}^0\theta = \left[\begin{array}{cc|cc} - & - & - & \cdot \\ \hline u_1 & 0 & u_2 & | \\ 0 & I & 0 & | \end{array}\right]$ and with $Y^{-1} = \begin{bmatrix} y_{-1} & y_0 \\ - & \boxed{-} \end{bmatrix}$

$$\widehat{T} = \begin{bmatrix} (\overline{A}'S_{L;0,0}' + \overline{B}')y_0' & \boxed{|} \end{bmatrix}, \tag{16.82}$$

and hence

$$\widehat{M} = (\overline{A}'S_{L;0,0}' + \overline{B}')y_0', \tag{16.83}$$

in which $S_{L;0,0}$ of dimensions $\eta^- \times (n-\eta^-)$ is contractive, but otherwise arbitrary. The important point is that $\overline{A}'S_{L;0,0}' + \overline{B}'$ generates (column) subspaces that contain degree η^- approximations to T. Such a subspace can be "tracked" when further data are inputted recursively, leading to a low-complexity algorithm for "subspace tracking," a topic that has been much studied (see the Notes). $S_{L;0,0}'$ picks out η^- columns out of $\overline{A}'$ in a contractive way.

This example was intended to show the mechanics of the recursive system's Hankel norm approximation. For the case of approximating a simple matrix M in norm with a low-rank model, a relatively simple direct procedure is indeed possible, fully conformal with the example just given, but omitting recursive details, which in this case reduce to just one step. The procedure requires $N := I - MM'$ to be nonsingular. The rows of $\begin{bmatrix} I & M \end{bmatrix}$ then generate a Krein space of signature $(\eta^+, -\eta^-)$ and a J-unitary matrix $\overline{\Theta}$ that satisfies

$$\overline{\Theta} J \begin{bmatrix} I \\ M' \end{bmatrix} = \begin{bmatrix} 0 \\ \overline{A} \\ \hline 0 \\ -\overline{B} \end{bmatrix} \tag{16.84}$$

is found by completion, just as before. The J-unitarity of $\overline{\Theta}$ results

$$I - MM' = \overline{A}'\overline{A} - \overline{B}'\overline{B}, \tag{16.85}$$

while the nonsingularity of $I - MM'$ ensures the nonsingularity of $\begin{bmatrix} \overline{A}' & \overline{B}' \end{bmatrix}$, and the result is again given by Eq. (16.83) now with $\overline{\Theta}$ replacing Θ. The difference between these two J-unitary matrices is technical (i.e., not essential) in the sense that $\overline{\Theta}$ embeds $\begin{bmatrix} I \mid M \end{bmatrix}$, while Θ embeds $\begin{bmatrix} Q & D \mid M \end{bmatrix}$ with $\begin{bmatrix} Q & D \end{bmatrix}$ unitary. Further processing allows for a judicial choice of S_L; see the literature for more information on the use of this procedure and further refinements as well.
End of the example.

16.5 Notes

1. The theory of this chapter appeared, to the best of the author's knowledge, for the first time in [32] and [70]. It was further elaborated in [29] and [69]. The novelty is the generalization of part of the Schur–Takagi [61, 66] and AAK [1] theory to the time-variant case. Generalizations in other directions, for example, to the nonlinear case, can be found in [8] and [9]. Although the results originally derived by Schur and Takagi seem to have a strong flavor of complex function calculus, it may come as a surprise that the mode of interpolation they introduced generalizes to the context of quasi-separable or time-variant systems, in which there is no obvious notion of poles or zeros.
2. Concerning the example of Section 16.4, the main references are [24, 77] and a chapter on Subspace Tracking in [29]. These references also contain a survey of various methods proposed for subspace tracking in the literature, as well as experiments evaluating the performance of the proposed method. There is an extensive mathematical literature on generalized QR factorization, where the unitary Q matrix is replaced by a J-unitary matrix Θ, also surveyed in the references cited.
3. The chapter leaves an important question open: What can be done when the condition of nonsingularity of Λ is not met? To our knowledge, this borderline case has not been treated so far for the general Schur–Takagi case. For the regular constrained interpolation case, an approach is presented in [2].

17 Isometric Embedding for Causal Contractions

Causal and bounded operators play an important role in signal processing, system theory and therefore also in numerical algebra. Finitely indexed matrices are automatically bounded, but their bound may be unwieldy, in particular when their dimensions are not a priori fixed, for example, when the largest or smallest index is allowed to grow recursively. It is therefore an important issue to consider matrices and operators that do have a precise bound, and, in particular, when that bound is 1, that is, when the matrix or operator is *contractive*. This chapter considers the causal case: first, we study how the realization of a contractive system can be extended to yield a realization of an isometric system with the same state degree, and next, how this effort also yields a canonical contractive realization for the original system. Once the extension is done, a further extension to unitary, still of the same degree, leads to the famous Darlington synthesis, which always exists in the finitely indexed case but not in the infinitely indexed case.

Menu

Hors d'oeuvre
Causal Isometric Embedding of a Contractive Causal System

Main Course
The Quasi-separable Case

Dessert
Darlington Synthesis

17.1 Introduction

In Chapter 13 we discovered that we could find interpolating matrices of a positive definite matrix $\frac{1}{2}(Y + Y')$, whereby the interpolation data consist of the diagonal entries (in the case we treated, normalized to I) and off-diagonal data defined on the complement of a staircase. The interpolation procedure was recursive, gradually emptying the staircase, to produce a J-unitary matrix, which defines an interpolating positive real (PR) matrix Y_a, and related Cholesky factors L_a and M_a. When the staircase has become empty, then an "embedding" of the original Y in a J-unitary matrix results, which can be given a causal form; see the notes of Chapter 13.

In this chapter, we investigate a more general situation, where, instead of a PR matrix Y, a "causal" or block lower contractive matrix S is given. S does not have to be square, while Y will automatically be square, because it has to match Y' dimension-wise, and so will be the resulting Cayley transform $(Y+1)^{-1}(Y-I)$, and we shall not even have to assume that S is strictly contractive.

17.2 Direct Isometric Embedding of Causal Matrices

To warm up for the treatment of the quasi-separable case, we explore an obvious elementary strategy to embed a block lower contractive matrix in a causal isometric matrix recursively, starting with the last column and moving backward to the first, thereby creating the necessary additional entries that produce the total isometry directly on the block entries, without assuming a state-space or quasi-separable model. Let S be the given matrix, assumed to be contractive (the procedure will allow us to determine whether that is indeed the case), and let us derive a (minimal) new block lower matrix T such that the combination $\begin{bmatrix} S \\ T \end{bmatrix}$ is isometric, that is,

$$S'S + T'T = I, \tag{17.1}$$

and such that T has a right causal inverse (equivalently, is "right-outer") for definiteness and ensuring minimal dimensions as well. The embedding procedure can be done recursively, starting from the last column, using a block-wise variant of the classical Cholesky factorization. Indeed, for any block lower matrix with the general form

$$\left[\begin{array}{cc} S_{1,1} & \\ S_{2,1} & S_{2,2} \\ \hline T_{1,1} & \\ T_{2,1} & T_{2,2} \end{array}\right] \tag{17.2}$$

to be isometric, it is necessary (and, of course, not sufficient) that $\begin{bmatrix} S_{2,2} \\ T_{2,2} \end{bmatrix}$ be isometric. So, knowing $S_{2,2}$, one can start by determining $T_{2,2}$ and then move on to the left and determine $T_{1,1}$ and $T_{2,1}$.

We show now that this can be done recursively, assuming that the contractive S has the form

$$S = \begin{bmatrix} \ddots & & \\ \ddots & S_{n-1,n-1} & \\ \cdots & S_{n,n-1} & S_{n,n} \end{bmatrix}. \tag{17.3}$$

Step 1: Let $S_{n,n}$ be the last block diagonal entry in S; then, to determine T, the first step is to find $T_{n,n}$ such that $T'_{n,n}T_{n,n} = I - S'_{n,n}S_{n,n}$. To achieve minimum dimensions of the result, we determine $T_{n,n}$ so that it has a right inverse $T^{+}_{n,n}$ (e.g., by Cholesky factorization, keeping the smallest number of rows, so that the rows of $T_{n,n}$ form a basis).

General step: The inductive hypothesis is as follows. A partial isometric embedding $\begin{bmatrix} s_{\rm in} \\ t_{\rm in} \end{bmatrix}$ with $t_{\rm in}$ right invertible has been determined. Given in the next step are a new diagonal block S_d and an S_i for which $\begin{bmatrix} S_d & \\ S_i & s_{\rm in} \end{bmatrix}$ is contractive, and we have to find a right invertible diagonal element T_d and a T_i so that

$$\left[\begin{array}{cc} S_d & \\ S_i & s_{\rm in} \\ \hline T_d & \\ T_i & t_{\rm in} \end{array}\right] \tag{17.4}$$

is isometric. Orthogonality of the two block columns gives immediately $S_i' s_{\rm in} + T_i' t_{\rm in} = 0$; hence, for any pseudo-inverse $t_{\rm in}^+$,

$$T_i' = -S_i' s_{\rm in} t_{\rm in}^+. \tag{17.5}$$

Remark: T_i' is uniquely determined, thanks to the nonsingularity of $t_{\rm in}$ so that $s_{\rm in} t_{\rm in}^+ = s_{\rm in} t_{\rm in}^\dagger$, with $t_{\rm in}^\dagger$ the Moore–Penrose inverse of $t_{\rm in}$ – the proof is an interesting exercise.

Next, T_d should be chosen so as to make the first block column isometric. Hence, we must have $T_d' T_d = I - S_d' S_d - S_i' S_i - T_i' T_i$, and the right-hand side of this expression has to be shown positive definite (not necessarily strict), given that the given S is contractive, and hence the partial $\begin{bmatrix} S_d & \\ S_i & s_{\rm in} \end{bmatrix}$ contractive as well. From this (assumed) contractivity follows

Lemma 17.1

$$\begin{bmatrix} I - S_d' S_d - S_i' S_i & T_i' \\ T_i & I \end{bmatrix} \geq 0 \tag{17.6}$$

and $I - S_d' S_d - S_i' S_i - T_i' T_i \geq 0$.

Proof

Since $\begin{bmatrix} S_d & \\ S_i & s_{\rm in} \end{bmatrix}$ is assumed contractive, we have

$$\begin{aligned} &\begin{bmatrix} I & \\ & t_{\rm in}^{+\prime} \end{bmatrix} \begin{bmatrix} I - S_d' S_d - S_i' S_i & -S_i' s_{\rm in} \\ -s_{\rm in}' S_i & I - s_{\rm in}' s_{\rm in} \end{bmatrix} \begin{bmatrix} I & \\ & t_{\rm in}^+ \end{bmatrix} \\ &\quad = \begin{bmatrix} I - S_d' S_d - S_i' S_i & -S_i' s_{\rm in} t_{\rm in}^+ \\ -t_{\rm in}^{+\prime} s_{\rm in}' S_i & I \end{bmatrix} \geq 0, \end{aligned} \tag{17.7}$$

which is Eq. (17.6), using Eq. (17.5). The Schur complement of the (2,2) block entry I evaluates to $I - S_d' S_d - S_i' S_i - T_i' T_i$ and is hence positive definite. □

The lemma gives the necessary positivity for $D_d' D_d$, and D_d can be chosen right invertible as well, thereby minimizing its dimensions.

The recursion amounts to a Cholesky factorization of $I - S'S = T'T$, in which T is block lower and right invertible. Just as is the case with any direct Cholesky factorization, the computational complexity increases at each step, since the relevant "elimination" matrices s_{in} and t_{in} increase in dimension at each step. $s_{\text{in}}t_{\text{in}}^{+}$ can, of course, also be recursively computed:

$$t_{\text{new}}^{+} := \begin{bmatrix} T_d^{+} & \\ -t_{\text{in}}^{+}T_iT_d^{+} & t_{\text{in}}^{+} \end{bmatrix}, \; s_{\text{new}}t_{\text{new}}^{+} = \begin{bmatrix} S_dT_d^{+} & \\ (S_i - s_{\text{in}}t_{\text{in}}^{+}T_i)T_d^{+} & s_{\text{in}}t_{\text{in}}^{+} \end{bmatrix}. \tag{17.8}$$

A similar (dual) procedure allows us to augment a causal contractive block matrix S with another causal contractive block T_f so that $\begin{bmatrix} S & T_f \end{bmatrix}$ is co-isometric: $SS' + T_fT_f' = I$, with T_f left invertible. This time, the recursion will start at the upper-left corner (first block *row*) and proceed in a forward manner.

17.3 Quasi-separable Isometric Embedding

The increase in the complexity of the direct isometric embedding at each step can be avoided when the system to be embedded has a quasi-separable or state-space realization, very much in the style that we have used for the Bellman and the Kalman filters. We assume that we are given a causal system, ending at some index point $n + 1$, and which has the following system description:

1. at the last stage: $y_{n+1} = C_{n+1}x_{n+1} + D_{n+1}u_{n+1}$,
2. for any $k < n$: $\begin{bmatrix} x_{k+1} \\ y_k \end{bmatrix} = \begin{bmatrix} A_k & B_k \\ C_k & D_k \end{bmatrix} \begin{bmatrix} x_k \\ u_k \end{bmatrix}$,

or, in terms of the matrix transfer operator:

$$S_1 = \begin{bmatrix} \ddots & & & & \\ \cdots & D_{1;k} & & & \\ \vdots & \vdots & \ddots & & \\ \cdots & C_{1;n}A^{>}B_k & \cdots & D_{1;n} & \\ \cdots & C_{1;n+1}A^{>}B_k & \cdots & C_{1;n+1}B_n & D_{1;n+1} \end{bmatrix}, \tag{17.9}$$

in which the notation "$A^{>}$" is used to indicate an appropriate (but not further specified) continuous product (e.g., $C_{1;n}A^{>}B_k = C_{1;n}A_{n-1}\cdots A_{k+1}B_k$, where $A^{n>k} := A_{n-1}\cdots A_{k+1}$), and in which we have anticipated that the second factor S_2 that is going to complete the embedding will share A's and B's with S_1, with the goal of achieving a realization for an isometric

$$\begin{bmatrix} S_1 \\ S_2 \end{bmatrix} = \begin{bmatrix} D_1 \\ D_2 \end{bmatrix} + \begin{bmatrix} C_1 \\ C_2 \end{bmatrix} (I - ZA)^{-1}ZB, \tag{17.10}$$

in which S_2 is right-outer (i.e., causal and causally right invertible). We shall see that the recursive procedure we derive achieves this efficiently using the same state sequence, and, as in Section 17.2, the recursion will start with the right most column and move backward (and upward). Notice that our last entry is now D_{n+1}.

Given a contractive S_1, we have to determine S_2. (At this point, we do not have a characterization of contractivity of S_1 in terms of the given state-space realization, but we shall be able to derive one as a consequence of the procedure, which will also produce an important result, known as the "BR lemma.") For simplicity, we assume that the realization given is minimal. This assumption is, strictly speaking, not necessary but does not restrict generality, since nonminimal realizations may be reduced to minimal.

Step 0 is the same as in Section 17.2: embedding $D_{1;n+1}$ in an isometry $\begin{bmatrix} D_{1;n+1} \\ D_{2;n+1} \end{bmatrix}$. For this, we must have

$$D'_{2;n+1} D_{2;n+1} = I - D'_{1;n+1} D_{1;n+1}, \tag{17.11}$$

and we choose a minimal factorization, that is, one for which the rows of $D_{2:n+1}$ form a basis (which then is also of minimal dimension). Hence, $D_{2;n+1}$ will have a right inverse $D^+_{2;n+1}$.

Step 1 aims at embedding the partial block $\begin{bmatrix} D_{1;n} & \\ C_{1;n+1}B_n & D_{1;n+1} \end{bmatrix}$ and, in the process, at obtaining an isometric realization for stage $n+1$. Step $n+1$ takes x_{n+1} as an input. Here is the critical observation: an *isometric* realization of this stage requires the incoming state (call it $\widehat{x}_{n+1}$) to be normalized $\|\widehat{x}_{n+1}\| = \|y_{n+1}\|$; hence, to achieve this and assuming x_{n+1} minimal, a state transformation $\widehat{x}_{n+1} = R_{n+1}x_{n+1}$ with square invertible R_{n+1} will be required. The strategy hence will be to determine the isometric realization in the process recursively. So, putting $R_{n+1}x_{n+1}$ with R_{n+1} to be determined as an input to the last stage, we have to satisfy

$$\left[\begin{array}{cc} D_{1;n} & \\ \widehat{C}_{1;n+1}\widehat{B}_n & D_{1;n+1} \\ \hline D_{2;n} & \\ \widehat{C}_{2;n+1}\widehat{B}_n & D_{2;n+1} \end{array}\right]' \left[\begin{array}{cc} D_{1;n} & \\ \widehat{C}_{1;n+1}\widehat{B}_n & D_{1;n+1} \\ \hline D_{2;n} & \\ \widehat{C}_{2;n+1}\widehat{B}_n & D_{2;n+1} \end{array}\right] = I, \tag{17.12}$$

with $\widehat{B}_n := R_{n+1}B_n$ and $\widehat{C}_n = C_n R_n^{-1}$; see the state diagram in Fig. 17.1 for a realization of the partial block considered at this step. From Step 0, the last column is already isometric. Next we want the realization of stage $n+1$ to be isometric; hence, we want to satisfy the following two conditions with R_{n+1} nonsingular:

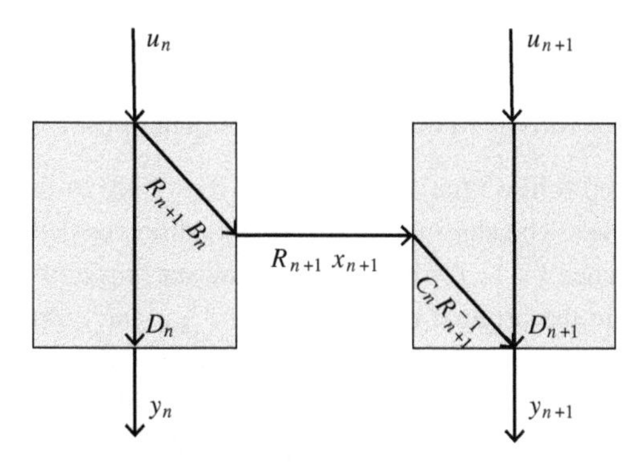

Figure 17.1 The realization of the isometric partial block considered in Step 1.

1. $(D'_{1;n+1}C_{1;n+1} + D'_{2;n+1}C_{2;n+1})R^{-1}_{n+1} = 0$;
2. $R^{-\prime}_{n+1}(C'_{1;n+1}C_{1;n+1} + C'_{2;n+1}C_{2;n+1})R^{-1}_{n+1} = I$.

Condition 1 will be satisfied by choosing

$$C_{2;n+1} = -D^{+\prime}_{2;n+1}D_{1;n+1}C_{1;n+1}, \tag{17.13}$$

while, and putting $M_{n+1} := R'_{n+1}R_{n+1}$, Condition 2 requires

$$M_{n+1} = C'_{1;n+1}C_{1;n+1} + C'_{2;n+1}C_{2;n+1}. \tag{17.14}$$

We assumed the realization given for S_1 to be minimal; hence, $C_{1;n+1}$ is *left*-invertible as being an observability operator at stage $n+1$ all by itself ($\mathcal{O}_{n+1} = C_{1:n+1}$ in the notation of Chapter 4), and M_{n+1} is a fortiori nonsingular as the observability Gramian at this stage. This determines R_{n+1} up to orthogonality and the realization at the index $n+1$. To guarantee the isometry of Eq. (17.12) and in preparation for the next step, we still have to determine $D_{2;n}$ minimal, to satisfy

$$D'_{2;n}D_{2;n} = I - D'_{1;n}D_{1;n} - \widehat{B}'_n\widehat{B}_n \tag{17.15}$$

(since $\widehat{C}_{n+1}$ is already isometric). Just as in Section 17.2, and starting from the contractivity of $\begin{bmatrix} D_{1;n} & \\ \widehat{C}_{1;n+1}\widehat{B}_n & D_{1;n+1}\end{bmatrix}$, we find, in sequence (repeating the argument of Section 17.2),

$$\begin{bmatrix} I - D'_{1;n}D_{1;n} - \widehat{B}'_n\widehat{C}'_{1;n+1}\widehat{C}_{1;n+1}\widehat{B}_n & -\widehat{B}'_n\widehat{C}'_{1;n+1}D_{1;n+1} \\ -D'_{1;n+1}\widehat{C}_{1;n+1}\widehat{B}_n & I - D'_{1;n+1}D_{1;n+1}\end{bmatrix} \geq 0, \tag{17.16}$$

using $D'_{2;n+1}D_{2;n+1} = I - D'_{1;n+1}D_{1;n+1}$, the right invertibility of $D_{2;n+1}$, premultiplying with $\begin{bmatrix} I & \\ & D^{+\prime}_{2;n+1}\end{bmatrix}$, postmultiplying with $\begin{bmatrix} I & \\ & D^{+}_{2;n+1}\end{bmatrix}$ and using $\widehat{C}_{2;n+1} = -D^{+\prime}_{2;n+1}D'_{1;n+1}\widehat{C}_{1;n+1}$,

$$\begin{bmatrix} I - D'_{1;n}D_{1;n} - \widehat{B}'_n\widehat{C}'_{1;n+1}\widehat{C}_{1;n+1}\widehat{B}_n & \widehat{B}'_n\widehat{C}'_{2;n+1} \\ \widehat{C}_{2;n+1}\widehat{B}_n & I\end{bmatrix} \geq 0, \tag{17.17}$$

and the result follows from the positivity of the Schur complement of the (2, 2) entry I.

General Step. The general step follows the same mode of reasoning as the previous step but is slightly more complex. The starting (recursive) situation consists of an isometric realization for the sequence $k+1: n$ with $k < n$ and the starting state $R_{k+1}x_{k+1}$, with x_{k+1} the $(k+1)$th state in the original model, inputs $u_{k+1:n}$ and outputs $y_{k+1:n}$. The key observation now is that the partial system (with indices $\ell \leq k$) that feeds into the state x_{k+1} and produces outputs $y_{:,k}$ from inputs $u_{:,k}$ itself will become isometric recursively. This partial system has the following unnormalized global realization (with $\ell < k$ some anterior index):

$$\mathbf{S}_k := \left[\begin{array}{ccccc} \ddots & & & & \\ \cdots & D_\ell & & & \\ \vdots & \vdots & \ddots & & \\ \cdots & C_{k-1}A^{>}B_\ell & \cdots & D_{k-1} & \\ \hline \cdots & C_k A^{>} B_\ell & \cdots & C_k B_{k-1} & D_k \\ \cdots & A_k A^{>} B_\ell & \cdots & A_k B_{k-1} & B_k \end{array}\right], \tag{17.18}$$

with the input of the last stage in this partial realization (corresponding to the last two rows) being $\begin{bmatrix} x_k \\ u_k \end{bmatrix}$ and the output being $\begin{bmatrix} y_k \\ x_{k+1} \end{bmatrix}$. We have now

$$\begin{bmatrix} y_k \\ x_{k+1} \end{bmatrix} = \mathbf{S}_k u_{:k} = \begin{bmatrix} C_k & D_k \\ A_k & B_k \end{bmatrix} \begin{bmatrix} x_k \\ u_k \end{bmatrix}, \tag{17.19}$$

where x_k summarizes the effect of $u_{:(k-1)}$ on k and the following stages.

The corresponding isometric system will end up having $\{A_k, B_k, C_k\}$ transformed to

$$\begin{bmatrix} \widehat{A}_k & \widehat{B}_k \\ \widehat{C}_k & D_k \end{bmatrix} = \begin{bmatrix} R_{k+1} A_k R_k^{-1} & R_{k+1} B_k \\ C_k R_k^{-1} & D_k \end{bmatrix}. \tag{17.20}$$

Comparing this with the previous step, we see $\begin{bmatrix} C_k \\ A_k \end{bmatrix}$ taking the role of C_{n+1} and $\begin{bmatrix} D_k \\ B_k \end{bmatrix}$ the role of D_{n+1} in Step 1. Having performed step $k+1$, we assume as an *induction hypothesis* that R_{k+1} and D_k have been determined and that $\begin{bmatrix} D_k \\ R_{k+1} B_k \end{bmatrix}$ is isometric, with $D_{2,k}$ right invertible. (Just replace k by n to obtain the correspondence with Step 1, but notice that the aspects of the C's and D's have changed.) The goal at step k is to replace the local realization with an isometric one, that is, making the map $\begin{bmatrix} R_k x_k \\ u_k \end{bmatrix} \mapsto \begin{bmatrix} y_k \\ R_{k+1} x_{k+1} \end{bmatrix}$ isometric. Hence, we want

$$\left[\begin{array}{cc} C_{1;k} R_k^{-1} & D_{1;k} \\ C_{2;k} R_k^{-1} & D_{2;k} \\ \hline R_{k+1} A_k R_k^{-1} & R_{k+1} B_k \end{array}\right] \tag{17.21}$$

to be isometric (see Fig. 17.2). $C_{2;k}$ and R_k are unknown, the other entries are known, and $D_{2;k}$ is right invertible by inductive assumption. Hence, the kth state transition of the embedding will be isometric if

$$C_{2;k} = -D_{2;k}^{+\prime}(D_{1;k}^{\prime} C_{1;k} + B_k^{\prime} M_{k+1} A_k) \tag{17.22}$$

and

$$M_k = A_k^{\prime} M_{k+1} A_k + C_{1;k}^{\prime} C_{1;k} + C_{2;k}^{\prime} C_{2;k}. \tag{17.23}$$

The second equation will always produce a strictly positive M_k, thanks to the observability of the system $\{A, C\}$ (an inductive proof is easy to give). To complete the general step, we still have to determine $D_{2;k-1}$ so that, with $\widehat{C}_k := C_k R_k^{-1}$ and $\widehat{B}_{k-1} = R_k B_k$,

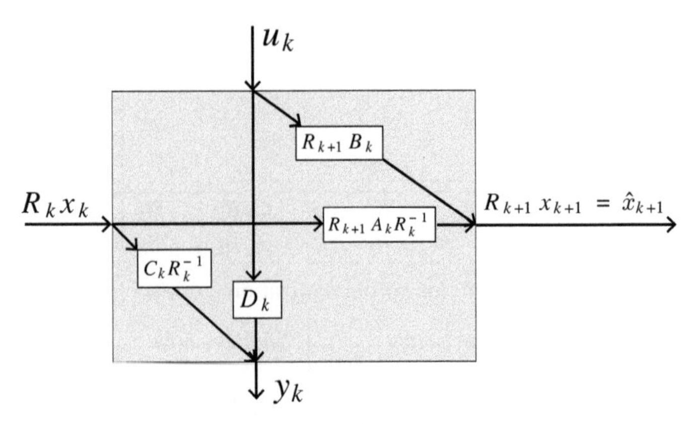

Figure 17.2 The realization of the isometric partial block considered at the General Step.

$$\left[\begin{array}{c} D_{1;k-1} \\ C_{1;k}B_{k-1} \\ \hline D_{2;k-1} \\ C_{2;k}B_{k-1} \end{array}\right] = \left[\begin{array}{c} D_{1;k-1} \\ \widehat{C}_{1;k}\widehat{B}_{k-1} \\ \hline D_{2;k-1} \\ \widehat{C}_{2;k}\widehat{B}_{k-1} \end{array}\right] \tag{17.24}$$

is isometric as well, thereby defining $D_{2;k-1}$ right invertible by

$$D'_{2;k-1}D_{2;k-1} = I - D'_{1;k-1}D_{1;k-1} - \widehat{B}'_{k-1}\widehat{B}_{k-1} \tag{17.25}$$

(very much as in Step 1). The proof will be complete if we show that the right-hand side is positive definite (not necessarily strictly), because then all the quantities needed to move to the next step have been determined, namely R_k, M_k and $D_{2;k-1}$. For this, we need the contractivity of $\left[\begin{array}{cc} D_{1;k-1} & \\ \hline \widehat{C}_{1;k}\widehat{B}_{k-1} & D_{1;k} \\ \widehat{A}_k\widehat{B}_{k-1} & \widehat{B}_k \end{array}\right]$, which maps $\begin{bmatrix} u_{k-1} \\ u_k \end{bmatrix} \mapsto \begin{bmatrix} y_{1;k-1} \\ y_{1;k} \\ \widehat{x}_{k+1} \end{bmatrix}$.

This follows from the (recursive) contractivity of the map $\begin{bmatrix} \widehat{x}_{k+1} \\ u_{k+1:n} \end{bmatrix} \mapsto y_{1;k+1:n}$, due to the isometry of the chain of realizations from the index $k+1$ on (achieved so far in the recursion), which maps $\begin{bmatrix} \widehat{x}_{k+1} \\ u_{k+1:n} \end{bmatrix}$ to $y_{k+1:n}$ isometrically, making the partial map $\begin{bmatrix} \widehat{x}_{k+1} \\ u_{k+1:n} \end{bmatrix}$ to $y_{1;k+1:n}$ contractive. Hence,

$$\begin{bmatrix} I - D'_{1:k-1}D_{1:k-1} - \widehat{B}'_{k-1}(\widehat{C}'_{1;k}\widehat{C}_{1;k} + \widehat{A}'_k\widehat{A}_k)\widehat{B}_{k-1} & -\widehat{B}'_{k-1}(\widehat{C}'_{1;k}D_{1;k} + \widehat{A}'_k\widehat{B}_k) \\ -(D'_{1;k}\widehat{C}_{1;k} + \widehat{B}'_k\widehat{A}_k)\widehat{B}_{k-1} & I - D'_{1;k}D_{1;k} - \widehat{B}'_k\widehat{B}_k \end{bmatrix} \geq 0, \tag{17.26}$$

and using the definitions of $\widehat{C}_{2;k} = -D^{+\prime}_{2;k}(D'_{1;k}\widehat{C}_{1;k} + \widehat{B}'_k\widehat{A}_k)$ and $D'_{2;k}D_{2;k} = I - D'_{1;k}D_{1;k} - \widehat{B}'_k\widehat{B}_k$, with $D_{2;k}$ right invertible, we find, after premultiplication with

$\begin{bmatrix} I & \\ & D_{2;k}^{+\prime} \end{bmatrix}$, postmultiplication with $\begin{bmatrix} I & \\ & D_{2;k}^{+} \end{bmatrix}$ and using $\widehat{A}_k'\widehat{A}_k + \widehat{C}_k'\widehat{C}_k = I$,

$$\begin{bmatrix} I - D_{1:k-1}'D_{1:k-1} - \widehat{B}_{k-1}'(I - \widehat{C}_{2;k}'\widehat{C}_{2;k})\widehat{B}_{k-1} & \widehat{B}_{k-1}'\widehat{C}_{2;k}' \\ \widehat{C}_{2;k}\widehat{B}_{k-1} & I \end{bmatrix} \geq 0. \tag{17.27}$$

The positivity of the Schur complement of the (2, 2) entry produces the positivity of $I - D_{1:k-1}'D_{1:k-1} - \widehat{B}_{k-1}'\widehat{B}_{k-1}$, very much as before in Step 1.

The construction just given shows that given a realization $\{A, B, C_S, D_S\}$, there is an isometric embedding T for S, a *local state transformation* R ($\widehat{x}_k = R_k x_k$) and a nonsingular diagonal $M := R'R$, such that the combined realization

$$\left[\begin{array}{cc} \widehat{A} & \widehat{B} \\ \hline \widehat{C}_S & D_S \\ \widehat{C}_T & D_T \end{array}\right] = \left[\begin{array}{cc} R^{\langle -1\rangle}AR^{-1} & R^{\langle -1\rangle}B \\ \hline C_S R^{-1} & D_S \\ C_T R^{-1} & D_T \end{array}\right] \tag{17.28}$$

is isometric. Moreover, M is strictly positive definite and satisfies

$$\begin{cases} A'M^{\langle -1\rangle}A + C_S'C_S + C_T'C_T &= M, \\ B'M^{\langle -1\rangle}A + D_S'C_S + D_T'C_T &= 0, \\ B'M^{\langle -1\rangle}B + D_S'D_S + D_T'D_T &= I. \end{cases} \tag{17.29}$$

This property (and its converse) is commonly known as the Kalman–Yakubovitch–Popov (or KYP) lemma:

Lemma 17.2 *A minimal state-space realization $\{A, B, C_S, D_S\}$ of a regular quasi-separable system S describes a contractive system if and only if there exists a sequence of strictly positive definite matrices M_k, a sequence of locally nonsingular matrices $\overline{D}$ and a bounded sequence of matrices $\overline{C}$ such that*

$$\begin{cases} M_k &= A_k'M_{k+1}A_k + C_k'C_k + \overline{C}_k'\overline{C}_k, \\ \overline{D}_k'\overline{C}_k &= -(D_k'C_k + B_k'M_{k+1}A_k), \\ \overline{D}_k'\overline{D}_k &= I - D_k'D_k - B_k'(C_k'C_k + \overline{C}_k'\overline{C}_k)B_k. \end{cases} \tag{17.30}$$

Proof

The "only if" part is by the construction given earlier, which shows the existence of the sequences of matrices $M, \overline{C}$ and $\overline{D}$ when the system described by $\{A, B, C, D\}$ is contractive: just substitute $C \leftarrow C_S, \overline{C} \leftarrow C_T$, etc.

The "if" part is immediate: the claimed properties allow us to factor $M_k = R_k'R_k$, with R_k square nonsingular, and produce an isometric realization defined as $\widehat{A}_k := R_{k+1}A_kR_k^{-1}$, $\widehat{B}_k := R_{k+1}B_k$, $\widehat{C}_{1;k} = C_kR_k^{-1}$, $\widehat{C}_{2;k} = \overline{C}_kR_k^{-1}$ and $\widehat{D}_{1;k} = D_k$ and $\widehat{D}_{2;k} = \overline{D}_k$ locally nonsingular, making also the resulting operator $\begin{bmatrix} S \\ T \end{bmatrix}$ isometric, and hence S contractive as a subsystem. □

Further properties

1. T obtained in the way presented earlier will be right-outer. Its finitely indexed subsystems for indices $[k : n]$ for all $k \leq n$ are all right-outer by construction.

Assuming inputs to be ℓ_2, this also makes the extended system right-outer (Theorem 8.1).

2. Let $\mathcal{O}_S$ and $\mathcal{O}_T$ denote the derived observability operators of S and T respectively for the realization $\{A,B,C,D\}$; then Eq. (17.23) shows that the global observability Gramian satisfies

$$\mathcal{O}'\mathcal{O} = \mathcal{O}_S'\mathcal{O}_S + \mathcal{O}_T'\mathcal{O}_T. \tag{17.31}$$

Positioning ourselves at some index $k < n$, and decomposing the operators in strict past and future, we have

$$\begin{bmatrix} S \\ T \end{bmatrix} = \left[\begin{array}{cc} K_S & \\ \mathcal{O}_S\mathcal{R} & E_S \\ \hline K_T & \\ \mathcal{O}_T\mathcal{R} & E_T \end{array}\right]_k \tag{17.32}$$

using the original $\mathcal{R}$ based on $\{A,B\}$. This equation is valid for all relevant k's, with $K_{\cdot,k}$ being the strict past to strict past operator and $E_{\cdot,k}$ the future to future operator, each with respect to k. Isometry gives,

$$[E_{S,k}'\mathcal{O}_{S,k} + E_{T,k}'\mathcal{O}_{T,k}]\mathcal{R}_k = 0. \tag{17.33}$$

Since $\mathcal{R}_k$ is nonsingular and E_T right-outer, we have

$$\mathcal{O}_{T,k} = -[E_{T,k}^\dagger]'E_{S,k}'\mathcal{O}_{S,k}. \tag{17.34}$$

Hence,

$$\mathcal{O}_k'\mathcal{O}_k = \mathcal{O}_{S,k}'\mathcal{O}_{S,k} + \mathcal{O}_{T,k}'\mathcal{O}_{T,k} = \mathcal{O}_{S,k}'(I + E_{S,k}E_{T,k}^\dagger[E_{T,k}^\dagger]'E_{S,k}')\mathcal{O}_{S,k}. \tag{17.35}$$

But $E_{T,k}'E_{T,k} = I - E_{S,k}'E_{S,k}$ by isometry, and hence (since these are all finite matrices) $E_{T,k}^\dagger[E_{T,k}^\dagger]' = (I - E_{S,k}'E_{S,k})^{-1}$ (which exists!), and finally, after a standard exchange,

$$M_k := \mathcal{O}_k'\mathcal{O}_k = \mathcal{O}_{S,k}'[I - E_{S,k}E_{S,k}']^{-1}\mathcal{O}_{S,k}, \tag{17.36}$$

where the inverse exists again. This fully characterizes the global M in terms of S, for the case where E_T is right-outer. Moreover, the observability Gramian of the right-outer T satisfies (with U being a diagonal unitary matrix) $\mathcal{O}_{T,k} = U[I - E_{S,k}E_{S,k}']^{-1/2}\mathcal{O}_{S,k}$ (actually valid for the Cholesky factor as well; there is, of course, some indeterminacy in the definition of the state: modulo a left diagonal unitary factor U). One can show (outside the scope of this book, but done in [29]) that this formula generalizes to the doubly infinite indexed case when the resulting $\widehat{A}$ with $\widehat{A}_k = R_{k+1}A_kR_k^{-1}$ and $M_k = R_k'R_k$ is u.e.s. Expression (17.36) gives a *closed-form solution to the recursive Riccati equation for M defined by* (17.30).

The Dual Case: Co-isometry

Starting out with a contractive S, we now wish to construct a complement T such that $\begin{bmatrix} S & T \end{bmatrix}$ is co-isometric. We assume S to be quasi-separable with realization

$\begin{bmatrix} A & B_S \\ C & D_S \end{bmatrix}$ and search for a realization for T, which shares A and C with S and has the realization $\begin{bmatrix} A & B_T \\ C & D_T \end{bmatrix}$, so that the transformed realization using a state transformation matrix R

$$\begin{bmatrix} \widehat{A} & \widehat{B}_S & \widehat{B}_T \\ \widehat{C} & \widehat{D}_S & \widehat{D}_T \end{bmatrix} = \begin{bmatrix} R^{\langle -1\rangle} A R^{-1} & R^{\langle -1\rangle} B_S & R^{\langle -1\rangle} B_T \\ C R^{-1} & D_S & D_T \end{bmatrix} \tag{17.37}$$

is co-isometric. Let $N := R^{-1}R^{-\prime}$; then we must find N, B_T, and D_T, so that

$$\begin{cases} N^{\langle -1\rangle} &= ANA' + B_S B_S' + B_T B_T', \\ 0 &= ANC' + B_S D_S' + B_T D_T', \\ I &= CNC' + D_S D_S' + D_T D_T' \end{cases} \tag{17.38}$$

(remark the shift in the index with respect to the previous case).

N is simply the Gramian of the overall reachability operator, which in this case equals $N_S + N_T$, where N_S is the reachability Gramian of S in the given realization. Now, if S is strictly contractive, then the partial operator $\widehat{K}_S$ is strictly contractive as well, and, of course, also $\widehat{K}_S^{\langle -1\rangle}$, while the completion T must be such that the combination of $\begin{bmatrix} K_S & K_T \end{bmatrix}$ and

$$\begin{bmatrix} K_S & K_T \end{bmatrix}^{\langle -1\rangle} = \left[\begin{array}{cc|cc} \widetilde{K}_S & 0 & \widetilde{K}_T & 0 \\ \widetilde{V}_S & [T_S]_{[0]} & \widetilde{V}_T & [T_T]_{[0]} \end{array}\right] \tag{17.39}$$

is co-isometric. Hence, to get a co-isometric completion, one must have

$$\begin{cases} \widetilde{K}_S \widetilde{K}_S' + \widetilde{K}_T \widetilde{K}_T' &= I, \\ \widetilde{V}_S \widetilde{K}_S' + \widetilde{V}_T \widetilde{K}_T' &= 0, \\ \widetilde{V}_S \widetilde{V}_S' + [T_S]_{[0]}[T_S]_{[0]}' + \widetilde{V}_T \widetilde{V}_T' + [T_T]_{[0]}[T_T]_{[0]}' &= I. \end{cases} \tag{17.40}$$

Canonical Contractive Realizations

The isometric or co-isometric embedding theory of this chapter allows us to define *canonical forms* for regular contractive operators. Such a transfer function S with realization $\{A, B, C_S, D_S\}$ can be embedded in a causal isometric system, and Eq. (17.29) shows that S has an observability Gramian given by the Riccati equation

$$\begin{aligned} M = {}& A_S' M^{\langle -1\rangle} A_S + C_S' C_S \\ & + (A' M^{\langle -1\rangle} B + C_S' D_S) \left[I - (B' M^{\langle -1\rangle} B + D_S' D_S)\right]^{\dagger} (B' M^{\langle -1\rangle} A + D_S' C_S). \end{aligned} \tag{17.41}$$

The realization will be isometric when it has the special property that $M = I$. Such a realization turns out to be "canonical" in the sense that all other minimal realizations for which the embedding observability operator is isometric are unitarily equivalent. We formulate this as a proposition.

Proposition 17.3 *The canonical realization $\{A, B, C, D\}$ of a contractive operator S is a partial realization of a minimal isometric embedding if and only if it satisfies*

$$I = \begin{bmatrix} A' & C' \end{bmatrix} \left(I - \begin{bmatrix} B \\ D \end{bmatrix} \begin{bmatrix} B' & D' \end{bmatrix} \right)^{\dagger} \begin{bmatrix} A \\ C \end{bmatrix}. \tag{17.42}$$

Proof

The first part of the proposition follows immediately from the Riccati equation, putting $M = I$ and the observation that (as utilized already many times before)[1] $I + X(I - X'X)^{\dagger}X' = (I - XX')^{\dagger}$ in general. The second part follows from the fact that, given minimal realizations and Expression (17.36), two realizations that have their observability Gramians $M = I$ must have their corresponding observability operators $\mathcal{O}$ equal, which defines the minimal state realization up to unitary equivalence. □

One obvious consequence is that contractive operators have contractive realizations. However, a contractive realization is not necessarily "canonical" in the sense that the realization is part of the realization of the corresponding isometry. Canonical realizations have the property that they can be realized using an algebraic minimum of elementary rotations (Givens transformations) – also this point is beyond our treatment here and is explored in [29].

17.4 Darlington Synthesis

After isometric or co-isometric embedding of a contractive transfer function, the next step is embedding in a causal unitary transfer operator, that is, a fully inner operator. This is what is known in the electrical engineering literature as *Darlington synthesis*, although, strictly speaking, Darlington synthesis is a more specific method of filter design (see the notes on this). So, starting from an isometric

$$\begin{bmatrix} S_{1,1} \\ \hline S_{2,1} \end{bmatrix} \sim_c \begin{bmatrix} A & B \\ C & D \\ \hline C_2 & D_2 \end{bmatrix}, \tag{17.43}$$

can we find the $S_{1,2}$ and $S_{2,2}$ causal operators such that $S = \begin{bmatrix} S_{1,1} & S_{1,2} \\ S_{2,1} & S_{2,2} \end{bmatrix}$ is unitary? The Darlington synthesis boils down to (1) an isometric (or, equivalently, a co-isometric) embedding of a given contractive $S_{1,1}$, and (2) a factorization

$$\begin{bmatrix} S_{1,1} \\ \hline S_{2,1} \end{bmatrix} = S \begin{bmatrix} I \\ 0 \end{bmatrix} \tag{17.44}$$

[1] The property is pretty obvious when $(I - XX')$ is nonsingular. If it merely has a Moore–Penrose inverse, the property follows by expressing the SVD of X. In case X is an infinite-dimensional operator, extra conditions have to be imposed to ensure the existence of the operators concerned.

in which S is inner. We immediately recognize (17.44) as an external factorization of the type $U_\ell \Delta_r'$, in which the causal Δ_r happens to be constant, that is, a unitary main diagonal (or the conjugate factorization in the co-isometric case). Hence, a simple external factorization of the isometric embedding gives the Darlington synthesis – and the latter will exist if and only if (1) the isometric embedding exists and (2) the external factorization leads to a unitary S. (Since the generalized Beurling theorem only guarantees S to be isometric, it might turn out not to be globally unitary, even though it is locally unitary: energy may "leak out" at $+\infty$.) It turns out that both conditions can easily be violated but that the synthesis will exist under broad regularity conditions, which are always satisfied in the finitely indexed case.

17.5 Discussion and Research Problems

1. Even though the embedding theory may seem computationally complex at first, it is perfectly elementary. The complexity has mostly to do with the number of different quantities (matrices or operators) present and the fact that each of them has at least four representative quantities in turn (the matrices $\{A, B, C, D\}$ in their realization). All these quantities are, taken together, much smaller in numerical complexity than the objects they represent – that is their main merit. The reduction comes at the prize of apparent but not real algebraic complexity! The state-space representation exemplifies the role various suboperators play: it introduces a new level of detail that makes operations more specific but somehow diminishes transparency.
2. It is nonetheless remarkable that the whole embedding theory rests on the single, absolutely elementary, recursive extension of a contractive causal matrix, as shown in the early sections of this chapter. The elementary embedding theory as we have presented it in this chapter, when translated to the state-space formalism in Section 17.3, quickly leads to the famous bounded real lemma, that is, the contractive version of the "KYP lemma" or "Kalman, Yacubovich, Popov lemma" that characterizes the state-space representation of a causal and contractive transfer function. This lemma is not so easy to prove in the LTI version, although in the LTV or matrix version, it is elementary (see, e.g., [5]). This is due to the extra difficulty the LTI situation introduces in the problem, namely that the recursion that determines the embedding has to start at infinity (in the case of the embedding that we have considered at $+\infty$), and remain stationary forever. The KYP lemma has therefore to posit the solution and then prove its validity under conditions that ensure stationarity (namely stability), rather than derive it from a recursion that necessarily starts at some finite time. Nonetheless, it is not really difficult to show that the method we have presented gives the desired solution even in the LTI case, assuming time invariance and stability, because the recursion converges to fixed values in that case. It is an interesting issue to further explore.
3. This chapter has been put entirely in the context of finite (or at most half-infinite) indexed matrices. The extension to full indexation requires a more powerful analytic approach at the cost of additional assumptions on the state-space model. The

study of quasi-infinite indexation provides for interesting cases where the embedding theory works recursively in one or the other direction, but not in both, and existence of an embedding is questionable. We invite the reader to explore such situations further.

17.6 Notes

Unitary embedding has a long history in mathematics, system theory and engineering. A straight unitary extension of a contractive matrix S appears to be an elementary problem in matrix theory, solvable by Cholesky factorization, followed by a basis completion (this is also how we approached the problem) – nothing extraordinary. The embedding issue becomes a little more interesting when the original matrix has some structure, and a similar structure is desired for the matrices that complete the embedding. We started out with requesting causality, translated into "block lower," and required the completion to consist of block lower matrices. Again, not much of a deal. The next structure to be considered is a dynamic system representation, and here the embedding issue becomes more interesting.

From the point of view of a dynamic structure, there have been two major approaches historically. One is based on the use of transform theory, specifically Laplace transforms, and considers the embedding of either a scalar or matrix transfer function that is contractive in the right half complex plane. The rational scalar case has been known as Darlington synthesis and was developed as one of the main methods to realize the desired contractive rational transfer function as a lossless circuit terminated in a unit resistor. More precisely, the Darlington synthesis required a cascade realization of so called sections, each realizing one "transmission zero," that is, a complex frequency such that any component of that frequency in the signal that is being transmitted is smothered, but this issue in selective filtering is often overlooked and the notion of Darlington synthesis assigned to any type of lossless embedding.

It was soon realized that the most effective formalism to handle Darlington synthesis is the scattering formalism, derived from the propagation of signals along lossless transmission lines. Instead of propagating voltages and currents that are intimately related along a transmission line, one considers propagating waves, which propagate independently from each other, one from right to left and the other from left to right along the line. These propagating waves have the property that the energy they carry is given by the square of their amplitude, so that energetic transmission or reflection properties can simply be expressed by contractive or unitary transfer functions or transfer matrices. When such a wave hits a device, it gets partly absorbed and partly reflected. The resulting mathematical framework for Darlington synthesis then turns out to be unitary embedding. Pioneering work in this area has been done in particular by Belevitch [11, 12].

The other approach is using state-space representations, pioneered for use in estimation and control theory by Kalman, but with many contributors to it in various fields, ranging all the way back to the work of Newton, and with ramifications in many areas of mathematics and mathematical physics. In the case of linear systems, as we have shown in the present book, systems are represented by a quartet of block diagonal matrices (A, B, C, D). These do not form a univocal representation, as we know from realization theory. The following issue arises: Which of these characterizes a contractive system or a lossless one when represented by any correct representation? In the early 1960s, papers appeared with such a characterization for the state-space representation of continuous-time LTI systems. This became the famous KYP lemma, which, given the knowledge at the time, was thought somewhat hard to prove. It was, from the beginning, clear that the lemma was closely related to spectral factorization, a topic that had been extensively studied by Wiener and Masani [54]: the lemma is, in fact, a reformulation of the existence of an appropriate spectral factorization and hence as difficult to prove as spectral factorization. (For a beautiful proof of spectral factorization in this context, see Lecture XI in the book of Helson on invariant subspaces [42].)

However, spectral factorization is only half the embedding story! At some point in the early history of this subject, it was thought that any causal contractive system could be embedded in a causal unitary one – this seemed like an almost evident proposition (and there are several papers in the literature claiming this). It is, however, not true. The catch is that additional structural properties like causality have to be respected in any meaningful embedding. This means that a Darlington synthesis is not possible in general, and in the LTV case, even very simple causal contractive systems do not have a Darlington embedding. However, finitely indexed systems do have unitary embeddings (as we showed in this chapter), and so it is for contractive rational matrix functions in the LTI case. There is a general criterion of existence valid for both cases, but that requires extending the theory to infinitely indexed systems [26].

Appendix: Data Model and Implementations

A.1 Introduction

To describe computational or time-variant systems comfortably, we need a data model and representations that are more flexible than the traditional signal flow diagrams, because the computing nodes (computers or computing units) we use may have to execute more than one function, connect to multiple sources of data and handle parallel computations. By the same token, we wish to have representations that are both close to the algebra of mathematical functions *and* allow easy mapping to computer architectures, including sequentialization, control generation and memory assignment. It turns out that there is such a model, which we conveniently call an *Applicative State Transition* or AST model. (Applicative computer languages have become quite common and have a long history as well, but they are especially well suited as executable descriptions of various types of system.)

A.2 The AST Model

The AST model is based on a simple *functional representation* of the proposed algorithm, defined as a directed graph consisting of nodes and edges with the following properties (semantics):

1. *nodes* execute algebraic functions, and data is transmitted along *edges*;
2. a node has *input ports* (indicated by incoming arrows) and *output ports* (indicated by outgoing arrows); output ports of one node connect to input ports of another, or to the external environment, which may also provide inputs to the system;
3. the behavior of the external environment is modeled just like any node in the system;
4. a node executes a sequence of computing cycles; at the beginning of a computing cycle of a given node, a specific function is installed by the node controller, the function reads (and consumes) the input data it needs from its relevant input ports, does the computation, puts (pushes) the output data on the relevant output ports (these ports may differ per function) and installs the next function (possibly depending on the data computed so far);
5. input and output data accumulate on the edges in a first-in first-out (FIFO) fashion; any function installed in a node waits until its relevant input ports have obtained the

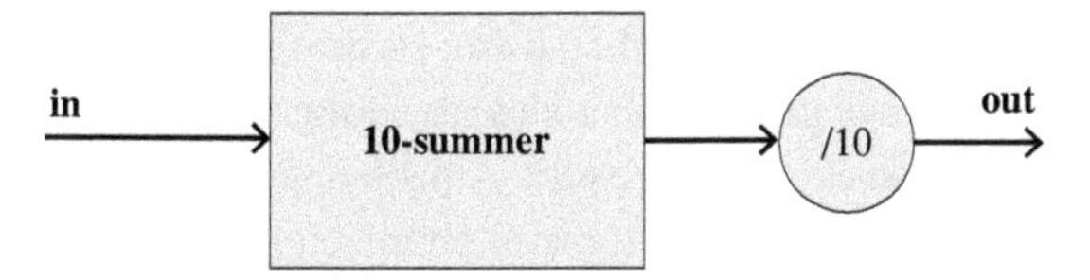

Figure A.1 The schema to average 10 items of data, output the average, and start all over again.

input data the function needs and pushes the output data it produces on its relevant output ports.

This means, in particular, that the edges must be equipped with the necessary memory and that the execution proceeds in a Petri-net fashion, that is, data items represent tokens that are selectively consumed by the nodes (called "colored tokens" in the Petri net language), and nodes "fire" when the selected input ports have tokens waiting.

Each node hence executes a *trace* of functions in the course of its history,[1] and each edge carries a sequence of data produced by the node at its input and consumed by the node at its output.

In the AST diagrams we use in this book, we make a further highly simplifying assumption: each function has specific input ports and specific output ports assigned to it, and the next function to be executed is also fixed. In other words, the whole operation of the graph is *not data dependent* (detailed operations may depend on parameters attached to the objects being computed, such as matrix dimensions, but the sequence of generic operations is defined by the graph). Admittedly, this extra assumption greatly reduces the flexibility of the representation, but the benefit is that it represents the type of algorithm we are considering with the least possible clutter of visible or explicit control.

Example : Average-10 Consider a device that takes in 10 data items, averages them and then outputs the average to start the cycle again. The diagram is shown in Fig. A.1. What the "10-summer" has to do is: read in 10 items, sum them and divide the result by 10 (i.e., consume 10 items from the input) and, after that, output the result (i.e., put the result on the output edge), reset its memory and start again.

With the conventions just defined, the whole network executes what may be called a data-independent precedence graph, which is, itself, the most elementary example of an AST network, whereby each node is used only once (and may then be thrown away) and the graph represents the data dependences of the operations in a single tree (per definition without a closed loop). An AST network is seen to be correct

[1] This data model is close, but not identical, to the Kahn network formalism, which is sometimes used in the literature. Rather, it adheres closely to the algebraic practice of composing functions and arguments. It also reduces to the traditional time-invariant signal flow diagrams in a straightforward way.

from the point of view of data processing, when it can be devolved into a precedence graph. Conversely, any precedence graph with nodes executing single functions can be partitioned and folded into a correct AST network – this being the task of the system designer. We only sporadically touch on the topic of architecture design (partitioning, timing, memory assignment, control generation, etc.); see, for example, Giovanni De Micheli's book [51] for an excellent introduction to the subject.

We may draw functional diagrams from right to left, in accordance with mathematical practice. (Example: in $y = Au$, u is fed in from the right and y gets out at the left side, and we draw $y \leftarrow \boxed{A*} \leftarrow u$, $y \leftarrow A \leftarrow u$ or even $y \overset{A}{\leftarrow} u$.) A split node and a join typically act as controlled switches. (There are other possibilities, of course, but these conventions respect "losslessness" of the individual nodes, meaning: energy out = energy in measured in square norms.)

The model we use cannot describe all or even most possible computer architectures, but it can handle all data-independent algorithmic schemes and many data-dependent algorithmic schemes as well (but by far not all). The limited ability of handling data-dependent sequences of computations is precisely due to a priori assignment of functions to nodes and data to edges. Advantages and disadvantages of this representation mode can already be seen in the simple example just shown: the functional schema is very simple, because it does not deal with detailed control issues, but it cannot handle more complex issues, like interpreting the input stream and then deciding on what to do next (e.g., having an input stream with tokens for "begin" and "end" and an arbitrary number of data in between to be averaged). A more general Petri net formalism is able to handle such adaptive cases, but at the cost of algorithmic transparency (which is our main goal here).

Simplifying Conventions

When there is no confusion, we retreat to the classical signal flow graph convention, where an edge is seen as a multiplication by a coefficient or a matrix marked on top of it, like in $y \overset{A}{\leftarrow} u$, which according to the functional data model, should be denoted as "$\leftarrow A \leftarrow$," since multiplication by A is an operation, to be executed. A "join" in a traditional signal flow graph actually represents a node, in which matrix addition is performed, but in our datamodel, the trace representing the sequence of data transmitted on an edge is often put on top of the arrow. The applicable situation is to be derived from the context!

A.3 Realizations Using Jacobi/Givens Rotors

As an example, let us look at orthogonalization using rotors. The simplest possible components for orthogonal transformations (often called *orthogonal filtering*) are elementary real or complex *rotors*. (The elementary rotor goes back to Jacobi, but in

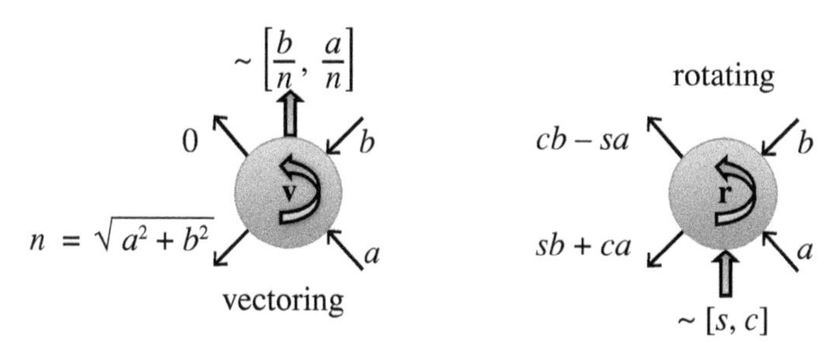

Figure A.2 The single Jacobi/Givens rotor in two modes: vectoring and rotating. In the vectoring mode, angle information is produced, labeled here as $[s, c]$, while in the rotating mode, the angle information is used to perform the rotation. In many cases, the angle information remains resident in the node and is not shown. Signal propagation is from SE to NW.

numerical algebra is known as a "Givens rotation.") Only the real case is considered here, but the rotors can easily be extended to the complex case.

Rotors

A rotor comes in two (switchable) versions: the *vectoring* mode, where a rotation matrix is first computed from the input data and then applied on the same input data; and the *rotating* mode, where a given rotation matrix is applied on new data; see Fig. A.2. In the *vectoring mode (v)* of the single rotor, a vector $\mathrm{col}(b,a)$ is inputted, the rotor calculates $n = \sqrt{a^2 + b^2}$, $c = a/n$ and $s = b/n$, and outputs a "control" output $\begin{bmatrix} s & c \end{bmatrix}$ and a "data" output $\begin{bmatrix} 0 & n \end{bmatrix}$ so that

$$\begin{bmatrix} c & -s \\ s & c \end{bmatrix} \begin{bmatrix} b \\ a \end{bmatrix} = \begin{bmatrix} 0 \\ n \end{bmatrix}. \tag{1.1}$$

In the *rotating mode (r)*, a control input $\begin{bmatrix} s & c \end{bmatrix}$ is given and applied to an input vector $\begin{bmatrix} b & a \end{bmatrix}$ to produce an output vector $\begin{bmatrix} d & c \end{bmatrix}$ so that

$$\begin{bmatrix} c & -s \\ s & c \end{bmatrix} \begin{bmatrix} b \\ a \end{bmatrix} = \begin{bmatrix} d \\ c \end{bmatrix}. \tag{1.2}$$

Figure A.3 shows a simple application in which the vectoring mode is used recursively to produce the norm of a vector. (In this example, the recursive edge is initialized to 0 and the trace is vvv ⋯.) There is an electronic device called CORDIC that implements the rotor using elementary (bitwise) rotations and that is used in high-frequency applications.[2] The Jacobi/Givens matrix is, in principle, unitary, but in finite arithmetic, the unitarity cannot be exact. A conservative solution to this problem ensures that the actual matrix is slightly contractive (as also used in standard wave digital filtering), so that no extra energy that could lead to instabilities is generated.

[2] See, for example, www.actel.com/ipdocs/CoreCORDIC_HB.pdf.

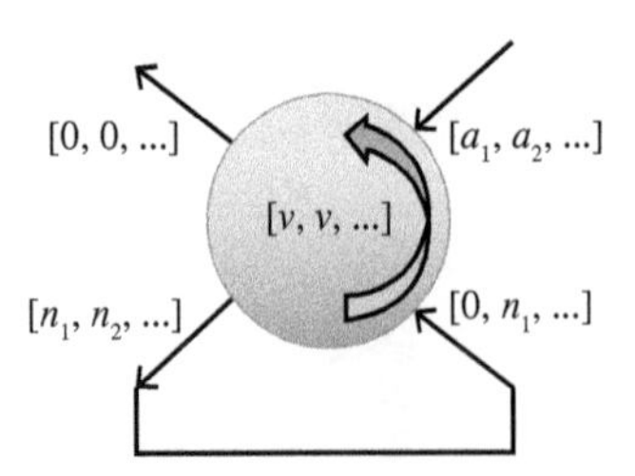

Figure A.3 The single rotor used as a norm generating filter (traces shown).

For a next example, let us look at the QR factorization of a 3×3 matrix a. In words, the procedure using Givens rotations goes as follows (see also the appendix):

1. using the first two rows of a, first vectorize $\begin{bmatrix} a_{1,1} \\ a_{2,1} \end{bmatrix}$ with a rotor $q_{1,2}$ and rotate the remainder entries $\begin{bmatrix} a_{1,2} & a_{1,3} \\ a_{2,2} & a_{2,3} \end{bmatrix}$ with this rotor; row 3 is passed unscathed to the next stage; this produces at the output two new rows 1 and 2 (which we can denote shortly as a^+, i.e., $a^+_{1,1}$, etc.) obtained by applying the partial rotation $\begin{bmatrix} q_{1,2} & \\ & 1 \end{bmatrix}$ to a;
2. next, do the same operation on a^+, using rows 1 and 3, producing a next update a^{++} and a new rotor $q_{1,3}$;
3. finally, a new rotor $q_{2,3}$ is applied on rows 2 and 3 to produce the new update a^{+++}, which is equal to the R factor;
4. to obtain the Q factor in the matrix form (instead of a computational schema), one just has to apply the three-dimensional unit matrix in the same order, but now using only rotations, and the Q flows out of the array.

Figure A.4 shows the data flow schema. The same schema (i.e., using the same $q_{i,j}$ rotors) produces the Q-factor in the form Q' (since $Q'a = r$), as shown in Fig. A.5. This operation can, of course, be performed right after the production of the R factor. It should be clear that the two flowcharts can be combined into one by just concatenating all the traces.

The QR-factorization just described is one instance of a parametrized collection of factorizations indexed by the dimensions of the input matrix (called a here). This means that when such a matrix is going to be offered as input to a processor, first its dimensions have to be read in by the processor, which is then able to set up the corresponding data flow schema and receive the matrix data to be converted to the output data consisting of parametrized generic matrices Q and R. These two output matrices partition in further sub-matrices representing, in the case of Q, a co-kernel and a range, and in the case of R, an echelon form with a specific echelon structure. All this has to happen in a parametrized way, only the generic sequence of events is determined by the data-flow diagram. Of course, the processor involved has a lot of detailed bookkeeping to do, which we thankfully ignore at the algorithmic level; the

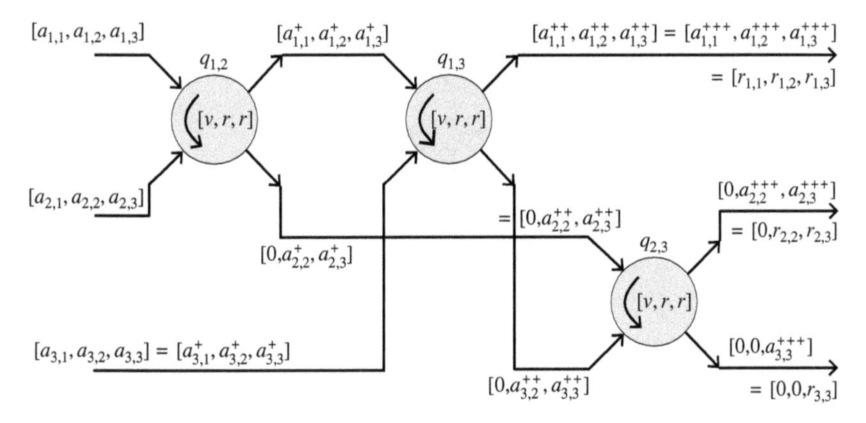

Figure A.4 QR factorization of the 3×3 matrix a using Givens rotors. The traces on the operation nodes and edges are indicated.

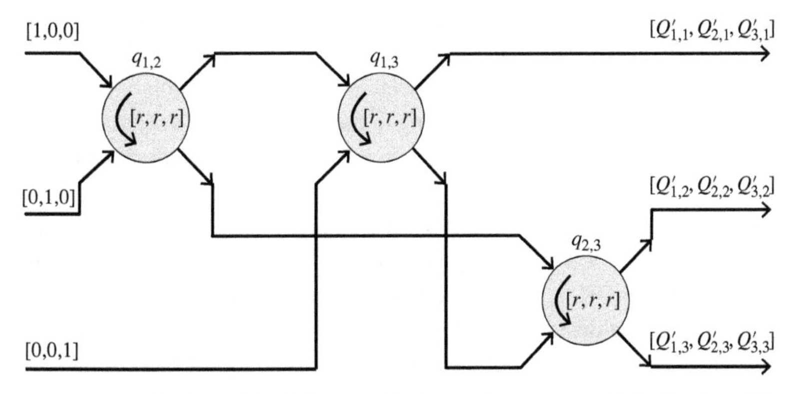

Figure A.5 Production of the Q factor with the previous array, with indication of the traces. Notice that the scheme actually produces Q'.

important point is that the processor obtains all the necessary information from the network and is able to convert and transmit the further necessary information to its successor nodes in an AST fashion.

A.4 Architecture Design

Once a functional representation of the algorithm has been made, it is ready to be mapped to an actual architecture. The latter consists of a list of devices (processors and memories) and an interconnection network. This is what one calls *architecture design*. It consists of the following:

- assignment of operations to processors;
- assignment of memory locations to data (both input/output and intermediate);
- generation of the local controllers;

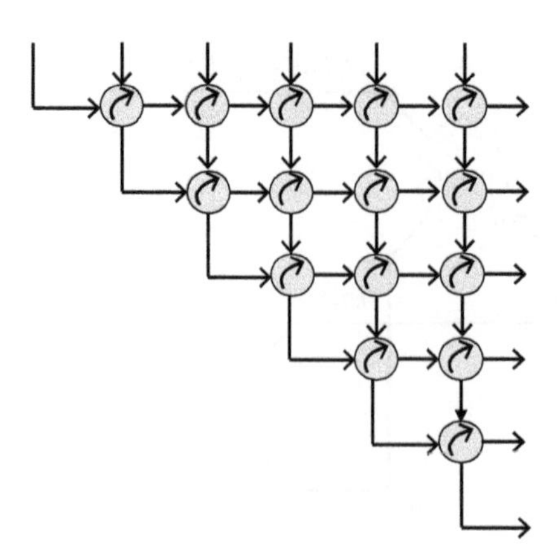

Figure A.6 Array structure of the QR factorization with Givens rotations: the "Gentleman–Kung array."

and all this with the aim to ensure correct and, if possible, efficient or even optimal operation of the overall system, taking into account physical limitations. The data flow graph we defined gives an architecture-free representation of the algorithm: each operation and each data item occurring in the algorithm has a separate identity and can be mapped on any adequate device in a proposed architecture, respecting causalities of course.

For example, the 3×3 QR factorization already discussed may be represented as a single-operation 3×3 QR taking the full matrix A as its input and outputting the full R factor. A 9×9 QR then looks very much like the original 3×3, but now done with 3×3 blocks. The designer (or the design software) will have to do the bookkeeping to get all the control right, but the bookkeeping is straightforward, and the data model remains the same.

The regularities of the QR computing schema can be exploited further. A little rearranging of the computing nodes, letting the input flow in from the top and the output flow out from the right, produces a schema like the one shown in Fig. A.6 and known as the "Gentleman–Kung array" for computing QR.

A great advantage (and sometimes great disadvantage) of our restrictive data flow description is its deterministic character that allows for hierarchical design on dedicated hardware. The flow graph can be partitioned in "tiles," which can be assigned to processors either in a parallel or sequential way, keeping the same data flow modeling method throughout, while assembling processing and data transfer traces according to the prescriptions of the original flow graph.

There are two main strategies one can follow after tiling, each one having a big impact on the memory handling of the consumed and produced data. Consider the tiling in Fig. A.7, whereby the tiles shown are taken out of a (much) larger schema:

- In one strategy, each tile is assigned to an individual processor, and the tiles are allowed to run in parallel; this strategy requires each processor to provide memory for its local data, since the local operations in the tile are executed sequentially. The strategy is called "LSGP" for "local sequential, global parallel."

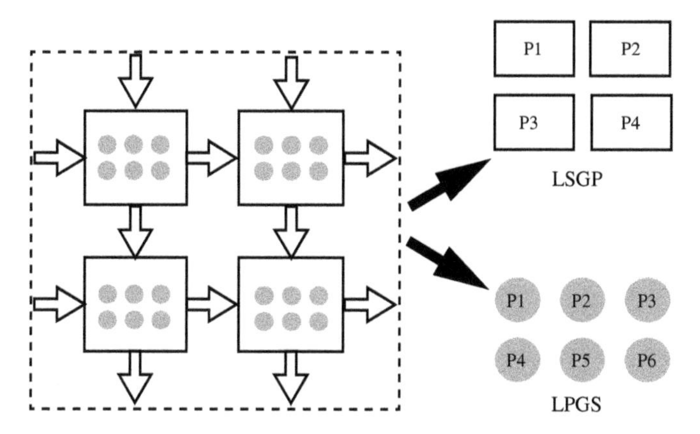

Figure A.7 Cutting out a set of tiles in a two-dimensional array and mapping to individual processors.

- In another strategy, each node in a tile is given a separate processor, and the tiles are run sequentially (in a causal order of course). This strategy requires the availability of a tile-attached tile-global memory capable of handling the data production of the tiles. The strategy is called "LPGS" for "local parallel, global sequential."

  Assuming a two-level hierarchy of tiling, the most optimal strategy of assigning memory will be to use LSGP first, up to the exhaustion of the local memory size, whereupon data produced by the tiles have to be handled by global memory in an LPGS fashion. The strategies can be mixed in more levels, and other considerations, like the efficient use of processing hardware may be considered as well.

  The two strategies can equally well be used with nonuniform partitioning.

Once such mappings have been done, the detailed architecture specification can be made: local controllers have to be generated to supply operations with the correct data, to schedule the operations and, after any operation, to channel the results to correct locations, ready for the next moves. Operations and memory may be distributed over a number of processing units (like in an array processor) in a regular or irregular fashion, and memory may be available locally (as in register files) or (partly) concentrated in background memory, in which case, local or global controllers have to request data when needed, or, conversely, channel results back to appropriate memory locations. The important point is that all these operations can be done using a unified data model. However, architecture design is not the subject of this book, and we have to refer to the literature for more information on this important topic (starting with [51], and see further for many partitioning examples and properties of the model, e.g., [15,21,23]).

References

[1] V. M. Adamjan, D. Z. Arov, and M. G. Krein. Analytic properties of Schmidt pairs for a Hankel operator and the Schur–Takagi problem. *Math. USSR Sbornik*, 15(1):31–73, 1971.

[2] D. Alpay, P. Dewilde, and D. Volok. Interpolation and approximation of quasiseparable systems: the Schur–Takagi case. *Calcolo*, 42:139–156, 2005.

[3] B. D. O. Anderson and J. B. Moore. *Optimal Filtering*. Prentice Hall, Englewood Cliffs, NJ, 1979.

[4] B. D. O. Anderson and J. B. Moore. *Optimal Control, Linear Quadratic Methods*. Prentice Hall, Englewood Cliffs, NJ, 1989.

[5] B. D. O. Anderson and S. Vongpanitlerd. *Network Analysis and Synthesis*. Prentice Hall, Englewood Cliffs, NJ, 1973.

[6] J. Anneveling and P. Dewilde. Object-oriented data management based on abstract data types. *Softw. Pract. Exp.*, 17(11):751–781, November 1987.

[7] W. Arveson. Interpolation problems in nest algebras. *J. Funct. Anal.*, 20(3):208–233, 1975.

[8] J. A. Ball and J. W. Helton. A Beurling-Lax theorem for the Lie group $U(m,n)$ which contains most classical interpolation theory. *J. Oper. Theory*, 9(2):107–142, 1983.

[9] J. A. Ball and J. W. Helton. Inner-outer factorization of nonlinear operators. *J. Funct. Anal.*, 104(2):363–413, 1992.

[10] H. Bart, I. Gohberg, and M. A. Kaashoek. Convolution equations and linear systems. *Integr. Equ. Oper. Theory*, 5(3):283–340, 1982.

[11] V. Belevitch. *Classical Network Theory*. Holden Day, San Francisco, CA, 1968.

[12] V. Belevitch. Elementary applications of the scattering formalism to network design. *IRE Trans. Circuit Theory*, 3(2):97–104, 1956.

[13] R. Bellman. The theory of dynamic programming. *Bull. Am. Math. Soc.*, 60(6):503–516, 1954.

[14] A. Ben-Artzi and I. Gohberg. Nonstationary inertia theorems, dichotomy and applications. *Proc. Sympos. Pure Math.*, 51(Part 1):85–95, 1990.

[15] J. Bu. *Systematic Design of Regular VLSI Processor Arrays*. PhD thesis, Delft University of Technology, Delft, the Netherlands, May 1990.

[16] S. Chandrasekaran, P. Dewilde, M. Gu, T. Pals, X. Sun, A. J. van der Veen, and D. White. Some fast algorithms for sequentially semi-separable representations. *SIAM J. Matrix Anal. Appl.*, 27(2):341–364, 2005.

[17] Th. M. Cover and J. A. Thomas. *Elements of Information Theory*. John Wiley & Sons, Inc., New York, 1991.

[18] G. Cybenko. A general orthogonalization technique with applications to time series analysis and signal processing. *Math. Comput.*, 40(161):323–336, 1983.

[19] W. N. Dale and M. C. Smith. Existence of coprime factorizations for time-varying systems – an operator-theoretic approach. In H. Kimura and S. Kodama, editors, *Recent*

Advances in Mathematical Theory of Systems, Control, Networks and Signal Processing I (Proc. Int. Symp. MTNS-91), pages 177–182. MITA Press, Japan, 1992.

[20] L. J. Castriota, D. C. Youla, and H. J. Carlin. Bounded real scattering matrices and the foundation of linear passive network theory. *IRE Trans. Circuit Theory*, 4(1):102–124, 1959. Corrections, *Ibid.*, 317 (September 1959).

[21] A. A. J. de Lange, A. J. van der Hoeven, E. F. Deprettere, and P. Dewilde. HiFi: an object oriented system for the structural synthesis of signal processing algorithms and the VLSI compilation of signal flow graphs. In L. Claesen, editor, *Applied Formal Methods for Correct VLSI Design*, pages 462–481. IFIP, 1989.

[22] E. Deprettere. Mixed-form time-variant lattice recursions. In *Outils et Modèles Mathématiques pour l'Automatique, l'Analyse de Systèmes et le Traitement du Signal*, CNRS, Paris, 1981.

[23] E. Deprettere and A. J. van der Veen. *Algorithms and Parallel VLSI Architectures*. Elsevier, Amsterdam, 1991.

[24] A.-J. Van der Veen. A Schur method for low-rank matrix approximation. *SIAM J. Matrix Anal. Appl.*, 17(1):139–160, 1996.

[25] P. Dewilde. On the LU factorization of infinite systems of semi-separable equations. *Indagationes Math.*, 23:1028–1052, 2012.

[26] P. Dewilde. Generalized Darlington synthesis. *IEEE Trans. Circuits Syst. I Fundam. Theory Appl.*, 45(1):41–58, January 1999.

[27] P. Dewilde and E. Deprettere. Approximative inversion of positive matrices with applications to modeling. In R. F. Curtain, editor, *NATO ASI Series, Vol. F34 of Modeling, Robustness and Sensitivity Reduction in Control Systems*, pages 211–238. Springer Verlag, Berlin, 1987.

[28] P. Dewilde, Y. Eidelman, and I. Haimovici. LU factorization for matrices in quasiseparable form via orthogonal transformations. *Linear Algebr. Appl.*, 502:5–40, 2016.

[29] P. Dewilde and A.-J. van der Veen. *Time-Varying Systems and Computations*. Kluwer, Dordrecht, 1998. [Out of print but freely available at ens.ewi.tudelft.nl.]

[30] P. Dewilde and A.-J. van der Veen. Inner-outer factorization and the inversion of locally finite systems of equations. *Linear Algebr. Appl.*, 313:53–100, 2000.

[31] P. M. Dewilde. A course on the algebraic Schur and Nevanlinna-Pick interpolation problems. In F. Deprettere and A. J. van der Veen, editors, *Algorithms and Parallel VLSI Architectures*. Elsevier, Amsterdam, 1991.

[32] P. M. Dewilde and A. J. van der Veen. On the Hankel-norm approximation of upper-triangular operators and matrices. *Integr. Equ. Oper. Theory*, 17(1):1–45, 1993.

[33] P. Van Dooren. A unitary method for deadbeat control. In P. A. Fuhrmann, editor, *Mathematical Theory of Networks and Systems. Lecture Notes in Control and Information Sciences*, vol 58. Springer, Berlin, Heidelberg, 1984.

[34] P. Van Dooren. Deadbeat control, a special inverse eigenvalue problem. *BIT*, 24:681–699, 1984.

[35] H. Dym and I. Gohberg. Extensions of band matrices with band inverses. *Linear Algebr. Appl.*, 36:1–24, 1981.

[36] Y. Eidelman and I. Gohberg. On a new class of structured matrices. *Integr. Equ. Oper. Theory*, 34(3):293–324, 1999.

[37] Y. Eidelman and I. Gohberg. A modification of the Dewilde-van der Veen method for inversion of finite structured matrices. *Linear Algebr. Appl.*, 343–344:419–450, 2002.

[38] I. A. Feldman and I. Gohberg. On reduction method for systems of Wiener–Hopf type. *Dokl. Akad. Nauk SSSR*, 165(2):268–271, 1965.

[39] A. E. Frazho and M. A. Kaashoek. Canonical factorization of rational matrix functions. a note on a paper by P. Dewilde. *Indag. Math.*, 23:1154–1164, 2012.

[40] G. H. Golub and Ch. F. Van Loan. *Matrix Computations*. John Hopkins University Press, Baltimore, MD, 1983.

[41] L. M. J. Hautus. A simple proof of Heymann's lemma. *IEEE Trans. Autom. Control*, 22(5):885–886, 1977.

[42] H. Helson. *Lectures on Invariant Subspaces*. Academic Press, New York, 1964.

[43] J. W. Helton. Orbit structure of the Möbius transformation semigroup acting on H_∞ (broadband matching). In *Topics in Functional Analysis*, volume 3 of *Advances in Mathematics: Supplementary Studies*, pages 129–133. Academic Press, New York, 1978.

[44] B. L. Ho and R. E. Kalman. Effective construction of linear, state-variable models from input/output functions. *Regelungstechnik*, 14(1–12):545–548, 1966.

[45] K. Jainandunsing and E. F. Deprettere. A new class of parallel algorithms for solving, systems of linear equations. *SIAM J. Sci. Stat. Comput.*, 10(5):880–912, 1989.

[46] T. Kailath. *Lectures on Linear Least-Squares Estimation*. Springer Verlag, CISM Courses and Lectures No. 140, Wien, New York, 1976.

[47] T. Kailath. A view of three decades of linear filtering theory. *IEEE Trans. Inform. Theory*, 20(2):145–181, 1974.

[48] R. E. Kalman. A new approach to linear filtering and prediction problems. *J. Basic Eng.*, 82(1):34–45, 1960.

[49] R. E. Kalman, P. L. Falb, and M. A. Arbib. *Topics in Mathematical System Theory*. International Series in Pure and Applied Mathematics. McGraw-Hill, New York, 1970.

[50] David G. Luenberger. *Introduction to Dynamic Systems*. John Wiley & Sons, 1979.

[51] G. De Micheli. *Synthesis and Optimization of Digital Circuits*. McGraw-Hill, New York, 1994.

[52] M. Morf and T. Kailath. Square-root algorithms for least-squares estimation. *IEEE Trans. Autom. Control*, 20(4):487–497, 1975.

[53] H. Nelis, P. Dewilde, and E. F. Deprettere. Inversion of partially specified positive definite matrices by inverse scattering. *Ope. Theory: Adv. Appl.*, 40:325–357, 1989.

[54] P. Masani P and N. Wiener. The prediction theory of multivariable stochastic processes. *Acta Math.*, 98 and 99:111–150 and 93–137, 1957 and 1958.

[55] M. Van Barel R. Vandebril and N. Mastronardi. *Matrix Computations and Semiseparable Matrices*. John Hopkins University Press, Baltimore, MD, 2008.

[56] J. K. Rice and M. Verhaegen. Distributed control: a sequentially semi-separable approach for spatially heterogeneous linear sysems. *IEEE Trans. Autom. Control*, 54(6):1270–1284, 2009.

[57] The New Oxford American Dictionary. [Edition available as application on macOS Mojave.]

[58] John. R. Ringrose. On some algebras of operators. *Proc. Lond. Math. Soc.*, 15(1):61–83, 1965.

[59] W. Rudin. *Real and Complex Analysis*. McGraw-Hill, New York, 1966.

[60] P. M. Van Dooren, R. V. Patel, and A. J. Laub. *Numerical Linear Algebra Techniques for Systems and Control*. IEEE Press, New York, 1994.

[61] I. Schur. Uber Potenzreihen, die im Innern des Einheitskreises beschränkt sind, I. *J. Reine Angew. Math.*, 147:205–232, 1917. Eng. Transl. *Oper. Theory: Adv. Appl.*, 18:31–59, 1986.

[62] L. M. Silverman and H. E. Meadows. Equivalence and synthesis of time-variable linear systems. In *Proceedings of 4th Allerton Conference on Circuit and System Theory*, pages 776–784, 1966.

[63] G. W. Stewart. *Matrix Algorithms*. SIAM, Philadelphia, PA, 1998.

[64] G. W. Stewart. *Matrix Algorithms, Vol. I: Basic Decompositions*. SIAM, Philadelphia, PA, 1998.

[65] G. W. Stewart. *Matrix Algorithms, Vol. II: Eigensystems*. SIAM, Philadelphia, PA, 2001.

[66] Teiji Takagi. On an algebraic problem related to an analytic theorem of Carathéodory and Fejér and on an allied theorem of Landau. *Japan J. Math.*, 1:83–93, 1924.

[67] N. P. van der Meijs. Accurate and efficient layout extraction. *PhD thesis*, Delft University of Technology, The Netherlands, 1992.

[68] Arjan van der Schaft. *L_2-Gain and Passivity Techniques in Nonlinear Control*. Springer, New York, 2000.

[69] A. J. van der Veen. *Time-varying system theory and computational modeling: realization, approximation, and factorization*. PhD thesis, Delft University of Technology, Delft, The Netherlands, June 1993.

[70] A. J. van der Veen and P. M. Dewilde. On low-complexity approximation of matrices. *Linear Algebr. Appl.*, 205/206:1145–1201, 1994.

[71] M. Verhaegen. Subspace model identification. Part III: Analysis of the ordinary output-error state space model identification algorithm. *Int. J. Control*, 58(3):555–586, 1994.

[72] M. Verhaegen and P. Dewilde. Subspace model identification. Part I: The output-error state space model identification class of algorithms. *Int. J. Control*, 56(5):1187–1210, 1992.

[73] M. Verhaegen and P. Dewilde. Subspace model identification. Part II: Analysis of the elementary output-error state space model identification algorithm. *Int. J. Control*, 56(5):1211–1241, 1992.

[74] J. H. Wilkinson and C. Reinsch. *Linear Algebra, vol. 2, Handbook for Automatic Computation*. Springer, New York, 1971.

[75] Jan Willems. The behavioral approach to open and interconnected systems. *IEEE Control Syst. Mag.*, 27(6):46–99, 2007.

[76] J. C. Willems. Dissipative dynamical systems – part i: general theory. *Arch. Ration. Mech. Anal.*, 45:321–351, 1972.

[77] Mu Zhou and Alle-Jan van der Veen. Stable subspace tracking algorithm based on signed URV decomposition. In *Proceedings of IEEE International Conference on Acoustic, Speech, and Signal Processing*, pages 2720–2723, Prague, Czech Republic, May 2011.

[78] Max Dehn. Über die Zerlengung von Rechtecken in Rechtecke. *Math. Ann.* 57:314–322, 1903.

[79] C. J. Bouwkamp, A. J. W. Duijvestijn, and P. Medema. *Table of c-nets of orders 8–19 inclusive*, 2 vols. Philips Research Laboratories, Eindhoven, The Netherlands, 1960; Unpublished available in UMT file of Mathematics of Computation.

[80] A. J. W. Duijvestijn. Simple perfect square of lowest order. *J. Comb. Theory. Ser. B.* (25):240–243, 1978.

Index

Milton Keynes UK
Ingram Content Group UK Ltd.
UKHW050724061124
450288UK00027B/51

9 781009 455626